Advances in
Earth Science
From Earthquakes to Global Warming

Royal Society Series on Advances in Science

Series Editor: J. M. T. Thompson *(FRS)*

Published

PHILOSOPHICAL
TRANSACTIONS
—— OF ——
THE ROYAL
SOCIETY
MATHEMATICAL, PHYSICAL & ENGINEERING SCIENCES

Royal Society Series on Advances in Science – Vol. 2

Advances in Earth Science

From Earthquakes to Global Warming

Editors

P R Sammonds
University College London, UK

J M T Thompson
University of Cambridge, UK

Imperial College Press

ICP

Published by

Imperial College Press
57 Shelton Street
Covent Garden
London WC2H 9HE

Distributed by

World Scientific Publishing Co. Pte. Ltd.
5 Toh Tuck Link, Singapore 596224
USA office: 27 Warren Street, Suite 401-402, Hackensack, NJ 07601
UK office: 57 Shelton Street, Covent Garden, London WC2H 9HE

British Library Cataloguing-in-Publication Data
A catalogue record for this book is available from the British Library.

Royal Society Series on Advances in Science — Vol. 2
ADVANCES IN EARTH SCIENCE
From Earthquakes to Global Warming

ISBN-13 978-1-86094-761-2
ISBN-10 1-86094-761-1
ISBN-13 978-1-86094-762-9 (pbk)
ISBN-10 1-86094-762-X (pbk)

Typeset by Stallion Press
Email: enquiries@stallionpress.com

Printed by FuIsland Offset Printing (S) Pte Ltd, Singapore

Preface

This welcome volume is a collection of articles principally adapted from articles published in the Philosophical Transactions triennial issue. It very largely reflects the views of younger scientists, and highlights how the Earth Sciences continue to delight us with new ideas and controversies. Many of the authors are research fellows, or former research fellows, of the Royal Society, and the editors have successfully assembled an entertainingly eclectic mix. The book is divided into three sections covering Environmental Change, the Dynamics of the Earth, and Applied Earth Science, and the topics range from costing climate change, to the properties of the Earth's core and objectively optimised Earth observation.

In costing climate change Dave Reay notes that economics and climate change have a great deal in common, in that they seek to predict the future on the basis of what has recently gone before. He examines existing cost-benefit analyses of greenhouse gas reduction policies and concludes that economics cannot provide a justification for political inaction on greenhouse gas emissions. Andy Ridgwell and Karen Kohfield illustrate the integrative thinking that is required in addressing the Earth system through the medium of dust. Dust is important globally because of the control it exerts on marine plant productivity and hence the uptake of CO_2 from the atmosphere. The current perturbation of the carbon cycle is so large and fundamental that it has been suggested that the Earth has entered a new geological epoch. Yadvinder Malhi reviews the likely causes of different carbon sinks and sources and highlights the limits to the amount of carbon that can be stored in natural vegetation. It may be that terrestrial carbon storage is unstable to significant global warming, and hence have the potential to accelerate rather than brake global warming. Finally in this section, Richard Twitchett explores what can be learnt from a better understanding of the largest mass extinction event in the last few 100 million years, at the end of the Permian.

In the Dynamics of the Earth, Cathyrn Mitchell explores how tomography has been developed from a medical tool into a technique for imaging the ionised plasma around the entire Earth. It may be a little early to achieve

'Ionospheric Weather' forecasting, but real-time movies now allow us to watch the result of the Earth's bombardment by solar wind during events known as storms. Eiichi Fukuyama changes scale sharply to show how the dynamic rupture process of real earthquakes can now be simulated, given the right information. A major development has been the ability to obtain improved information on the rates of natural processes from the geological record. This is illustrated by Simon Turner in his review of the application of short-lived U-series to investigate the time scales of the magmatic processes that occur beneath island arc volcanoes. New continental crust is generated at island arcs, and Tim Minshull discusses how such crust is subsequently broken apart in response to plate tectonic processes. Some continental margins have considerable volumes of igneous rocks associated with continental break-up, and others do not. More research is required on paired continental margins, and in the development of computer models that can handle the transition from continental deformation to sea-floor spreading and the formation of a new ocean basin. There is continuing interest in the Earth's core, its composition and when it was formed. Francis Nimmo and Dario Alfe focus on the properties of core-forming materials, how core motions generate the Earth's magnetic field, and the evolution of both the core and the dynamo. They also briefly review the current state of knowledge for cores and dynamos on other planetary bodies.

Chris Kilburn starts the Applied Earth Science section with a reappraisal of the hazards from large landslides. Their size and speed prevent effective hazard mitigation after collapse, and so the emphasis is on advance warning of collapse and how on far individual landslides may travel. The earthquake cycle remains poorly understood, but Tim Wright explains the exciting advances that have been made using radar interferometry with data from satellites. Detailed maps of the warping of the earth surface can now be obtained for the first time, and they provide remarkable observations of the earthquake cycle. Dominik Weiss, Malin Kylander and Matthew Reuer review the environmental and human impact of lead. Lead has been mined since ancient times, but by 1983 human activities accounted for $\sim 97\%$ of the global mass balance of lead. The amounts may have decreased since then, but the release of lead into the environment has also provided a geochemical tracer providing new insights into its fate and transport within marine and terrestrial systems. The move to clean up automobile emissions has resulted in a considerable demand for platinum in the manufacture of catalytic converters, and Hazel Pritichard reviews likely sources of platinum and palladium. They remain rare in the rocks of the Earth's surface, but significant amounts are accumulating in our cities from where it may be possible to recycle them. The volume culminates with a major vision for the future, an objectively optimised earth observation system with integrated scientific analysis. David Lary and Anuradha Koratkar show how this would dynamically adapt the what, where, and when of the observations made in an online fashion. It might change some of the ways in which we do science, and be used for a wide range of earth and environmental science observations, even including perhaps the sites of likely malaria outbreaks.

Chris Hawkesworth, FRS
Professor of Earth Sciences,
University of Bristol, England

Profiles

Editors

Peter Sammonds

Department of Earth Sciences, University College London, England

Peter Sammonds has been Professor of Geophysics at UCL since 2001. His research aims are to investigate the mechanics of the Earth's crust and ice sheets by studying the fundamental physics and mechanics of geological materials, particularly directed towards studying the impacts of climate change and natural hazards. He was a Royal Society University Research Fellow from 1992 to 2001 and he is on the Editorial Board of the Philosophical Transactions of the Royal Society.

J. Michael T. Thompson, FRS

Department of Applied Mathematics & Theoretical Physics,
Cambridge University

Michael Thompson was born in Yorkshire in 1937, and attended Hull Grammar School. He graduated from Cambridge with 1st class Honours in 1958 [ScD (1977)]. He was a professor at University College London and was appointed Director of the Centre for Nonlinear Dynamics in 1991. His fourth book, *Nonlinear Dynamics and Chaos* [2nd edn, Wiley (2002)], has sold 14 000 copies. Michael was elected a Fellow of the Royal Society in 1985, and served on the Council. He won the Ewing Medal (Inst. Civil Engineers) in 1992, the IMA Gold Medal for mathematics in 2004, and was a Senior SERC Fellow. Since 1998, Michael has been Editor of *Phil. Trans. R. Soc.* Michael is Emeritus Professor (UCL) and a Fellow at *DAMTP*, Cambridge. Married with 2 children and 8 grandchildren, he enjoys astronomy with his grandchildren, wildlife photography and badminton.

Authors

Dario Alfe   *The Earth's Core and Geodynamo*

Department of Earth Sciences, University College London, England and INFM DEMOCRITOS, National Simulation Centre, Trieste, Italy

Dario Alfe was born in 1968 in Napoli, Italy. He is married with two children. He graduated in physics from the University of Trieste, Italy in 1993 and gained a PhD at the International School for Advanced Studies, Trieste. He became a Royal Society University Fellow at University College London in 2000. He won the Philip Leverhulme Prize for outstanding young scientist in 2002.

Eiichi Fukuyama   *Rupture Dynamics of Earthquakes*

National Research Institute for Earth Science and Disaster Prevention, Tsukuba, Japan

Eiichi Fukuyama is a senior researcher at National Research Institute for Earth Science and Disaster Prevention working on *in-situ* stress measurements near the earthquake fault zone. He obtained his bachelor's, master's and PhD from Kyoto University. Much of his research has been on seismic waveform inversion and more recently on modelling dynamic earthquake rupture. He has worked in France, Italy and the USA.

Chris Kilburn   *Giant Catastrophic Landslides*

Benfield Hazard Research Centre, Department of Earth Sciences,
University College London, England

Christopher Kilburn is a Senior Research Fellow and Deputy Director of the Benfield Hazard Research Centre. He is a specialist in modelling geophysical hazards, notably the emplacement of sturzstroms and lava flows, as well as forecasting volcanic eruptions and improving the awareness of hazards among vulnerable populations.

Karen Kohfeld   *Dust in the Earth System*

School of Earth & Environmental Sciences, Queens College, NY, USA

Karen Kohfeld is assistant professor at Queens College of the City University of New York. Her research has focused on using global palaeoenvironmental datasets with Earth system models to understand the dominant processes and feedbacks controlling glacial-interglacial climate change. A believer in moderation, she only has two cats.

Anuradha Koratkar   *Objectively Optimised Earth Observation*

NASA Goddard Space Flight Center, MD, USA

Anuradha Koratkar obtained her PhD in Astronomy form the University of Michigan in 1990. She has worked as in instrument scientist at the Space Telescope Science Institute on the Hubble Space Telescope.

Malin Kylander   *Global Geochemical Cycle of Lead*

Department of Earth Science & Engineering, Imperial College, London, England

Malin Kylander grew up in Montreal and Vancouver, Canada. She attended McMaster University in Hamilton, Ontario where she received an Honours BSc in Biology in 1999. She earned an MSc in Applied Environmental Techniques from Chalmers University of Technology, Göteborg, Sweden in 2002. In 2003 she moved to Imperial College London where she is currently completing her PhD in Environmental Geochemistry. Her work looks at lead isotope in peat bogs and examining natural and anthropogenic forcing of isotopic signals. She enjoys dogs, hiking and planting trees.

David Lary   *Objectively Optimised Earth Observation*

NASA Goddard Space Flight Center, MD, USA

David Lary is a senior research scientist at the NASA's Global Modelling and Assimilation Office. He previously held a Royal Society University Research Fellowship in chemical data assimilation at the Centre for Atmospheric Science, University of Cambridge and an Alon Fellowship and a Senior Lectureship at Tel-Aviv University, Israel. He has developed advanced photochemical schemes for inclusion in atmospheric models, most recently in data assimilation models.

Yadvinder Malhi   *Carbon in the Atmosphere*

Department of Geography & Environment, University of Oxford, England

Yadvinder Malhi read Natural Sciences at the University of Cambridge and studied for a PhD in meteorology from the University of Reading. His research interest in tropical forests began as a researcher at the University of Edinburgh. Currently he is a Royal Society

University Research Fellow at the Oxford University Centre for the Environment. His research focuses on how the physiology, structure, biomass and dynamics of tropical forests are controlled by climate and soils, and how these features of the forest may respond to ongoing atmospheric change. He is co-founder of the RAINFOR project in South America and Africa and co-ordinator of the EU programme which trains students from across Amazonia in ecological field science techniques. He is co-editor of the book *Tropical Forests and Global Atmospheric Change.*

Tim Minshull   *The Break-Up of Continents*
National Oceanography Centre, Southampton, England

Tim Minshull has a degree in physics from the University of Cambridge and an MSc in geophysics from the University of Durham. He completed his PhD in marine geophysics at Cambridge in 1990. After a short period as a lecturer in geophysics in Birmingham, he returned to Cambridge in 1991, where he spent a further eight years, first as a research associate and then as a Royal Society University Research Fellow. In 1999, he moved to the National Oceanography Centre, Southampton, where he joined the academic staff. His main research interests are in the deep structure of rifted continental margins and in methane hydrates in marine sediments; in pursuit of these interests he has led or participated in fifteen research cruises since 1992.

Cathryn Mitchell   *Space-Plasma Imaging*
Department of Electrical Engineering, University of Bath, England

Originally from Staffordshire, Cathryn Mitchell studied at the University of Wales Aberystwyth, where she was awarded her PhD in 1996 in 'Tomographic Imaging of Ionospheric Electron Density'. She received the *Sir Granville Beynon Prize* from the University and the *Blackwell Prize in Geophysics* from the Royal Astronomical Society for her PhD work. Subsequent research involved the application of this work to HF communications systems. She was appointed to her first lectureship in September 1999 at the University of Bath where she has set up a new research area using GPS satellite signals to image the troposphere, ionosphere and plasmasphere. In 2003 she was awarded an EPSRC Advanced Research Fellowship to study the effects of the ionized atmosphere on navigation satellite signals. Out of work she enjoys walking and horse riding in the Wiltshire countryside.

Francis Nimmo   *The Earth's Core and Geodynamo*

Department of Earth Sciences, University of California
Santa Cruz, CA, USA

 Francis Nimmo is an assistant professor in Earth Sciences at the University of California Santa Cruz. He obtained his PhD from Cambridge University on the volcanic and tectonic evolution of Venus in 1996. Thereafter he was a Junior Research Fellow at Magdalene College, Cambridge; a Royal Society University Research Fellow at University College London; and an Adjunct Assistant Professor at UCLA before taking up his current position. His primary interests are the thermal and orbital evolution of solid solar system bodies, including the Earth.

Hazel Prichard   *Platinum and Palladium Occurrences*

School of Earth, Ocean & Planetary Sciences, Cardiff University, Wales

 Hazel Prichard graduated in geology and physical geography at Hull University; was awarded a PhD from the University of Newcastle and gained an MBA from the Open University in 1996. She was awarded a Royal Society University Fellowship to study platinum in ophiolite complexes in 1986. She was appointed as lecturer in Earth Sciences at Cardiff University in 1996. She was awarded a 4-year Royal Society Industrial Fellowship in 2000 for which the host companies were MinMet and Rio Tinto and applied academic models for Pt and Pd concentration to practical exploration with these companies.

Currently she is studying the transport and concentration of precious metals in the urban environment, funded by the Royal Society Senior Brian Mercer Award for 2004.

Dave Reay   *Price of Climate Change*

*Institute of Atmospheric & Environmental Science,
University of Edinburgh, Scotland*

 Dave Reay was born in Fleet, Hampshire, in 1972. He studied marine biology at Liverpool University and gained a PhD at Essex University studying the response of Southern Ocean algae to temperature change. He then studied the impact of land-use on the soil methane sink. In 2001 he moved to Edinburgh University to investigate greenhouse gas emissions from agriculture. He is author of Climate Change Begins at Home: Life on the Two-way Street of Global Warming (Macmillan) and editor of the leading climate change website GreenHouse Gas Online (www.ghgonline.org). His loves include Test Match Special, writing stories for his daughter and composting.

Matthew Reuer   *Global Geochemical Cycle of Lead*

*Department of Earth Science & Engineering, Imperial College,
London, England*

 Matthew K. Reuer recently joined the Environmental Science Program of the Colorado College, where he guides undergraduate research projects and teaches courses in Environmental Science. He received his doctoral degree from the MIT/ WHOI Joint Program in Oceanography in 2002, focusing on the environmental geochemistry of anthropogenic lead under the supervision of Edward Boyle. In 2002 he received a Harry Hess Postdoctoral Fellowship from Princeton University, studying oxygen triple isotopes with Michael Bender. Matt's research interests include aquatic environmental chemistry and the development of novel isotopic and elemental tracers; his teaching interests

include undergraduate Biogeochemistry and Environmental Chemistry. In his spare time Matt greatly enjoys mountaineering, alpine skiing and trail running.

Andy Ridgwell   *Dust in the Earth System*

Department of Earth & Ocean Sciences, University of British Columbia, BC, Canada

Andy Ridgwell is an assistant professor and is 'Canada Research Chair in Global Process Modelling'. Although in practice spending most of his time tending to the every need of 5 cats, his research addresses fundamental questions surrounding the past and future controls on atmospheric CO_2 and the role of feedbacks in the climate system. His weapon of choice in this endeavour is an Earth System Climate Model.

Simon Turner   *Magmatic Processes Occurring Beneath Island Arc Volcanoes*

Department Earth & Planetary Science, Macquarie University, NSW, Australia

Born in Melbourne, Australia, Simon Turner studied at the University of Adelaide where he graduated in Earth Sciences in 1986, and obtained his PhD in 1991. He moved to the Open University in England in 1992 as a NERC researcher, investigating continental flood basalts and the dynamics of the Tibetan plateau. In 1995 he was awarded a Royal Society University Research Fellowship and moved to the University of Bristol in 2000. In 2002 he was awarded the Lyell Fund by The Geological Society of London and received a Federation Fellowship from Australia. He is currently Federation Fellow and Professor of Geochemistry at Macquarie University in Sydney.

Richard Twitchett   *Late Permian Mass Extinction*

Department of Earth & Planetary Science, University of Tokyo, Japan

Richard Twitchett graduated in geology and biology from Bristol University in 1993. He went on to complete a PhD on Early Triassic marine palaeoenvironments at Leeds University, based primarily on studies of the Permian-Triassic record of northern Italy. Since then, he has continued to study the facies, fauna, ecology and environments of the Permian-Triassic extinction-recovery interval during research positions in the UK, USA, Netherlands and, most recently, Japan. He gained a permanent position at the University of Plymouth in 2003. Specific research interests include the question of size change through extinction events. Alongside, he has learnt to play craps in the casinos of Las Vegas, snorkeled among the stromatolites of Shark Bay and climbed Mt. Fuji. He is an occasional spin bowler, life long bridge player and aspiring kendoist.

Dominik Weiss   *Global Geochemical Cycle of Lead*

Department of Earth Science & Engineering, Imperial College, London, England

Dominik Weiss was born in Basle, Switzerland and graduated with a degree in natural sciences at the ETH. He completed his PhD work under the supervision of William Shotyk at the University of Berne reconstructing atmospheric deposition of Pb in Europe using peat bogs archives. After three months climbing the mountains of East Africa, he spent one and a half years at MIT working as post-doctoral research assistant with Edward Boyle on the Pb isotope geochemistry of ocean surface waters. In 2000, he was appointed as a lecturer in Environmental Geochemistry at Imperial College London. Recent work has been focussing on the low temperature isotope geochemistry of trace metals, mainly Cu, Zn and Fe, and on the geochemistry of trace metals with respect to air, water and soil quality.

Tim Wright   *Remote Monitoring of the Earthquake Cycle*
Department of Earth Sciences, University of Oxford, England

Born in 1974, Tim Wright graduated from Cambridge University in 1995 in natural sciences. After spending a year working in a day centre for adults with learning difficulties, he returned to university, obtaining an MSc in remote sensing from the University of London (intercollegiate) in 1997. From 1997 to 2000 he completed his DPhil at Oxford University on InSAR studies of active tectonics in Turkey. He was then awarded a NERC postdoctoral research fellowship, also at Oxford, to study continental shear zones. He has been a Royal Society University Research Fellow since October 2004.

CONTENTS

Introduction

Peter Sammonds

Department of Earth Sciences, University College London, England

The earth sciences are enjoying a renaissance. Global issues in the earth sciences, such as building a tsunami warning system or burning of fossil fuels, are discussed at meetings of world leaders; there is a strong level of popular interest as witnessed by the public response to the acclaimed BBC series, "Walking with Dinosaurs"; while debate about the Permian extinction amongst intellectuals has not been so intense for over a century. Elsewhere the earth sciences are not seen in a positive light, caught in a political storm as "intelligent design" is pitted against Darwinism in the educational boards of the USA and are the target of environmentalists, because of the damage caused by the mining and oil industries. The earth sciences deal with the dynamics and evolution of Earth's crust and the life it supports, its interactions with the ocean-atmosphere system and the Earth's deep interior, and the Earth's near-space environment. It is because the earth sciences deal so directly with our "life support system" they are at the centre of intellectual and political controversy. Of course this is not new. Charles Lyell, one of the founders of geology in the nineteenth century, was not only embroiled in the controversies over the science of the evolution of the Earth and of life, but also the politics. What is new today is the urgency which some of these issues need to be addressed.

The modern earth sciences cover a huge subject range — from earthquakes to global warming. But where are the advances being made and which topics do we need to keep abreast of? A generation ago the principal paradigm driving research in the earth sciences was plate tectonics. The construction of plate tectonic theory was surely one of the great intellectual achievements of the twentieth century. The idea of a dynamic Earth has made as big an intellectual impact as any scientific discovery. Indeed, the triumph of plate tectonic theory seems so complete is there anywhere else for it to go? Some scientists have argued that it is to the terrestrial planets of the Solar System we need to look for breakthroughs of the same significance. However in this book we see that studying the dynamics of the Earth is still a key research area where advances are being made. But perhaps the biggest issues driving research in the earth sciences are about understanding the complexities of environmental change and environmental hazards. This shift is reflected in this book where these issues feature prominently.

One of the key developments in the earth sciences has been a move away from a reductionist approach, where the earth sciences can be reduced in their supposed basic components of the disciplines and sub-disciplines of physics, chemistry and biological. What drove this was a perceived failure of the reductionist approach to deliver on its promises of understanding the complexity at the Earth's surface; examples of which are earthquake prediction and safe disposal of nuclear waste. In the 1990s we saw the collapse of the US and Japanese earthquake prediction programmes and the UK government's refusal to sanction the building of a new underground radioactive waste repository. These can be seen as consequences of the inherent complexity in the mechanical and physical behaviour of the crust, which cannot be solved by classical physics. What has however arisen is an appreciation that the Earth is a complex system, which has to be treated in a holistic way. This is an earth system science approach that is as interested in the interaction between processes as much as in the processes themselves. In the climate system these are called feedbacks — but in these feedbacks, such as cloud formation, that are the principal controls. This change in perception in the earth sciences is seen in this book. Alongside the rise of the treatment of the earth as a complex system, have been the development of the tools that have allowed earth scientists to do this: Improvements in earth observation and particularly satellite remote sensing; improvements in computing power to allow ever more detailed simulations; improvements in the resolution of laboratory analytical techniques. The rise of the internet, ever lower travel costs and improved infrastructure have allowed global inter-comparisons to be made ever more readily. The globalisation of science has also brought us successful large international collaborations such as the ocean-drilling programme (IODP), on a scale which no one country could fund, but can be accessed by scientists worldwide.

The articles in this book have been written by earth scientists from a broad range of backgrounds specialising in a diverse range of research subjects. Our key criterion has been to accept articles only from world-class scientists. But a volume of this nature cannot hope to be comprehensive in capturing all the advances in the earth sciences. The contributing authors are mostly younger scientists, at the forefront of their subjects: They are indeed the future of the subject. The articles address the key areas of advances in the earth sciences in:

- Environmental change
- Dynamics of the Earth
- Applied earth sciences

Editors to a certain extent are at the mercy of the scientists who choose to contribute, or at least to those whom the editors have managed to persuade to break from their research to explain their field to a broader readership. The geographical distribution of the authors does reflect the provenance of this book in the articles originally published in the Philosophical Transactions of the Royal Society, London — the world's longest running scientific journal. There are some obvious gaps: An example is physical volcanology. Volcanic eruptions have been predicted and evacuations carried out following prediction by scientists. There has been a huge increase in the understanding of physical volcanology to facilitate this. The very recent development of landscape evolution and modelling as a subject area is missing. Satellite remote sensing of the cryosphere is not covered — although this might be seen as being more directly linked to meteorology. We do not have an article on Japan's huge computer, the "Earth Simulator". One of the new hot topics of Eocene climate change, 30 millions years ago, is not dealt with. We have no report on how laboratory experiments are transforming our understanding of the Earth's mantle. However we believe the book does give a flavour of the advances currently being made in research on the earth system.

Environmental Change

Environmental change is now one of the key drivers of research in the earth sciences. Geologists have of course always studied environmental change. "The present is the key to the past" was the dictum of James Hutton in the eighteenth century. This dictum was taken up by his successors such as Lyell who could demonstrate that sedimentary rocks were deposited gradually in similar environments to those of today: Old red sandstones originated from deserts; limestones which might cover them were laid down in shallow seas; whilst sandy-clay layers were the run-out of giant submarine flows bringing material from the continental shelves into the oceans, triggered by tectonic activity. The changing environment is recorded in the geological record. What is new is that it is no longer just geologists, but physical geographers, ecologists, meteorologists and oceanographers too who now work on environmental change. The resurrection of a nineteenth century idea that carbon dioxide in the atmosphere is a determinant of our climate, along with evidence of rapid past climate change from ice and ocean sediment and cores, satellite measurements of the global temperature distribution and ice extent and the availability of sophisticated computer programs written to predict the "nuclear winter", created the intellectual environment in which concerted efforts could be made to predict what the future climate holds for us, and its consequences. This research cuts across traditional scientific boundaries, but undeniably forms the most dynamic part of the today's earth science.

The book opens with an article by Dave Reay on the price of climate change. It is probably not possible to start with a more contentious or political scientific issue. He argues that the combined uncertainties in both the science and the economics of climate change are so large that a limitless range of outcomes is possible. However he examines existing cost-benefit analyses and concludes that there are host of abatement strategies that are able to deliver significant carbon dioxide emission reductions at little or no net cost when the full economic impacts of climate change are considered. Yadvinder Malhi examines carbon in the atmosphere and terrestrial biosphere. He argues that the anthropogenic perturbation of the global biogeochemical cycle is so large, that understanding and managing its effects are amongst the most pressing issues of the twenty-first century. One of the key issues is how much carbon dioxide is absorbed by vegetation — the "terrestrial carbon sink". He proposes that controlling deforestation and managing forests has the potential to play a significant role in stabilising atmospheric carbon dioxide concentrations. Andy Ridgwell and Karen Kohfeld, continuing this theme of the need to treat the earth system as a whole, investigate the biogeochemical linking of the land, air and sea. Specifically they examine the role of dust in the earth system. The atmospheric transport of mineral dust is a key pathway for the delivery of nutrients essential to plant growth not only on land, but also more importantly in the oceans. The stimulation of plant productivity by these nutrients controls carbon dioxide take-up from the atmosphere, so the whole system is linked. Finally Richard Twitchett discusses the Late Permian mass extinction. This was a biological catastrophe in a greenhouse world and a salutary reminder of what has happened in earth history.

Dynamics of the Earth

Research into the evolution and dynamics of the Earth is a major research area in the earth sciences, however this is not confined to the solid Earth. Indeed some of the most active research is studying the Earth in its near-space environment. But the widespread acceptance of plate tectonic theory has not meant diminished interest in the solid Earth. Within the broad plate tectonic framework there is the need to understand the details of the rifting of continents and the formation of ocean basins; the ascent of magma in the formation and eventual eruption of volcanoes; the dynamics of the Earth's deep interior and how it is coupled to the Earth's surface evolution. Nowhere has research been more promising than advances in understanding the Earth's iron core. Its enigmatic behaviour is at last giving way to the application of new models of magneto-hydrodynamics coupled with a far better understanding of the core's composition through computation mineral physics. There is also a societal need to understand earth dynamics driven by the need to assess and mitigate earthquake hazard. "Is this even possible?" is a question that drove much theoretical research in crustal dynamics at the end of last century. The problem is not so much that crustal dynamics cannot be modelled, but that the expectation of being able to predict behaviour of the crust during a tiny time interval, of far less than a human lifespan, and in tiny area, covering that of a suburb, is probably unrealistic, when the driving forces operate on geological time and spatial scales. But even here, new techniques are coming to bear on this problem, which may make some resolution possible.

This section starts with a review by Cathryn Mitchell of research on the Earth's dynamics in relation to its environment in space and in particular the Earth's ionosphere. Echoing the need for the Earth to be treated as an integrated system, she says it is becoming clear that to produce "space weather" forecasts new research projects are needed to link together models of the entire solar-terrestrial system, including the Sun, solar wind, magnetosphere, ionosphere and thermosphere. Moving to the solid Earth, Eiichi Fukuyama argues that we are now able to simulate the dynamic rupture process of real earthquakes, once the fault geometry, stress field applied to the fault, and friction law on the fault surface have been provided. Simon Turner looks at the processes of magma formation, ascent and storage in shallow magma chambers prior to eruption beneath island arc volcanoes. The details of these processes can be followed by high resolution dating of between 100 to 10 000 years ago using radioactive isotopes. Tim Minshull examines the new theories on the the break-up of continents and the formation of new ocean basins necessitated by observations of mantle rocks at continental margins. Francis Nimmo and Dario Alfe review recent advances in understanding the properties and evolution of the Earth's core and geodynamo. They focus on the properties of the core-forming materials and how core dynamics generates the Earth's magnetic field (the geodynamo). This article then links back to the first in this section.

Applied Earth Science

The earth sciences have always had a strong applied side. Indeed the world's first geological map was prepared by William Smith who earned his living as a surveyor for constructing canals. Geologists and geophysicists are central to the mining and petroleum industries, which underpin our modern society, but nowadays applied earth scientist are as likely to involved in mitigating natural hazards and controlling pollution. In a complex system, making assessments, which satisfy public and political expectations, is testing. For instance, even if we can understand the dynamics of a major fault, earthquakes unfortunately continue to occur on previously unrecognised faults.

Chris Kilburn describes a new understanding of one of the most devastating natural hazards: Giant landslides. They are caused by the collapse of whole mountainsides, which feed giant landslides that travel kilometres within minutes. Both their size and speed prevent effective hazard mitigation after collapse. Tim Wright reports on one of the most exciting advances in the earth sciences, that of using satellite radar interferometry for remote monitoring of the earthquake cycle. For the first time, detailed maps of the deformation of the Earth's surface during the earthquake cycle can be obtained with a spatial resolution of a few tens of meters and a precision of a few millimetres. In his article, he reviews some of the remarkable observations of the earthquake cycle already made using radar interferometry and speculates on breakthroughs that are tantalisingly close. Dominik Weiss, Malin Kylander and Matthew Reuer address the human influence on the global geochemical cycle of lead. Human activity dominates this cycle as a result of large lead consumption over human history and accounts for an estimated 97% of the global mass balance of lead. The overall burden of anthropogenic lead emissions has decreased but new pollution sources have become important meaning it is still a global problem. Lead is not biodegradable and finds its way into the ecosystem. Hazel Prichard examines other heavy metals, platinum and palladium and their natural and artificial occurrences worldwide. The catalytic converters used to reduce poisonous exhaust emissions from cars use platinum and palladium, which are now accumulating in our cities and approaching concentrations found in natural deposits. In rounding off the volume, David Lary and Anuradha Koratkar look forward to an objectively optimised earth observation system, which will dynamically adapt the what, where, and when of the observations made in real time to maximise information content and minimise uncertainty. They describe a prototype system applied to atmospheric chemistry. An example of its application might be the remote identification of sites of likely malaria outbreaks, the early identification of potential breeding grounds for mosquitoes and sites to apply larvicide and insecticide. Optimising the response would reduce costs, lessen the chance of developing pesticide resistance and minimise the damage to the environment. They describe in effect the practical application of earth system science.

SECTION 1

ENVIRONMENTAL CHANGE

The Price of Climate Change

David S. Reay

Institute of Atmospheric and Environmental Science
University of Edinburgh
Crew Building, West Mains Road, Edinburgh, EH9 3JN
United Kingdom
David.Reay@ed.ac.uk

Economics and climate change science have a lot in common. Both rely on sound predictions of what the future will bring, these being largely based on what has gone before. Just, though, as you can only make an educated guess at what the housing market will do next year, you cannot be wholly sure how emissions of greenhouse gas will increase in years to come, and exactly how the planet's climate will then react. Bring all these unknowns together, by attempting on the one hand to calculate the economic impacts of climate change and on the other the costs of climate change mitigation, and the range of possible outcomes is almost limitless. Given such uncertainty, both the environmental lobby and the oil lobby can use economic arguments to justify their differing stance on climate change mitigation.

Existing cost-benefit analyses of greenhouse gas reduction policies are examined, with a view to establishing whether any such global reductions are currently worthwhile. The potential for, and costs of, cutting our own individual greenhouse gas emissions is also assessed. I find that a host of abatement strategies are able to deliver significant emission reductions at little or no net cost when the full economic impacts of climate change are considered. Additionally, I find that there is great potential for individuals to simultaneously reduce their climate impact and save money. I conclude that the use of economics to excuse political inaction on greenhouse gas emissions is not justified.

1. Introduction

Debate over how, when, and even whether, human-made greenhouse gas emissions should be controlled has grown in its intensity even faster than the levels of greenhouse gas in our atmosphere. Many argue that the costs involved in reducing emissions outweigh the potential economic damage of human-induced climate change. Exaggerated claims and forecasts of climate meltdown in the media have naturally given rise to both fervent belief in, and hardened scepticism of, measures to limit greenhouse gas (GHG) emissions.

Many of those who argue against the need to reduce GHG emissions cite economic analyses as proof that such measures would not be 'cost-effective'. Though such economic arguments might sometimes be dismissed on moral or ideological grounds [Brown (2003)], we live in a world where the importance of money cannot be ignored. Cost-benefit analyses of climate change are prone to numerous pitfalls. Firstly, the economic time-horizons applied are commonly of the order of years or decades, whilst the climate change impacts, driven by the enhanced greenhouse effect, are likely to continue and intensify for centuries. Secondly, the true 'externalities' of each tonne of greenhouse gas emitted — the costs of a flooded home in Bangladesh, a failed harvest in the Sudan or deaths from a heatwave in the South Eastern US — are difficult to quantify. Nevertheless, cost-benefit analyses continue to inform governmental policy on climate change mitigation around the world. They are also beginning to take more account of the time-horizon and externality issues.

Here I examine some of these 'cost-benefit' aspects of global warming and GHG abatement, to assess whether climate change mitigation can ever be cost effective, and how the full inclusion of externalities can affect the balance between the impact and mitigation costs. In addition to discussing costs on a national and international scale, I will also look at the costs and benefits of GHG reductions in our own day-to-day lives. How much GHG do most of us produce in a lifetime? Can we easily make large reductions? And, if so, will these reductions cost us money? Two theoretical Londoners, one who lives a comparatively 'GHG ignorant' life, and the other who lives a more 'GHG aware' life are compared.

2. The Rise and Fall of Mitigation Costs

In 1997 over 160 nations came together in Japan to discuss the intensifying problem of climate change and the burgeoning greenhouse gas emissions apparently to blame. The result of this was the Kyoto Protocol — a set of emission targets for developed-world nations designed to cut global GHG emissions by over 5% compared to emissions in 1990. Four years later though, and George W Bush withdrew the US from Protocol, throwing its survival into doubt. In justifying their abandonment of the US commitment to the Kyoto protocol George Bush's spokesman, Ari Fleischer, stated "...it is not in the United States' economic best interest" [Kleiner (2001)]. As the world's biggest GHG polluter, it is vital that the US be involved sooner rather than later, in efforts to tackle global climate change. However, certain cost-benefit analyses do seem to bear out some of the US administration's objections to GHG cuts.

Some studies have indicated that Kyoto Protocol-like GHG emission limits may have potentially large economic costs [e.g. Nordhaus (1994); Lutter (2000); Nordhaus and Boyer (2000)]. With more extensive reductions leading to an ever-escalating cost per tonne of GHG saved [Lomborg (2001)]. There are also many cost analyses of specific GHG abatement strategies, which have shown that cuts are often possible at little or no cost. These include new energy technologies [Morthorst (1998); Brown *et al.* (1998)], solid waste treatment [Ayalon *et al.* (2001)], biogas use [Smith *et al.* (2000)], afforestation and reforestation [De Cara and Jayet (2000); Baral and Guha (2004); van Kooten *et al.* (2004)] and land-management [De Jong *et al.* (2000)]. Even in the US, significant reductions in GHG emissions are possible at essentially no net cost to the US economy [Brown (2001)].

The implementation of large-scale GHG reduction schemes here in the UK is already underway. Nationally the government aims to replace 10% of our energy requirements with energy from renewable resources by 2010. Cost-energy analyses of such schemes in Scotland have indicated great potential for wind, wave and tidal power at costs of around 3 p/kWh [Scottish Executive (2001)] — a price comparable to that for fossil fuel-powered energy generation.

Technological development has always been a key area to consider in economic analyses of GHG abatement [Nordhaus (1994)]. Future technological developments can be very difficult to predict and consequently so can their impact on abatement costs. One aspect of such technological change is that of the abatement policy itself driving further technological development. In his analysis of this so-called 'induced innovation' for wind power in Denmark, Rasmussen [2001] showed that such 'added value' may significantly reduce abatement costs.

The financial costs of implementing the Kyoto protocol may also be significantly reduced by the use of a 'multi-gas control' strategy [Reilly *et al.* (1999)]. John Reilly and his group at MIT showed the potential savings possible using such a multi-gas approach, the reduction cost per tonne of carbon equivalent being markedly lower when a range of GHGs are targeted, rather than just CO_2.

Greater flexibility in the timetable of GHG cuts under Kyoto may lower overall abatement costs [Toman (1999)]. While 'fringe benefits' of GHG reduction strategies may also lead to reduced implementation costs. Cutting coal combustion, for instance, will not only reduce GHG emissions but will also lead to savings in public health costs arising from air pollution [Butraw *et al.* (1999); De Leo *et al.* (2001)].

It is clear that the cost of climate change mitigation can be reduced in a host of ways. There remains, though, the 'bottom-line' question: Is the final cost of mitigation likely to be lower than the cost of climate change without mitigation?

The answer appears to be yes.

In their analysis of implementation of the Kyoto protocol in Italy, De Leo *et al.* [2001] demonstrated that, where costs incurred in rectifying damage to human health, material goods, agriculture and the environment (the externalities of GHG emission) are included with those of energy production, the cost argument for inaction breaks down. As they state, "...the social and environmental costs of GHG emissions are not included in company balance sheets, but must be included in national balances". Even in the high-emitting US, such inclusion of externalities in economic analyses indicate that the net costs of Kyoto for the nation's economy are likely to be insignificant (less than 1% GDP) [Barker and Ekins (2004)].

The question of costs incurred as a direct result of human-made greenhouse gas emissions has sparked an intense debate in recent months. Late in 2004, Peter Stott and his co-authors [Stott *et al.* (2004)] took the brave step of suggesting a direct link between the devastating effects of the southern European heatwave in 2003 and human-made greenhouse gas emissions. Assuming our ability to link emissions and specific climate change impacts will increase, some of the world's largest emitters of greenhouse gas may face a litigious and very costly future.

In January 2004 Friends of the Earth (FOE) published a report on the 'Climate footprint' of perhaps the planet's largest corporate emitter of greenhouse gas: Exxon [FOE (2004)]. This report fell short of specifying a financial cost of Exxon's emissions over its 140 year history, and so the total amount any damage claims made against them might run into. There are, though, some published estimates of the cost of damage caused by the emission of each tonne of greenhouse gas. These estimates may be broad, given the uncertainties in climate change impacts and time horizons, but they recognise the social costs of greenhouse gas emission in terms the economists are familiar with: Money.

The range of estimates of this type goes from around to £3 to £80 per tonne of carbon emitted today. In a review of such estimates, the UK Government Economic Service [Clarkson and Deyes (2002)] cites the value calculated by Eyre *et al.* [1999] — £70 for each tonne of carbon emitted (at year 2000 prices) — as representing the most sophisticated. This estimate goes beyond many of the others in that it encompasses a much wider range of climate change impacts, and means £19 of damage for every tonne of CO_2 that is currently being emitted.

Combining this estimate with data available on the greenhouse gas emissions of major corporations, we can therefore get some idea of the climate impact costs to which some of our biggest corporations may one day be held liable:

Table 1   Annual CO_2 emissions of five leading corporations during 2000 and the associated costs of these emissions based on an impact cost of £19 per tonne CO_2.

Corporation	Emissions (MMT CO_2) in 2000	Cost (million £)
Exxon	122.9	2335
Shell	101	1919
BP	83.7	1590
Ford Motors	9.3	176
IBM	3.1	57

Using the estimate of £19 of damage per tonne of CO_2 emission it is apparent that annual costs arising from the emissions of these corporations may run into many millions, and in some cases billions, of pounds worth of damage globally.

With such emissions litigation a possibility, significant potential to reduce mitigation costs, and an increasing number of studies indicating these costs are outweighed by the costs of inaction, the immediate implementation of GHG cuts is hard to dismiss [Howarth (2001)]. While some critics of the Kyoto protocol argue that the GHG reductions it proposes are woefully inadequate, the United Nations Framework Convention on Climate Change (UNFCCC) maintains that the Kyoto protocol is only the first step in the process of tackling global warming. The eventual size of GHG reduction will need to be many times that outlined in the current Kyoto protocol if the most severe climate change impacts are to be avoided. However, the very high costs predicted for such wide-reaching GHG emission cuts [McKibbin and Wilcoxen (2004)], do not mean that immediate action, albeit on a relatively moderate scale of the Kyoto Protocol, should not be taken now.

3. Place Your Bets

As I've said, the key difficulty faced in predicting both the economic costs of global warming and the costs of GHG reduction strategies is the, often large, degree of uncertainty inherent in such predictions. On a time scale of hundreds of years, predictions involve a significant amount of guesswork, but such time-scales are short in terms of global climate dynamics. Even the most convincing economic argument against cuts in GHG emissions is essentially gambling on our future, betting against the possibility of catastrophic climatic events caused by global warming.

The UNFCCC promotes action on global warming in spite of the uncertainty surrounding its precise extent and impacts, based on what is called the 'precautionary principle' [see Kuntz-Duriseti (2004)]. This principle basically allows the international implementation of GHG reduction strategies like the Kyoto Protocol before there is absolute scientific certainty, based on avoidance of serious or irreversible damage to the environment [UNFCCC (2001)]. The economist William Nordhaus asserts that, though "a massive effort to slow climate change would be premature" we must be alert to the possibility of "catastrophic and irreversible changes" [Nordhaus (1994)]. It seems then, that we are faced with two options: Either we do nothing to reduce GHG emissions, and so gamble on the resulting effects being within those predicted by existing models, OR, we insure ourselves and future generations against the possibility of catastrophic climate change by cutting GHG emissions.

Overall, a start to the reduction of global GHG emissions seems not only to be economically viable, but also vital as a basis for any future international response to GHG driven climate change. The use of the UNFCCC's 'precautionary principle' appears entirely correct given the magnitude of the catastrophe climate change may bring about, not only for us, but also for our descendants.

As I write the Kyoto Protocol has just recently come into force (February 16th 2005). Russia's ratification late in 2004 meant that the two key criteria for the protocol's survival — that it involved at least 55 nations and that these nations were responsible for at least 55% of global GHG emissions — were met. The US though, itself representing 25% of global emissions, remains opposed the Kyoto Protocol. Kyoto then, remains a rather faltering first step in international action to tackle climate change. Nonetheless, it is a step in the right direction.

It may be frustrating to know how large a threat climate change poses but see an apparently very slow response by the international community. Each of us though, can take direct steps to fight climate change while we're waiting for the politicians to take concerted action. As members of a global society we each have a stake in, and responsibility for, the global environment. The UNFCCC emphasises the need to educate individuals about climate change, to try and change the way we think about our impact on the environment, both now and for future generations. Let's therefore examine such individual environmental impacts, the GHG emissions of a lifetime and, since economic matter as much to individuals as nations, the savings possible.

4. Greenhouse Gas Budgets for Individuals

It is in our own lifestyles that many of the most cost-effective reductions in GHG emissions can be made. If we add to this the 'bottom-up' effect such lifestyle changes could have on community, business and eventually government GHG policy then the huge importance of individual GHG emissions on a global scale is clear. To assess what kind of GHG emissions reductions might be possible, and the monetary cost of these reductions for typical westerners, let us consider two people born in modern day London. For the purposes of this comparison both our Londoners will live for 75 years, one living a 'GHG ignorant' life and the other a 'GHG aware' life. To avoid confusion we will call the two subjects Mr Black and Mr Meyer, respectively. GHG emission results of these analyses are shown in Fig. 1.

They are based on the five main sources — transport, holiday, household, food, and waste. Similarly, the associated financial costs over a lifetime for our two subjects are shown in Fig. 2. For details on GHG emission and monetary calculations for each stage of our subjects' lives, together with any assumptions made, see the Appendix section.

Though, for most of their respective childhood's, our two Londoners will not be able to determine their own lifestyles and so GHG emissions, we will initially explore how their parents can affect the GHG budgets of their children.

(a) Baby (Aged 0–2)

The first big 'GHG sensitive' decision our subjects' parents face is that of whether to use disposable or real nappies? Mr and Mrs Black decide to go

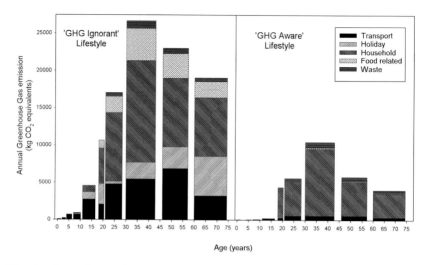

Fig. 1 Annual greenhouse gas emissions over a 75 year lifespan, for a greenhouse gas ignorant and greenhouse gas aware lifestyle.

for disposable nappies for their son, while Master Meyer's parents opt for 'real nappies'. The high energy cost for manufacture of disposable nappies, relative to that of both manufacture and cleaning of reusable nappies, leads to Master Black's nappies causing around 12 kg more GHG emission than the 'real' nappy option each year. The added environmental problems disposable nappies pose, due to their sheer volume and slow break down times, also results in increased local authority costs through refuse transport and landfill maintenance.

(b) Toddler (Aged 2–4)

After nappies, the next major 'GHG sensitive' decision our two sets of parents make is that of transport to nursery school. While the Meyers opt for taking their son to the nearby nursery by bike, the Blacks use their large four-wheel drive for all of these short journeys. Consequently, while going to nursery costs nothing in terms of GHG produced or fuel bought for the Meyers, the Blacks have to pay £55 extra on petrol and produce an extra 211 kg of GHG each year.

(c) Infant (Aged 4–7)

Now our two 'GHG guinea pigs' are old enough for infant school, they both qualify for a free bus service. However, the Blacks decide against using this

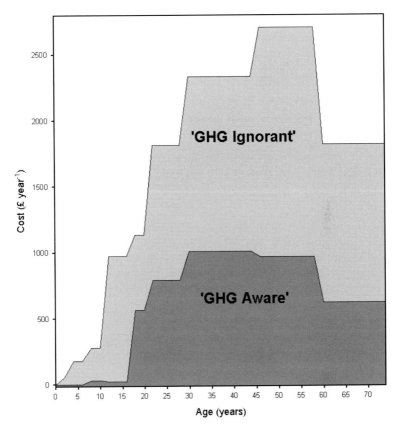

Fig. 2 **Annual financial cost associated with greenhouse gas emissions over a 75 year lifespan, for a greenhouse gas ignorant and greenhouse gas aware lifestyle.**

service and opt to continue using their four-wheel drive, despite these infant school trips (10 km) being around twice the distance of those to the nursery. This decision costs the Blacks an extra £176 on petrol and produces 677 kg of GHG each year. The Meyers do make use of the free bus service and so pay no extra money. The bus only produces about 53 kg of GHG to carry the young Meyer to and from infant school over the course of the year, a saving of more than half a tonne of GHG.

(d) Junior (Aged 7–11)

Now our two subjects are getting older their parents decide they are old enough to start going on summer holiday. As both families live on the outskirts of London they have relatively easy access to all major road, rail and air routes. The Meyers decide to spend their annual holiday in Plymouth, UK, while the Blacks opt for holidays in Paris. Travelling by train from London to Plymouth and back with the young Master Meyer costs an additional £29 and produces about 12 kg of GHG each year. Meanwhile, the Blacks clock up an extra 100 kg of GHG and a bill of around £94 to fly young master Black to Paris and back every summer.

During term-time the Blacks continue to use their four-wheel drive for school trips, while the Meyers still make use of the free bus service, so widening the divide in both GHG produced and associated costs between the two families.

For the first time in the lives of our young subjects begin to become directly responsible for some of their climate impact. Initially this takes the form of Master Black leaving the television, DVD player and video games console on for hours at a time when he isn't using them. This extra energy use adds up to an extra 120 kg of GHG each year and costs his parents about £6 each year. Young Master Meyer, on the other hand, usually switches off these appliances.

(e) Senior (Aged 11–18)

In this last period, before our two subjects will gain complete control of their lifestyle and its associated climate impact, they can already make quite an impact on their GHG emissions. Master Black now routinely leaves his computer, stereo, TV and DVD player on while they are not being used, so clocking up an extra 160 kg of GHG at a cost of £10 each year. He has also got into the habit of turning the home heating up to full blast instead of putting on a jumper. His use of an electric radiator in his room for an extra 2 hours each day during much of the year. Compared to Master Meyer, results in 700 kg of extra GHG emissions at a cost of £35 to his parents every year. For holidays during this period the Meyers travel to Chester each year by train, at a cost of £22.20 and producing 17.6 kg of GHG each time. The Blacks, on the other hand, fly to Cairo every summer, with the flights for our teenage Master Black costing £225 and producing over 1 tonne of GHG.

Both our subjects have also moved on to senior school in this period. Master Meyer again takes advantage of the free bus service to school, a journey that produces 211 kg of GHG over the course of each year. The

Blacks persist with using their car for these longer school runs, at an extra annual cost of some £706 and producing 2707 kg of GHG.

(f) Student (Aged 18–21)

Having left their family homes our two subjects have now become very much more responsible for their personal GHG emissions and climate impact. One of the first and most important decisions they make is their form of transport while students in London. Mr Meyer opts to use his bike and an all year public transport pass costing £264 and leading to annual GHG emissions of around 260 kg. Meanwhile, Mr Black chooses to buy and use a 7 year old hatchback for his transport. Purchase and maintenance costs aside, this option costs him £407 a year in fuel and produces over tow tonnes of GHG. For their holidays, Mr Black now travels each year to Bangkok at a price of £427, with the flights producing a massive 2745 kg of GHG each holiday. Mr Meyer instead travels by train to Bath at a cost of £26 — a trip producing only 10.5 kg of GHG.

The GHG emission and monetary savings possible in their rented student accommodation are fairly limited for both our subjects. As relatively 'low' energy users in London, each would normally produce around 4750 kg of GHG from 'household' sources, at a cost of nearly £300 in energy bills each year. However, Mr Meyer saves over £14 and around 300 kg of GHG simply by setting his home PC to 'sleep mode' for those times when he's not using it. He saves a further £10 a year in electricity bills and about 200 kg of GHG by replacing the two light bulbs in his bedroom with 'energy efficient' light bulbs.

(g) Young adult (Aged 21–30)

Having finished their studies our two subjects have now taken on the responsibility of their first jobs and homes. Together with transport, holidays, and food they are now directly responsible for the climate impact of their homes. Consequently, it is during these years that a very large divergence between their respective GHG emissions becomes evident. As 'medium' household energy users in London they both would normally produce about 9200 kg of GHG at a cost of £519 each year. Mr Meyer though, cuts his household emissions by 4.6 tonnes and his energy bill by £154 each year by making use of a range of energy saving strategies around the house. These include using energy efficient appliances, lighting, shower and water heating. Improved house insulation and energy-aware home design complete the savings. Mr Meyer is able to save a further 125 kg of GHG each year at no cost to himself, by simply leaving his old newspapers out for recycling instead of putting them straight in the bin — avoiding waste going into landfill, by recycling or even aerobically composting it, can reduce the amount of methane produced per kg of rubbish, with methane being around 20 times more potent a GHG as carbon dioxide this can make a big difference.

Our subjects' increasing affluence enables them to be more selective about their diet. Mr Black does his food shopping at a large out of town supermarket, the food he buys is fairly varied with quite a lot of organic fruit and vegetables. However, he is unaware of the 'food miles' many of the items he buys have clocked up. His weekly shopping basket of 16 kg of goods is responsible for 2184 kg of GHG over the year. Taking advantage of the local Farmer's market Mr Meyer is able to buy most of his family's food from local sources, though there is little monetary saving, Mr Meyer's shopping basket of 16 kg is responsible for only 74 kg of GHG over a year.

For their holidays during these years Mr Black travels by plane to Madrid each summer at a cost of £134 and produces 361 kg of GHG. Mr Meyer opts to spend his annual holiday in Manchester, travelling by rail at a cost of £66 and producing just 19 kg of GHG.

For transport as young adults, Mr Meyer again combines use of his bicycle with London's public transport system. His tickets cost £380 and his annual transport-related emissions total 528 kg. Mr Black trades in his old runabout for a brand new sporty hatchback. As a result his petrol costs alone rise to £580, with related GHG emissions reaching 4773 kg each year.

(h) Older adult (Aged 30–45)

With ever-greater spending power, the opportunity to increase GHG emissions through 'energy rich activities' tends to grow, while the relative financial incentives for cost effective emission reduction fall. As such, with higher incomes and children of their own, our two subjects are now faced with more and more GHG sensitive decisions. The difference in their GHG emissions (Fig. 1) and associated monetary costs (Fig. 2) now widens even further. Mr Black sells his sporty hatchback and instead buys a new family-sized estate car. This new car costs £673 in fuel and creates 5444 kg of GHG each year. Mr Meyer sticks to biking (with baby seat) and public transport at the annual cost of £380, and with an associated GHG emission of just 528 kg.

With growing families, the food purchases of both our subjects rise to 33 kg of goods each week, with Mr Black continuing to buy without regard to 'food miles' the GHG arising from the transport of his food goes up to 4368 kg. Meanwhile, the transport of Mr Meyer's food, sourced locally, gives rise to only 147 kg of GHG over the course of a year. The increase in family size leads to Mr Black's household waste related GHG emissions rising to more than a tonne. Mr Meyer limits this increase by continuing to recycle, leading to a saving of 240 kg of GHG a year.

The energy use of both Mr Black's and Mr Meyer's family could be expected to rise into the 'high user' category at this point in their lives. Indeed, Mr Black's household does just that. Energy related GHG emissions rising to over 13 tonnes a year at a cost of £739. Using various energy saving strategies, Mr Meyer is able to cut emissions by over 4.5 tonnes below this level and save £154 in energy costs at the same time.

During this period Mr Meyer opts to take his holidays in Skegness each year, travelling by rail at a cost of £43.20 and giving rise to around 13 kg of GHG each time. Mr Black flies each year to Seattle at a cost of £249 and produces 2229 kg of GHG on each round trip.

(i) Pre-retirement (Aged 45–60)

Until now we have examined only those activities of our two subjects which have an impact on their GHG emissions at home and while travelling. However, as holders of senior posts at work, both Mr Black and Mr Meyer now have responsibility for GHG-sensitive decisions at work. With 70 employees, work place energy use and associated GHG emission is large. By replacing the 200 lights in his block of offices with energy efficient lighting and ensuring waste paper is recycled wherever possible, Mr Meyer is able to cut GHG emissions by over 20 tonnes per year and save over £1400 in energy costs. (These office based GHG savings are so large that they have been left out of the lifetime comparison figures to improve clarity).

At home, our subjects' children have moved out and their energy usage would normally drop back to the 'medium' user bracket. However, Mr Meyer's energy saving strategies around the house also maintain the previous cuts in household energy related GHG emissions and costs. Similarly, his sourcing of locally grown food and recycling of household waste continues to reduce his personal climate impact.

Both Mr Black and Mr Meyer continue to clock up around 18 000 km in day-to-day travel each year. With no young children and more expendable income Mr Black decides to buy a large engined saloon, which leads to annual GHG emissions of nearly 7 tonnes at a fuel cost of over £1700. In comparison, Mr Meyer's chosen combination of public transport and bicycle continues to produce only 528 kg of GHG and cost only £380 each year (even without the savings on gym fees).

For his annual holidays Mr Meyer decides to stay in London and visit the various museums, galleries and shows the capital has to offer. By making use of his annual public transport pass he is able to travel around London for free, the 25 trips he clocks up each summer producing only 7 kg of GHG. Mr Black flies each year to Lima at a cost of £459, producing almost 3 tonnes of GHG on each round trip.

(j) Retirement (Aged 60–75+)

Free from the need for daily trips to work and with grown up children, our two subjects should see substantial drops in their energy use, GHG emissions and energy costs. Indeed, both could be expected to drop into the 'low' household energy user bracket, though with Mr Meyer making further cuts in household GHG emissions and energy costs as outlined previously. However, Mr Black negates much of this post-retirement decrease in climate impact by buying himself a petrol guzzling classic car which, even though only used for 9000 km of travelling each year, produces over 3 tonnes of GHG and costs £863 in fuel.

Worse still for Mr Black's annual GHG budget is his decision to now travel to Auckland for his annual holiday, a flight which costs £655 and creates over 5 tonnes of GHG on each round trip. Meanwhile, Mr Meyer continues using public transport to get around and opts for annual holidays in Aberdeenshire. The return ticket to Aberdeen costs £94 and the round trip produces 27 kg of GHG.

(k) The final bill

The final bill, both in terms of their lifetimes' GHG emissions and its associated costs is, as you've no doubt guessed, vastly greater for Mr Black than for Mr Meyer (Table 2). Through relatively modest changes in lifestyle, Mr Meyer succeeded in cutting his total GHG emissions by 70% and saved himself around £80 000 compared to Mr Black. It is clear that, if such reductions are extrapolated to scales of thousands or millions of individuals, huge GHG cuts are possible.

Table 2   Cumulative lifetime greenhouse gas emissions
and associated costs for two theoretical Londoners.

	Greenhouse Gas (tonnes)	Cost (£ sterling)
Mr Black	1251	131 000
Mr Meyer	370	48 845

From a political perspective, the promotion and subsequent incorporation of such 'individual' cuts into national GHG budgets would seem extremely attractive. Certainly, recent years have seen increasing government interest in this area. The UK government is promoting domestic energy efficiency via better information, financial incentives and tighter regulations. Indeed, they predict that implementation of these strategies could cut UK carbon emissions by around 5 million tonnes by 2010 [DETR, (2000)]. As the UK's Deputy Prime Minister has said "We have a responsibility to take action, but it is also in our own interests to do so. Measures to reduce greenhouse gas emissions can be good for the economy, for businesses and for our communities."

Other governments too, are active in the promotion of GHG emissions cuts at the individual level. The Australian Greenhouse Office, for instance, last year launched 'Cool Communities' which not only provides detailed information on how individuals might reduce their own GHG emissions, but also provides funds for communities to implement these GHG reduction strategies.

In pure 'cost per tonne GHG reduction' terms, no definitive figures exist for reductions via the 'increased public awareness' route, but there seems little doubt that this option has huge potential in industrialised countries like the UK. Were a million people with a 'Mr Black' type lifestyle to switch to a 'Mr Meyer' lifestyle, the annual reduction in UK GHG emissions would be more than 5 million tonnes, with a monetary saving of around £1billion. If we consider the likely spread of 'GHG awareness' of individuals to their place of work, choice of business suppliers, and ultimately their political representative. . . well, you see the power that is individual action.

5. Conclusion

Economic concerns have been used by some as an excuse for inaction on climate change. Though cost-benefit analyses of global warming have often been dismissed on ethical grounds, it seems that in many situations GHG abatement strategies can in fact be implemented at no net cost. Indeed, the possibility of catastrophic climate change would seem to justify GHG abatement even where significant short-term costs are incurred. Contrary to the assertions of the current US administration, the cuts proposed under the Kyoto Protocol appear both economically viable and vital as a basis for future international GHG abatement.

An area with huge potential for cost-effective GHG abatement is that of personal emissions. In industrialised countries, like the UK, implementing relatively modest lifestyle changes can make large savings in both the GHG emissions and energy costs of individuals. Indeed, if only one or two of the lifestyle changes outlined here were implemented on a wide scale, significant reductions in national GHG emissions are possible. The key to successfully realising this huge potential for GHG abatement is, ultimately, increased public awareness.

Appendices

Financial costs of nappy and food purchase were not included in these analyses. GHG emissions represent CO_2 equivalents, unless otherwise stated. GHG emission data obtained from non-UK datasets (US and Australia) is assumed to be valid for UK. Where cost analyses required conversion of US or Australian dollars to pounds sterling, conversion factors of 1.6:1 and 2.5:1 have been used, respectively. Emissions and cost analyses take no account of possible future technological, economic and social variability in the UK. For household related GHG emissions and energy costs, Mr Black and Mr Meyer are assumed to be ultimately responsible for total emissions even where the presence of a spouse/children is inferred (i.e. household related emissions may not always be per capita).

Baby (0–2 years)

Calculation assumes 'high' ecological foot print [Best Foot Forward, http://www.bestfootforward.com], electricity use at 3.6 MJ/kWh, and a GHG emission rate of 1 kg GHG/kWh electricity (Australian Institute of Energy, http://www.aie.org.au).

Toddler (2–4 years)

Assumes 100 trips to nursery per year, 5 km round trip. GHG emission data for the four wheel drive data are for a 2001 Isuzu Trooper 4WD 3.5l (US Department of Energy, http://www.fueleconomy.gov) based on the Greet model (Argonne National Laboratory, http://www.transportation.anl/ ttrdc/greet/index). Fuel cost data were obtained from UK Vehicle Certification Agency (http://www.vcacarfueldata.org.uk/). No account was taken of car or bike purchase and maintenance costs.

Infant (4–7 years)

Calculations for the four-wheel drive are as described above, but this time for a 10 km round trip and 160 trips per year. GHG emissions from bus journeys are based on the assumption of 33 g GHG emission per km per person carried [Australian Greenhouse Office (2001)].

Junior (7–11 years)

Flight GHG emissions were derived from IPCC [1999]. Flight prices were for 2002 and were obtained from 'CheapFlights.co.uk' (http:// www.cheapflights.co.uk). Train-journey GHG emissions were based on the assumption of 33 g GHG per km per person carried [Australian Greenhouse Office (2001)]. Train ticket costs were obtained from 'The Trainline.com' (http://www.thetrainline.com) and assume 'Saver return' tickets.

Household GHG emission data based on a television, video recorder and games console being left on standby, rather than being switched off [Australian Greenhouse Office (2001)].

Senior (11–18)

GHG emissions and related costs calculated as previously stated, assuming 160 trips school trips per year at 40 km each time.

Student (18–21)

Student transport assumes car to be a 1995 Ford Focus (21) and annual distance travelled to be 9000 km. For Mr Meyer, 8000 km per year via public transport, 1000 km per year by bike. Public transport fare obtained from 'Transport for London' (http://transportforlondon.gov.uk) assuming purchase of annual pass to travel in London zones 1–4 and with 30% discount for 'Youth' pass.

Housing energy costs calculated on the basis of 'low' energy use: 10 000 kWh gas and 1650 kWh electric; 'medium' energy use: 19 050 kWh gas and 3300 kWh electric; 'high' energy use: 28 000 kWh gas and 4950 kWh electric. These data were based on London Electric 'dual fuel' at standard credit [Energywatch, December (2001), http://www.uSwitch.com]. GHG emission estimates for energy used calculated using 1 kg GHG per kWh electricity and 0.31 kg per kWh for gas [Australian Greenhouse Office (2001)].

Adult (21–30)

Distance and transport type data for theoretical 'shopping trolley good' were obtained from Sustain (http://www.sustainweb.org/). GHG emissions were calculated for air-freight using IPCC [1999]. Lorry transport emissions were derived from UK Department of Transport, Local Government and the Regions [2000] assuming 33 ton twin axle articulated lorry in 45 mph speed bracket. Van emissions were calculated for Volkswagen Multi-Van (UK Vehicle Certification Agency, http://www.vcacarfueldata.org.uk/) travelling 100 km.

Recycling data assume total newspapers amounting to 50 kg in weight each year. Savings of 2.5 kg GHG per kg of recycled newspaper, relative to landfill. [US Environment Protection Agency (1998)].

Transport data calculated assuming 18 000 km per year on public transport and 1000 km on bike for Mr Meyer (annual public transport pass now without 30% 'youth' reduction in price and 18 000 km per annum for Mr Black, 2001 Volkswagen Golf GTi 1.8l (GHG emissions and fuel costs obtained as described above).

Older adult (30–45)

Mr Black's car was a 2001 Volvo V70 (2.4l), 18 000 km per year. GHG emissions and related costs calculated as described above.

Pre-retirement (45–60)

GHG and monetary savings form lighting policy calculated using 'Work Energy Smart Lighting Calculator' (http://www.energysmart.com. au/WESlight.shtml) assuming a change of 200 lights from 80 Watt Flouro halo T8 to 12 Watt Compact (CFL). Total operating hours per year 1500, leading to total reduction in GHG emissions of 18.7 tonnes per annum. Savings in sterling, £1400, calculated using an electricity price of 7 pence per kWh. Waste recycling saving based on recycling of 500 kg of paper per year, with an associated GHG saving of 2750 kg relative to landfill [US Environment Protection Agency (1998)]. Other data obtained as described above.

Retirement (60–75)

Mr Black's car was a 2001 Jaguar XJ8 (4l), 9000 km per year. This and other data obtained and calculated as described previously.

References

Australian Greenhouse Office (2001) *A Home Guide to Reducing Energy Costs and Greenhouse Gases*, Australian Greenhouse Office, Canberra, Australia.

Ayalon, O., Avnimelech, Y. & Schlecter, M. (2001) Solid waste treatment as a high-priority and low-cost alternative for greenhouse gas mitigation. *Environmental Management* **27**(5), 697–704.

Baral, A. & Guha, G. S. (2004) Trees for carbon sequestration or fossil fuel substitution: The issue of cost vs. benefit. *Biomass & Bioenergy* **27**(1), 41–55.

Barker, T. & Ekins, P. (2004) The costs of Kyoto for the US economy. *Energy Journal* **25**(3), 53–71.

Brown, D. A. (2003) The importance of expressly examining global warming policy issues through an ethical prism. *Global Environmental Change — Human and Policy Dimensions* **13**(4), 229–234.

Brown, M. A., Levine, M. D., Short, W. & Koomey, J. G. (2001) Scenarios for a clean future. *Energy Policy* **29**(14), 1179–1196.

Brown, M. A., Levine, M. D., Romm, J. P., Rosenfeld, A. H. & Koomey, J. G. (1998) Engineering-economic studies of energy technologies to reduce greenhouse gas emissions: Opportunities and challenges. *Annual Review of Energy and the Environment* **23**, 287–385

Buttraw, D., Krupnick, A., Palmer, K., Paul, A., Toman, M. & Bloyd, C. (1999) *Ancillary Benefits of Reduced Air Pollution in the US from Moderate Greenhouse Gas Mitigation Policies in the Electricity.* Resources for the Future, Discussion Paper, 99–51. Washington, DC.

Clarkson, R. & Deyes, K. (2002) *Estimating the Social Cost of Carbon Emissions.* Government Economic Service Working Paper 140. (http://www.hmtreasury.gov.uk/media/209/60/SCC.pdf).

De Cara, S. & Jayet, P. A. (2000) Emissions of greenhouse gases form agriculture: The heterogeneity of abatement costs in France. *European Review of Agricultural Economics* **27**(3), 281–303.

De Jong, B. H. J., Tipper, R. & Montoya-Gomez, G. (2000) An economic analysis of the potential for carbon sequestration by forests: Evidence from southern Mexico. *Ecological Economics* **33**(2), 313–327.

Department of Transport, Local Government and the Regions (2000) *NERA Report on Lorry Track and Environment Costs.* DTLR, UK.

Department of Environment, Transport and the Regions (2000) *Delivering Emission Reductions. Climate Change: The UK Programme.* DETR. London UK (http://www.defra.gov.uk/environment/climatechange/cm4913/index.htm).

Environmental Protection Agency (1998) *Greenhouse Gas Emission from Management of Selected Material in Municipal Solid Waste.* Environmental Protection Agency, US.

Eyre, N., Downing, T. E., Hoekstra, R., & Tol, R. (1999) *Global Warming Damages.* Final Report of the ExternE Global Warming Sub-Task (September 98), DGXII, European Commission, Brussels.

Friends of the Earth (2004) *Exxon's Climate Footprint: The Contribution of Exxonmobil to Climate Change Since 1882.* Friends of the Earth report: (http://www.foe.co.uk/campaigns/climate/resource/exxonmobil_climate_footprint.html#reports).

Howarth, R. B. (2001) Intertemporal social choice and climate stabilization. *International Journal of Environment and Pollution* **15**(4), 386–405.

Intergovernmental Panel on Climate Change (1999) *Air Transport Operations and Relation to Emissions.* Aviation and the Global Atmosphere, IPCC.

Kleiner, K. (2001) *Heat is On. New Scientist Online* (http://www.newscientist.com/news/news.jsp?id=ns9999566).

Kuntz Duriseti, K. (2004) Evaluating the economic value of the precautionary principle: Using cost benefit analysis to place a value on precaution. *Environmental Science and Policy* **7**(4), 291–301.

Lomborg, B. (2001) *The Skeptical Environmentalist.* Cambridge University Press, UK.

Lutter, R. (2000) Developing countries' greenhouse emissions: Uncertainty and implications for participation in the Kyoto Protocol. *Energy Journal* **21**(4), 93–120.

McKibbin, W. J. & Wilcoxen, P. J. (2004) Estimates of the costs of Kyoto: Marrakesh versus the McKibbin-Wilcoxen blueprint. *Energy Policy* **32**(4), 467–479.

Morthorst, P. E. (1998) The cost of reducing CO_2 emissions — Methodological approach, illustrated by the Danish energy plan. *Biomass & Bioenergy* **15**(4–5), 325–331.

Nordhaus, W. D. (1994) *Managing the Global Commons: The Economics of Climate Change.* MIT Press, Cambridge, MA.

Nordhaus, W. D. & Boyer, J. (2000) *Roll the DICE Again: Economic Models of Global Warming.* Cambridge, MA, MIT Press.

Rasmussen, T. N. (2001) CO_2 abatement policy with learning-by-doing in renewable energy. *Resource and Energy Economics* **23**, 297–325.

Reilly, J., Prinn, R., Harisch, J., Fitzmaurice, J., Jacoby, H., Kicklighter, D., Melillo, J., Stone, P., Sokolov, A. & Wang, C. (1999) Multi-gas assessment of the Kyoto protocol. *Nature* **401**, 549–555.

Scottish Executive (2001) *Scotland's Renewable Resource 2001 — Executive Summary.* Scottish Executive, UK.

Smith, K. R., Uma, R., Kishore, V. V. N., Zhang, J. F., Joshi, V. & Khalil, M. A. K. (2000) Greenhouse implications of household stoves: An analysis for India. *Annual Review of Energy and the Environment* **25**, 741–763

Stott, P. A., Stone, D. A. & Allen, M. R. (2004) Human contribution to the European heatwave of 2003. *Nature* **432**, 610–614.

Toman, M. A., Morgenstern, R. D. & Anderson, J. (1999) The economics of "when" flexibility in the design of greenhouse gas abatement policies. *Annual Review of Energy and the Environment* **24**, 431–460.

United Nations Framework Convention on Climate Change (2001) *Understanding Climate Change* (http://www.unfccc.de/resource/beginner.html).

Van Kooten, G. C., Eagle, A. J., Manley, J. & Smolak, T. (2004) How costly are carbon offsets? A meta-analysis of carbon sinks. *Environmental Science and Policy* **7**(4), 239–251.

Carbon in the Atmosphere and Terrestrial Biosphere in the Early Anthropocene

Yadvinder Malhi

Oxford University Centre for the Environment
University of Oxford, United Kingdom
ymalhi@ouce.ox.ac.uk

The release of carbon dioxide from fossil fuel combustion and land use change has caused a significant perturbation on the natural cycling of carbon between land, atmosphere and oceans. The perturbation of the carbon cycle, and other aspects of global biogeochemical cycle, is so large and fundamental that it has been suggested the Earth has entered a new geological epoch, the Anthropocene, characterised by overwhelming human perturbation of global biogeochemistry. Understanding and managing the effects of this perturbation are likely to be amongst the most pressing issues of the twenty first century. However, the present-day carbon cycle is still poorly understood. One remarkable feature is that an increasing amount of atmospheric carbon dioxide appears to be being absorbed by terrestrial vegetation.

In this paper I review the recent evidence for the magnitude and spatial distribution of this "terrestrial carbon sink", drawing on (i) current research on the global atmospheric distribution and transport of carbon dioxide, oxygen and their isotopes (ii) direct measurement of CO_2 fluxes above various biomes, and (iii) inventories of forest biomass and composition. I review the likely causes of these carbon sinks and sources, and their implications for the ecology and stability of these biomes.

Finally, I examine prospects and key issues over coming decades. Controlling deforestation and managing forests has the potential to play a significant but limited part in reaching the goal of stabilising atmospheric CO_2 concentrations. However, there are likely limits to the amount of carbon storage possible in natural vegetation and in the long term terrestrial carbon storage may unstable to significant global warming, with the potential to accelerate rather than brake global warming.

1. Introduction

It takes us a 24-hour boat ride through isolated backwaters and an hour's trek through the majestic ancient forest to reach the observation tower at Caxiuaná, near the eastern coast of the Amazon rainforest. Climbing up the tower, you catch your breath at the spectacle of almost pristine Amazonian forest stretching to the horizon in all directions. To the west it stretches for over 3000 km until it meets the rain-drenched wall of the Andes mountains. Here, at the heart of the Amazon, away from constant deforestation that gnaws away at the edges, it can seem at first that you have reached a world barely touched by human influences, where the cycles of nature still run to prehistoric rhythms. Yet the sensor on top of the tower tells a different story. It measures the concentration and vertical flow of the greenhouse gas carbon dioxide (CO_2) above the forest. Every year since these measurements began in Amazonia, the measured concentration of CO_2 has crept up: it was 350 ppm (parts per million by volume) in 1987, and has risen relentlessly to reach 380 ppm in 2005. This rise started in the late 18th century, when concentrations stood at about 280 ppm (a value that has remained fairly constant since the end of the last Ice Age), and is a fingerprint of global human activity, a consequence of the combustion of coal, oil and gas that drives modern economies, and the widespread clearance of forests. By the end of the 21st century it will reach values between 500 and 1000 ppm, values not experienced on Earth for 20 to 40 million years [(Prentice (2001); IPCC (2002)]. It is now generally agreed by the scientific community that this CO_2 increase is driving global warming, but there are other consequences. Atmospheric CO_2 is the raw material from which plants manufacture the sugars and starches and other compounds that provide them with energy stores and build their biomass, and ultimately provide the energy supply for almost all life. As the concentration of CO_2 increases, what will this do to the growth, biomass and ecology of these forests? The measurements of the flow of carbon dioxide at the tower provide a clue. They measure a net long-term flow of CO_2 into the forest, suggesting that the growth of forests is being stimulated. How large is this stimulation, what are the consequences of this effect on the global atmosphere and climate, what are the consequences of this effect on the ecological balance and composition of these diverse tropical forests, and how long into the future will these effects last? These are questions that will need to be answered in the 21st century, and that have shaped my research agenda for the last ten years. In this chapter I will describe some of this work, but also take a wider perspective, examining the current state of our understanding of the terrestrial carbon cycle, and how this understanding may improve over coming decades.

2. The Global Carbon Cycle Today

2.1. *The natural carbon cycle*

The major components of the natural global carbon cycle are illustrated in Fig. 1 [from Prentice *et al.* (2001)]. The major reserves of carbon are, in decreasing order of size, the rocks and sediments of the lithosphere (a vast but relatively inactive store of carbon), the oceans (38 000 Pg C; 1 Pg C = 1 pedagram of carbon = 10^{15} g), the soils (1500 Pg C) and plants (500 Pg C) of the terrestrial biosphere, and the atmosphere (730 Pg C). Although a small pool, it is the carbon stocks in the atmosphere, mainly in the form of carbon dioxide (CO_2), but also more potently in the form of methane (CH_4), that provide the direct link with climate change.

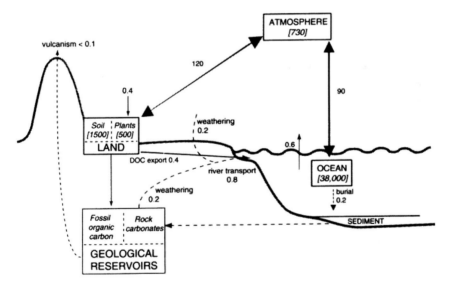

Fig. 1 Main components of the contemporary natural carbon cycle. The thick arrows represent gross primary production and respiration by the biosphere and physical sea-air exchange. The thin arrows denote natural fluxes which are important over longer timescales. Dashed lines represent fluxes of carbon as calcium carbonate. The units for all fluxes are Pg C/yr, the units for all compartments are Pg C. 1 Pg C is equal to 10^{15} g, or 1 gigatonne (Gt). Reproduced from IPCC [2001].

There is tight coupling between the atmospheric, terrestrial and oceanic carbon pools. Every year 120 Pg C (16% of the atmospheric stock) are transferred from the atmosphere to the biosphere via the process of plant photosynthesis. This is termed the gross primary production, GPP, of the terrestrial biosphere. About half of this amount is directly returned to the atmosphere through plant respiration, as the plants metabolise their manufactured sugars. The remaining amount (the net primary production, or NPP, of the terrestrial biosphere) is eventually transferred to the soil carbon pool as plants shed organic material (through litter-fall, root death, and, eventually, total plant death), with a small fraction being transferred to herbivores and other animals. Almost all this carbon is eventually returned to the atmosphere via decomposition or animal metabolism. By dividing the terrestrial biospheric carbon stock (2000 Pg C) by the gross photosynthesis (120 Pg C year^{-1}), we can estimate that the mean residence time for an atom of carbon in the terrestrial biosphere is about 17 years, with about 12 years of this being in the soil. The mean is an amalgam of a few months' residence time for leaves and fine roots, to thousands of years for some long-lived and some components of soil carbon.

CO_2 is also exchanged with the oceans, primarily by simple physical dissolution, with the net flux to and from any ocean surface depending on the gradient in the partial pressures of CO_2 in atmosphere and water. Once in solution, 99% of the CO_2 reacts chemically water to form bicarbonate, HCO_3^- and carbonate CO_3^{2-} ions. It is in this form that carbon is used by marine organisms, either for the photosynthesis of organic matter or for the synthesis of carbonate shells. Almost all of this activity occurs in the light-rich upper hundred metres of the ocean, where there is rapid recycling of carbon as organisms consume and die. A fraction of the organic carbon and carbonate, however, "rains" down into the deep ocean, where it enters the slow circulation of the abyss, to eventually re-enter contact with the atmosphere in regions of deep ocean upwelling (such as the equator, or western margins of continents). An even smaller fraction of carbonate (about 0.2 Pg C year^{-1}) and organic matter (0.01 Pg C year^{-1}) is buried in deep ocean sediments, and enters the lithosphere. Overall, about 90 Pg C are transferred each way to and from the ocean each year. The mean residence time carbon molecules in the oceans is therefore about 400 years, but this figure is misleading as most carbon is recycled within a year in the photic zone, whereas a small fraction resides for over a thousand years in the deep ocean.

Therefore, in total, 210 Pg of atmospheric CO_2 (more than a quarter of the atmospheric carbon stock) are cycled between the atmosphere, land and oceans every year, and the mean residence time for a CO_2 molecule in the atmosphere is only 3–4 years. Any human perturbations to the cycle are enmeshed within this huge natural cycle.

On a longer timescale, interactions with the geological carbon cycle become important. Some of the carbon locked in organic matter or removed from the atmosphere by land surface erosion is buried in deep ocean sediments and slowly transferred into the earth's mantle by the slow drift and subduction of the continental plates, or returned to the atmosphere by the uplift and exposure of buried sediments. Eventually, this carbon cycles back to the atmosphere through volcanic outgassing. The residence times for carbon in the lithosphere are millions of years: this slow, geological cycle is irrelevant on the timescales of human history, but has been the most important cycle on geological time scales of tens to hundreds of millions of years.

2.2. *The carbon disruption and the Anthropocene*

Human civilisation has always been associated with a modification of the local carbon cycle through the fire management and clearing of natural ecosystems [Perlin (1999)]. It is only since the 18th century, however, that the scale of this modification has accelerated such as to have a significant effect at the global scale. This "carbon disruption" has been driven partially by accelerated rates of deforestation, especially in the tropics, resulting in a net transfer of carbon from the terrestrial biosphere to the atmosphere and oceans. The most important agent, however, has been the hundredfold acceleration of the geological carbon cycle through the combustion of old biospheric carbon trapped in the lithosphere, the "fossil" fuels of coal, oil and natural gas. Figure 2(a) illustrates the rate of carbon transfer to the atmosphere from fossil fuel combustion since 1750, whilst Fig. 2(b) illustrates the cumulative emissions since the dawn of human history. In recent decades the carbon disruption has accelerated spectacularly. I was born in 1968. In my lifetime more CO_2 has been released to the atmosphere through human activity than in all of previous human history combined.

2.3. *The contemporary global carbon balance*

The fate of the carbon dioxide currently being released into the atmosphere by human activities is illustrated in Fig. 3. The values shown are estimates

(a)

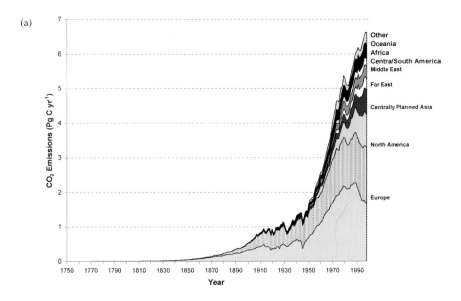

(b)

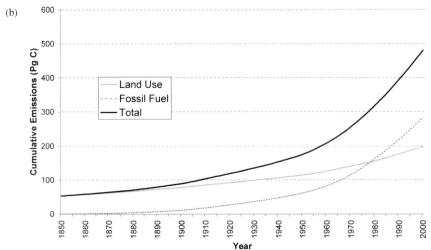

Fig. 2   (a) Total carbon emissions from fossil fuel combustion and cement production since 1750, divided by region. Data from Marland (2001), (b) Estimated cumulative carbon emissions since the start of human history, plotted since 1850. Data from Marland (2001) and Houghton (pers. comm.). Figure derived from one in Malhi *et al.* (2002).

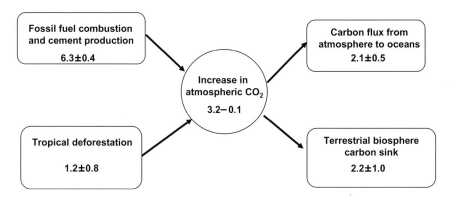

Fig. 3 An estimated carbon budget for the 1990s. The source from fossil fuels and deforestation is partially offset by carbon uptake in oceans and land. All units are in $Pg\,C\,year^{-1}$. Details are discussed in the text.

for the period 1990–1999. The source from fossil fuel combustion and cement production can be accurately estimated from economic data, and was $6.3 \pm 0.4\,Pg\,C\,year^{-1}$ for the 1990s (cement production accounts for only 2% of this total). The other main source is land use change, which encompasses deforestation, forest degradation, logging, reforestation and maturation of regrowing forests. Land use change is often quoted to be a carbon source of $1.7 \pm 0.5\,Pg\,C\,year^{-1}$. [IPCC (2002)], but there is considerable uncertainty in this value. The rates of forest clearance in the IPCC values are based on data from the Food and Agriculture Organisation (FAO). A number of recent studies [e.g. Achard *et al.* (2002); DeFries *et al.* (2002)] have suggested that the FAO may be significantly overestimating deforestation. Moreover, there is still a surprising uncertainty in even the biomass per unit ground area, with recent estimates varying by 30%, between $100\,Mg\,C\,ha^{-1}$ and $129\,Mg\,C\,ha^{-1}$ [Houghton (2005)]. In addition, a number of processes such as illegal logging and forest degradation are difficult to estimate. They may be a source of around $0.4\,Pg\,C\,year^{-1}$ [Nepstad *et al.* (1999)] and are not included in these calculations, but also must not be double-counted if degraded areas are subsequently deforested. In all, the net carbon source from land use change is likely to lie in the region of $1.2 \pm 0.8\,Pg\,C\,year^{-1}$.

Because CO_2 is relatively well-mixed in the global atmosphere, it is possible to measure the global rate of accumulation with high precision. In the 1990s CO_2 accumulated in the atmosphere at a rate of 3.2 ± 0.1 t C $year^{-1}$, corresponding to a rate of increase of CO_2 concentration of about $1.5\,ppm\,year^{-1}$. The discrepancy between the total carbon source

$(7.5 \pm 0.9 \, \mathrm{Pg\,C\,year}^{-1})$ and the rate of carbon accumulation in the atmosphere implies that a substantial amount of carbon is being removed from the atmosphere by other processes. The two most obvious "sinks" for carbon are the oceans, and the terrestrial biosphere.

The partitioning between these land and ocean sinks can be calculated from the fact that the exchange of CO_2 with the biosphere (through photosynthesis, respiration or combustion) results in an equivalent transfer of oxygen (O_2) in the other direction, whereas the dissolution of CO_2 is ocean water is not matched by a release of O_2. The difference between changes in atmospheric concentrations of O_2 and CO_2 can be used to quantify the ocean carbon sink, which is currently estimated to be $2.1 \pm 0.5 \, \mathrm{Pg\,C\,year}^{-1}$, a value that agrees with results from ocean carbon cycle models ($2.3 \pm 0.5 \, \mathrm{Pg\,C\,year}^{-1}$), and from studies of carbon isotopes ($2.1 \pm 0.9 \, \mathrm{Pg\,C\,year}^{-1}$). There is still significant uncertainty about the future magnitude of the ocean carbon cycle, as poorly understood processes such as changes in ocean circulation become increasingly important at longer timescales.

Therefore, fossil fuel emissions, the atmospheric CO_2 increase, and the modern day ocean carbon sink are all well constrained. We can therefore have high confidence that the remaining unknown factor, the net carbon sink into the terrestrial biosphere, had a value of $1.0 \pm 0.6 \, \mathrm{Pg\,C\,year}^{-1}$ in the 1990s. When we begin to tease apart this number into its component processes, however, we run into greater uncertainties. As described above, we know that land use change is a net source of $1.2 \pm 0.8 \, \mathrm{Pg\,C\,year}^{-1}$. Therefore, there is a likely carbon *sink* into the remaining terrestrial biosphere (away from tropical deforestation areas) of magnitude $2.2 \pm 1.0 \, \mathrm{Pg\,C\,year}^{-1}$ to balance the carbon budget.

3. The Causes of the Carbon Sink in the Terrestrial Biosphere

Where is the terrestrial carbon sink? The most likely suspects are forests, which hold about 50% of terrestrial biosphere carbon stocks, and account for about 50% of terrestrial photosynthesis. What could be causing forests to be net sinks of carbon? There are a number of possibilities.

The simplest possibility is that forests are simply *recovering from past disturbance* and are changing in age structure. In the temperate regions of North America and Europe, there has been a substantial abandonment of agricultural land over the 20th century. The basic change of land from agriculture to forest is already included in the estimate of net carbon emissions by land use change, but more subtle variations in the age and density of forests as forest management changes, or as dense vegetation encroaches into abandoned areas, are not included.

Another possibility is that some human-induced agent of global change is causing an enhanced rate of forest growth, or an expansion of forest area. The prime suspect is the very agent that is disappearing into the biosphere: *The enhanced carbon dioxide concentration in the atmosphere.* As explained in the introduction, the "CO_2 fertilisation" of photosynthesis is likely to be enhancing the growth of forests. The majority of lab studies of tree growth under high CO_2 concentration have shown enhanced growth rates, with on average a 60% increase in plant productivity for a doubling of atmospheric CO_2 concentrations [Norby *et al.* (1999)].

In natural, mature ecosystems (as opposed to young or small plants growing in controlled lab conditions), it has been less clear whether this growth enhancement with be significantly constrained by other limitations, such as the supply of essential nutrients like nitrogen and phosphorus (both essential to make new plant biomass), by water supply, or by whole system-level constraints such competition between species for limited light resources [(Norby *et al.* (2001)].

Another route by which increased atmospheric carbon dioxide may affect plant growth is through *increased water use efficiency*. Most of a leaf surface is waxy and relatively impermeable to CO_2, and most of the CO_2 to reaches the plant enters through small pores in the leaf surface known as stomata. These entry points for CO_2 are also exit points for water vapour, and in dry climates the loss of water through stomata is a serious problem. In response, water-stressed plants reduce water loss by closing their stomata, which consequently limits CO_2 uptake and the potential for photosynthesis. In a higher CO_2 atmosphere, a greater amount of CO_2 can enter for the same degree of water loss, thus increasing the amount of photosynthesis possible for a fixed supply of water.

In many regions human activities have greatly increased *the supply of nitrogen* in natural ecosystems, either though by-products of fossil fuel or biomass combustion (nitrogen oxides), or else through application of fertilisers (ammonium compounds and nitrates). The rate of supply of available nitrogen to the biosphere has increased from a pre-industrial value of $100\,\mathrm{Tg}$ N year^{-1} ($1\,\mathrm{Tg} = 1\,\mathrm{teragram} = 10^{12}\,\mathrm{g}$), to a 1990s value of $240\,\mathrm{Tg}$ N year^{-1} [Schlesinger (1998)]. This enhanced nitrogen supply is likely to have a number of consequences that are harmful to ecosystems, but one potential side effect may be a fertilisation of tree growth, either on its own or in combination with the CO_2 fertilisation effect [Norby (1998)].

Finally, *ongoing climate change* is also likely to affect vegetation carbon balance, although the direction of this effect is not clear. Warming trends have been most severe at high northern latitudes, where higher temperatures are likely to lengthen growing seasons at high latitudes, thus increasing the carbon stores in plant biomass, but may also accelerate the decomposition of vast reserves of carbon held in boreal forest and tundra soils. In tropical ecosystems, changes in precipitation are likely to have a greater effect than temperature changes, affecting, for example, whether an area can support a tropical rain forest or a savanna, but the expected regional pattern of precipitation changes in the face of climate change is far from clear.

Therefore, a wide variety of factors could be leading to a carbon sink in terrestrial vegetation. Which factor is likely to be the most important? The answer is likely to vary with region. In boreal latitudes (the forests and tundras of Canada and northern Eurasia), global warming may be the most important factor, leading to longer growing seasons and an increase in forest biomass, although this effect may be offset by the release of carbon from the soil. In temperate regions, the recovery from disturbance is likely to be most important, perhaps enhanced by nitrogen and CO_2 fertilisation. In tropical savannas and forests, responsible for over half of the earth's photosynthesis, CO_2 fertilisation may be most important. In arid and seasonally arid regions, the enhanced water use efficiency may be the most important factor.

How can we distinguish between these various causal effects? Perhaps by looking at the spatial pattern of the carbon sources and sinks.

4. Looking for the Terrestrial Carbon Sink

A variety of approaches have been employed to search for the terrestrial carbon sink. I will briefly review these techniques, moving down in scale from the global atmosphere to site-specific studies. Perhaps the biggest problem presented by the search for the carbon sink in the terrestrial biosphere is its spatial heterogeneity. In contrast to the oceans, the carbon dynamics of the terrestrial surface vary greatly from point to point, as an intact forest stand sits next to a natural treefall gap, as wet forest grades to dry forest to savanna to grassland, and as climatic regimes, topography and soil conditions vary. In each of these different surface types different carbon cycling mechanisms may predominate, and to derive a mechanistic understanding of the processes involved it is necessary to work at the local, site-specific scale. On the other hand, to understand how these processes affect the global atmosphere, it is necessary to scale up from individual sites, and here again the spatial heterogeneity of the biosphere presents problems. Measurements of carbon balance at continental or global scales can help us understand whether the scaling is successful, and whether we have captured all the processes involved. The study of the carbon sink benefits from research at a wide range of scales.

4.1. *The global atmosphere*

Although CO_2 is well mixed in the global atmosphere, it is not perfectly mixed, and there are small horizontal gradients in CO_2 concentration that are driven by the spatial pattern of sources and sinks for carbon dioxide. For example, CO_2 concentrations are higher in the northern hemisphere (the source of most fossil fuel combustion) than in the southern hemisphere. In principle, if the transport of CO_2 in the atmosphere can be accurately simulated by atmospheric transport models, it should be possible to use observations of atmospheric CO_2 concentrations to derive the spatial pattern of CO_2 sources and sinks. As the spatial pattern of CO_2 sources from fossil fuel combustion can be accurately predicted from economic data, the remainder will be the spatial pattern of biosphere CO_2 sources and sinks (including the effects of land use change). Further information can be derived by also including observations of the stable isotopes of CO_2 (biosphere exchange has a different isotopic signature than fossil fuel combustion), and observations of O_2 concentrations.

A number of research groups are using this "inverse modelling" approach. The primary requirements for these studies are atmospheric transport models, which developed in the 1980s with the expansion of computing power, and a global network of observation stations. However, most of these observation stations are in North America and Europe, with very little coverage over the oceans and no coverage over tropical land masses. This dearth of observations is the main problem that currently constrains this approach.

A number of research groups around the world are currently applying this atmospheric inversion technique. There is significant disagreement between the results, but some points of agreement. The TRANSCOM3 experiment compared the results from applying 16 different transport models to the same dataset [Gurney *et al.* (2002)]. The scientists found that results were relatively robust for the northern and southern extratropics, but very poorly constrained in the tropics. There appeared to be a significant carbon sink uniformly distributed across northern land regions, with a total northern land sink of $2.3 \pm 0.7 \, \mathrm{Pg \, C \, year^{-1}}$. The study suggests that the tropical lands are probably a net source of carbon, of magnitude $+1.0 \pm 1.3 \, \mathrm{Pg \, C \, year^{-1}}$, implying that the source from tropical deforestation more than compensates any carbon sink in intact forests. However, some more recent analyses [Rodenbeck *et al.* (2003)], with more sophisticated use of monthly weather data, estimates tropical carbon balance of $-0.8 \pm 1.3 \, \mathrm{Pg \, C \, year^{-1}}$, with a high probability of being outside the range of the TRANSCOM values, implying that the tropics are a net carbon sink, and that the intact tropics must therefore be a very strong carbon sink of about $2 \, \mathrm{Pg \, C \, year^{-1}}$. The uncertainties in this method are clearly still large, and primarily caused by the sparseness of data and problems in modelling the turbulent land-atmosphere interface.

4.2. Measurements of the vertical flux of CO_2 above a surface

Atmospheric studies have the advantage of covering large areas of a heterogeneous landscape, but it is still necessary to understand the fluxes at a more detailed spatial scale to arrive at a mechanistic understanding of processes. Over the last 15 years, towers between 10 and 200 m tall have been springing up over a range of forests and other vegetation types, hosting instrumentation that directly measures the vertical exchange of carbon dioxide between the surface and the atmosphere (Fig. 4a). The most commonly used instrumentation is based on the technique of eddy covariance. This technique is based on the fact that the CO_2 is transported by means of turbulent eddies, and that if both the vertical wind velocity and concentration of CO_2 at a point can be measured with sufficient frequency to adequately capture these eddies, the covariance of these two measurements will correspond to the vertical flux of carbon dioxide at that point. For example, in daytime over a forest, air carried out of the forest canopy will be depleted in CO_2 when compared with air sinking into the forest canopy. The greater the depletion, the larger the net carbon flux. The vertical wind velocity is usually measured with a sonic anemometer, which measures the difference in transit time between upwards and downwards pointing pulses

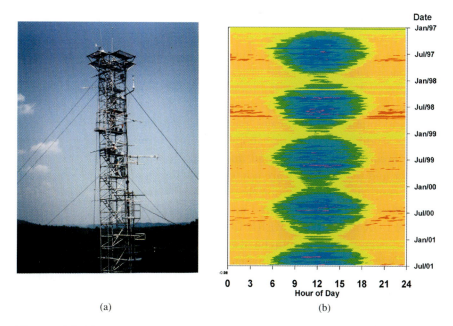

Fig. 4 (a) Measuring the flows of carbon in and out of a forest. A fully instrumented tower, measuring turbulent flows of carbon dioxide (photo: D. Baldocchi) (b) Watching the forest's breath: Four years of measurements from a sitka spruce plantation forest near Aberfeldy, Scotland. The horizontal axis is time of day and the vertical axis is time of year. Dark colours indicate periods of carbon uptake and light colours periods of carbon release. The maximum rates of carbon release and uptake are in early summer. The effect of varying day length on carbon uptake is clear. Overall, this plantation is taking up carbon at a rate of $7\,\mathrm{Mg\,C\,ha^{-1}\,year^{-1}}$ (data provided by R. Clement).

of ultrasound over a fixed distance. Both vertical wind speed and CO_2 concentration measurements are usually taken at frequencies of 1 once per second or greater, and then averaged over half-hour or one-hour periods. The derived flux usually represents an average for a region between 100 m and 5 km upwind of the tower, depending on tower height and local meteorological conditions

The diurnal and seasonal variation of carbon uptake over a sitka spruce forest in Scotland is illustrated in Fig. 4(b). At night-time, respiration and microbial decomposition are the only carbon cycling processes active, and there is a steady net efflux of carbon dioxide out of the forest (light colours). In the daytime, photosynthesis dominates over respiration, and there is a

net uptake of carbon by the forest (dark colours). The net carbon balance
of the forest on any particular day is the difference between this night-
time loss and daytime uptake. This net balance varies with meteorological conditions, with stage of vegetation development, and with season.
Such observations are powerful tools for understanding the specific mechanisms that are controlling the uptake and release of carbon at daily, seasonal and interannual timescales, particularly when combined with detailed
process studies in the forest canopy or soil. It is possible to witness the
daily "breath" of the forest, and understand how that breath varies with
weather and season. There are now over 100 flux towers set up around the
world, continuously monitoring the breath of the earth's terrestrial ecosystems. Most of these are clustered into regional networks (such as CARBO-
EUROFLUX in Europe, AMERIFLUX in North America, LBA in Brazil),
which cluster under the umbrella of a global flux tower network, FLUXNET
(http://www-eosdis.ornl.gov/fluxnet; [Baldocchi *et al.* (2001)].

The net carbon fluxes estimated by the tower measurements appear
too large to be consistent with other measurements of carbon uptake in
biomass and soils, with model expectations, or with global expectations of
the magnitude of the carbon sink.

What could be causing such a large overestimate of carbon uptake?
A number of causes have been hypothesised, including a bias in study
site selection towards growing forests recovering from natural disturbance,
unmeasured loss of carbon in the form of dissolved organic carbon in river
water, or volatile hydrocarbon emissions from leaves, but these do not seem
sufficient. The most favoured explanation, from the point of view of this
author at least, has to do with the nature of air movement at night. On
calm nights the condition of continuous, spatially homogeneous turbulent
transfer that is the basic requirement for eddy covariance measurements
no longer applies. Instead, air moves in a much more complex manner,
draining sluggishly along pressure and topographic gradients, occasional
generating turbulence at trigger points in the landscape, or being buffeted
and scooped up by occasional turbulence reaching down from the higher
atmosphere. Under such spatially heterogeneous conditions, it is extremely
unlikely that a measurement at a single tower is likely to capture the spatial
heterogeneity of vertical carbon flow, and in fact is likely to underestimate
the efflux, with the tower more likely to be located in a meteorologically
benign surface region rather than a region of active flux transfer. An underestimate of night-time efflux results in an overestimate of total 24-hour
carbon uptake, and therefore an overestimate of the net carbon sink.

So, whilst flux measurements can provide detailed mechanistic understanding of the processes controlling carbon uptake and release, the complex nature of night-time surface meteorology means they often fail to accurately determine the net carbon balance. Moreover, flux towers are relatively high-cost technology, and therefore are too few in number to sample the spatial variations within an ecosystem. To get this information, we need to directly measure and monitor the stocks of carbon in vegetation biomass and soils.

4.3. *Biomass and soil carbon inventories*

Most forests hold between 100 and $500\,\mathrm{t\,C\,ha^{-1}}$ in the form of biomass or soil organic carbon. The division between these pools varies with latitude. For example, in a typical boreal black spruce forest, $390\,\mathrm{t\,C\,ha^{-1}}$ are stored in the soil, but only $60\,\mathrm{t\,C\,ha^{-1}}$ in biomass [Malhi *et al.* (1999)]. In contrast, in a central Amazonian tropical forest, $240\,\mathrm{t\,C\,ha^{-1}}$ are stored in biomass, and $200\,\mathrm{t\,C\,ha^{-1}}$ in biomass. The primary reason for this difference is the effect of temperature on plant and soil metabolic activity: At high latitudes low temperatures restrict the decomposition of soil organic matter, allowing for the build-up of soil carbon but also restricting the recycling and availability of nutrients that are necessary for new growth. At tropical temperatures the soil organic matter is broken down ten times faster, allowing for more rapid biomass growth but resulting in a smaller soil carbon pool [Malhi *et al.* (1999)].

If forests are net sinks of atmospheric carbon, this carbon must be accumulating in either the soil or vegetation biomass. If a forest holds $400\,\mathrm{t\,C\,ha^{-1}}$, a rate of increase of carbon storage of $1\,\mathrm{t\,C\,ha^{-1}\,year^{-1}}$ should become detectable after about ten years. There are complications, however. Most mature forests have their own dynamic of local tree death (through storms, disease, fires etc.) and subsequent regrowth, and it is necessary to measure over a sufficiently wide area to include ensure adequate sampling of these events. Secondly, how do we actually measure plant biomass? For simple forests consisting of a few tree species, it is usually sufficient to harvest a few trees, determine the relationship between tree diameter, tree height and wood volume for each species, and then rely on tree diameter and height measurements to estimate the above ground biomass. If a few trees are completely excavated out of the soil, the relationship between tree diameter and root biomass can also be used to estimate below-ground biomass. In tropical forests, which typically host 200–300 different tree species in each hectare, the situation becomes more complicated, and usually an average (non-species specific) relationship between tree diameter and tree biomass is used [Chave *et al.* (2005)].

Tracking changes in soil carbon content can be more difficult, because the distribution of carbon within the soil depends on previous positions of now-dead trees, and is therefore extremely patchy. This spatial heterogeneity makes it very difficult to monitor long-term trends without a very intensive sampling network.

In temperate and boreal latitudes, the long history of commercial use of forests has resulted in extensive records of forest inventories. Thus is it possible to estimate biomass changes not only in mature forests, but also across the patchwork of forested landscapes affected by natural and human disturbance to arrive at a comprehensive estimate of biomass changes of the biome as a whole. Current estimates from forest inventory data suggest that the northern temperate and boreal forests are accumulating $0.7 \pm 0.2 \, \mathrm{Pg\,C\,year^{-1}}$ [Dixon *et al.* (1994); Houghton (1996)].

In tropical regions the data do not yet exist to compile a similar systematic inventory across the whole landscape. Instead, what work there has been has focussed on mature, undisturbed tropical forests. These are systems that should, on average, be in carbon balance in a stable atmosphere, and therefore provide a test bed for looking for the effects of global atmospheric change. This is an area of research that I have been pursuing with my colleagues Oliver Phillips and Timothy Baker from the University of Leeds since 1997. Initially, we compiled a database of mature tropical forest sites, mainly in Amazonia, that had been inventoried at least twice, and looked for evidence of changes in biomass. We concluded that there was large variability between plots, due to the natural dynamics of tree death and regrowth, but that *the majority of forest plots do seem to be accumulating carbon.* This was tantalising evidence that these pristine natural ecosystems were responding to global climate change. This work has subsequently been updated with an expanded analysis which reinforced the original conclusion. We estimated that South American tropical forests were accumulating carbon at a mean rate of $0.6 \pm 0.2 \, \mathrm{Mg\,C\,ha^{-1}\,year^{-1}}$, implying a total South American tropical forest biomass carbon sink of $0.6 \pm 0.3 \, \mathrm{Pg\,C\,year^{-1}}$ [Phillips *et al.* (1998)]. If we include potential rates of accumulation and turnover in dead wood and soils, estimated to be $0.2\text{--}0.4 \, \mathrm{Mg\,C\,ha^{-1}\,year^{-1}}$ [Telles *et al.* (2003)], the total Amazonian carbon sink could climb to $0.9 \pm 0.3 \, \mathrm{Pg\,C\,year^{-1}}$. Since this initial study, we have been working on building a more extensive tropical forest data set,

(a)   (b)

Fig. 5   Measuring carbon sinks the traditional way. Photos are from the
RAINFOR project, which is trying to quantify carbon uptake in old-growth
rainforests across Amazonia. (a) Quantifying above-ground biomass in a dry
forest in Noel Kempff National Park, Bolivia, May 2001. (b) Measuring soil
carbon and nutrient content in a wet rainforest at Sucusari, northeast Peru
(photos Y. Malhi.)

and on improving and standardising data collection and statistical analysis. We have developed a project, RAINFOR (the Amazon forest inventory
network: www.geog.leeds.ac.uk/projects/rainfor/), which involves revisit-
ing forest plots in key locations across Amazonia, remeasuring all trees to
a standard methodology, and analysing soil and leaf nutrient and carbon
content to build up a standardised database (Fig. 5). Our hope is that
these study sites will become reference sites that will be visited regularly in
the future for monitoring how 21st atmosphere change is affecting tropical
ecosystems. The latest RAINFOR results indicate that the rates of biomass
increase and biomass growth show no relation to climate, but are highest on
the more fertile soils of western Amazonia [Baker *et al.* (2004); Malhi *et al.*
(2004)]. We have also begun similar data compilation in tropical Africa,
where initial results suggest that carbon is accumulating in biomass at a
similar rate (S. Lewis, pers. comm.). If this pattern is repeated in the Asian
tropics, this implies that, in total, tropical forests are a carbon sink of about
$2.0 \, \mathrm{Pg} \, \mathrm{C} \, \mathrm{year}^{-1}$ [Malhi and Grace (2000)].

There are some methodological issues to be resolved, however. Is there a bias in where foresters choose to locate their forest plots? How large a sample is required to ensure that natural forest dynamics are adequately covered, and what is the bias if they are not adequately covered? Can changes in soil carbon be measured? Will the mean wood density of these changing tropical forests remain the same, or will it decrease? These are issues that we plan to tackle over the coming years.

4.4. *The overall distribution of the sink — reconciling different approaches*

A number of different approaches are suggesting the presence of carbon sinks in forests. How consistent are they with each other? After discounting deforestation, we are looking for a terrestrial carbon sink of $2.2 \pm 1.0 \, \mathrm{Pg\,C\,year^{-1}}$ (see 2.3 above). Extensive inventories are suggesting a sink of 0.6–$0.7 \, \mathrm{Pg\,C\,year^{-1}}$ in temperate and boreal forests, of which only 30% is in live biomass [Goodale *et al.* (2002)], and a sink of between 1.0 and $2.0 \, \mathrm{Pg\,C\,year^{-1}}$ in the tropics (depending on pending results from Africa and Asia). Atmospheric inversion studies carry perhaps the greatest uncertainty of all, but the recent analysis by Rödenbeck *et al.* [2003] suggests an extratropical sink of $\mathrm{Pg\,C\,year^{-1}}$, and a tropical net sink of $0.8 \pm 1.3 \, \mathrm{Pg\,C\,year^{-1}}$, or $2.0 \pm 1.5 \, \mathrm{Pg\,C\,year^{-1}}$ after factoring out land use change. The uncertainties are still astonishingly large but these values are roughly what we would expect. It is likely that the carbon sink is distributed across many biomes, and perhaps driven by different processes: Climatic warming in boreal latitudes, forest regrowth and nitrogen fertilisation in temperate latitudes, and CO_2 fertilisation and increased water use efficiency in the tropics.

5. Implications of a Biospheric Carbon Sink for the Biosphere

As outlined above, there is now clear evidence that terrestrial ecosystems are accumulating carbon from the atmosphere, and thereby are performing a global service by slowing the projected rate of climate change. However, it is unclear whether this is necessarily beneficial for the ecosystems themselves. Superficially, it may seem that a greater supply of a limiting factor (CO_2) may be a good thing, but in fact many of the richest and most diverse ecosystems, such as coral reefs or tropical forests, thrive in nutrient-poor habitats. A paucity of nutrients encourages evolutionary innovation,

and a variety of strategies to access sufficient nutrients. As nutrient supply increases, it is possible that a few species will be poised to exploit
this new abundance, increasingly dominating over other species. This is a
phenomenon frequently observed on fertilised grasslands. It is interesting to
speculate whether rising atmospheric CO_2 may lead to such a phenomenon.
In tropical forests, for example, it is possible that fast-growing plant types,
such as lianas, trees that fix their own atmospheric nitrogen, trees that
exploit gaps in the forest canopy, may be better positioned to exploit rising
CO_2. One of the aims of our RAINFOR project is to look for evidence that
the composition of undisturbed tropical forests is shifting. This is a field
that has barely been explored in the past, but the early results are tantalising. For example, it appears that in Amazonia lianas may have doubled
in abundance in the last ten years, as would perhaps be expected [Phillips
et al. (2002); Wright *et al.* (2004)]. More recently, a meticulous forest plot
survey in central Amazonia suggests that fast growing canopy tree species
may be increasing in abundance at the expense of slow-growing (and biodi-
verse) understorey species [Laurance *et al.* (2004)]. As the weight of lianas
has a strong influence on the likelihood of tree death, this may have significant effects on forest dynamics and forest structure. There are likely to
be many other shifts in forest composition that have simply not yet been
looked for. The signals of global change are probably sitting there in the
Amazonian forests. We simply have to look for them and know how to
interpret them.

6. Prospects for the 21st Century

6.1. *Scenarios for the 21st century*

As outlined above, it is clear that human activities are fundamentally altering the global atmosphere, and in turn this alteration is affecting the terrestrial biosphere. This alteration is just beginning and is set to continue and
probably accelerate for much of the coming century as human industrio-
economic activity and population increases. Over the coming century, what
will be the nature of the interaction between the biosphere and atmosphere,
and how will we be able to measure and understand it?

The Intergovernmental Panel on Climate Change (IPCC) has published
a range of scenarios for anthropogenic CO_2 emissions over the 21st century.
Their "business-as-usual scenario" projects that 1400 Pg C will released into
the atmosphere over the coming century, but with a range of values varying from 2100 Pg C if humanity follows a fossil fuel intensive strategy to

800 Pg C if it follows is an environment-conscious, low emissions strategy. Whichever scenario is followed, a lot of CO_2 is going to be pushed into the atmosphere.

How much of this CO_2 remains in the atmosphere depends primarily on the behaviour of the global carbon cycle. Both ocean and terrestrial carbon sinks are likely to increase in magnitude over the coming decades, but that the rate of increase will slow, as the sinks begin to "saturate". The sink capacity in the oceans is chemically limited by biocarbonate-carbonate chemistry, whereas the CO_2 fertilisation effect on land is probably limited by plant physiology and by structural considerations and ecosystem-level feedbacks that constrain how much biomass a forest can hold. Moreover, warming surface temperatures will reduce both the solubility of CO_2 in ocean water and the residence time of carbon in soils.

A number of global carbon cycle models have been applied to this problem. Using a wide range of carbon cycle models and potential economic scenarios, IPCC [2001] predicts that by 2100 atmospheric CO_2 concentration will have risen from its current value of 368 ppm to between 500 and 1000 ppm, and that global mean temperatures will rise by between 1.5°C and 5.8°C. A more recent uncertainty analysis [Stainforth *et al.* (2005)] suggest that the upper limit of possible change this century is about 10°C. Very approximately, these model outputs suggest that every 3 Pg of C emissions will result in a 1 ppm increase in atmospheric CO_2 concentrations by 2100, which in turn will increase global mean temperatures by about 0.01°C. It is now inevitable that in this century we will face (and probably already are facing) a fundamentally altered atmosphere and climate.

6.2. *The Kyoto protocol and carbon politics*

The evidence outlined above suggests that the terrestrial biosphere is already playing a clear role in slowing the rate of atmospheric and climatic change. This has led to the debate over whether management of the terrestrial biosphere could be employed as a tool to further slow atmospheric change. This management could take the form of enhanced reforestation, reduced forest degradation through logging, slowed down tropical deforestation, and sequestration of carbon in soils through "no-till" agricultural practices [Royal Society (2001)]. These options would also have a number of beneficial environmental side-effects, such as protection of biodiversity, watershed protection, and reduced soil erosion. Another potentially exciting option is the growing of bio-fuels to replace fossil fuels as an energy source.

In the last few years, the potential use of biosphere sinks has moved from academic debate to the forefront of international politics. Since the adoption of the United Nations Framework Convention on Climate Change (UNFCCC) in 1992, world political leaders have moved towards recognising the need to act to prevent dangerous levels of climate change. In 1997, the Kyoto Protocol of the UNFCCC was signed, committing the industrialised countries to reduce their CO_2 emissions to various targets below 1990 levels by 2010. Since then, the Kyoto Protocol has faced a number of difficulties, in particular the withdrawal of the USA in 2001, but it finally came into force in 2005. One of the major points of contention has been whether human-induced carbon sinks should be included as a CO_2 emissions reduction strategy, and in particular deliberate planned carbon sinks generated by appropriate forest and land management. The debate centres on two points: Whether these sinks really make a long-term contribution to climate change mitigation, and whether these sinks can be reliably measured.

It has been estimated that throughout human history about 190 Pg C have been lost from the biosphere to the atmosphere by the clearing of forests for settlement and food production, about 10% of the total carbon content of the biosphere. An intensive and active deforestation reduction and reforestation program could at maximum return 75 Pg C back to the biosphere by 2050. Such a program could make a significant contribution to reducing net carbon emissions for the next few decades, but over the century becomes increasingly irrelevant compared to the 1400 Pg C emissions projected by 2100 under the IPCC "business-as-usual" scenario. Some countries (particularly in Europe) and many NGOs have argued that because these sinks can only make a small contribution to the long-term solution, they are a dangerous opt-out from the primary challenge, which is to develop technology and restructure energy supply and energy use to build low carbon emission societies. Other countries (the USA, Canada, Japan and Australia) have argued that the biosphere carbon sinks can be a viable component of an overall CO_2 emissions reductions program. The debate still rages, but as of 2001 biosphere carbon sinks are included within the Kyoto Protocol, albeit in a rather arbitrary way.

Whatever the rights and wrongs of the debate, it is increasingly clear that the global carbon cycle will need to be closely monitored for political and economic as well as scientific reasons. The evidence outlined above shows that the scientific community is making progress in understanding the carbon cycle, but we are still somewhat fumbling in the dark. What is lacking is the right observational tool, a "macroscope" that can tell us how much carbon is being emitted from a particular place at a particular time. Over the coming decade, there is the exciting possibility that this macro-scrope will be developed, and that the sharp light of comprehensive global observation will finally be shed on the obscure secrets of the global carbon cycle. That macroscope will be satellite observations of CO_2 concentration in high spatial and temporal detail [Houweling *et al.* (2004)].

6.3. *Surprises in the biosphere*

Thus far it appears that the terrestrial biosphere has been absorbing a sig-nificant fraction of anthropogenic CO_2, and thereby slowing down the rate of global climate change. Thus, from the point of view of the atmosphere at least, it can be viewed as a "friend", a negative feedback that favours stability in the climate system. However, there are some concerns as to how fickle this "friendship" might be, and whether, if pushed too far, the biosphere might become a net source of CO_2, thereby accelerating climate change and causing a positive feedback loop.

Attention has focused on two possibly unstable biosphere carbon reserves. One is the reserve of soil organic carbon that has been accumulat-ing in boreal regions (forest and tundra) since the end of the last ice age. This holds 600 Pg of carbon, carbon that has largely accumulated because of the slow rates of microbial decomposition in sub-zero temperatures. Because of a positive feedback between warming temperatures, retreating ice, dark-ening land surface and increased absorption of sunshine, these northern regions are predicted to warm far more than any other region of the earth, with IPCC estimating that by 2100 temperatures will 8 to 12°C warmer. As the permafrost thaws and these cold reserves of carbon heat up, there is considerable concern as to whether these cold stores of carbon will oxidise back into the atmosphere, and to what extent this release would be com-pensated for by enhanced plant growth. A release of 300–400 Pg of boreal carbon (compared to the 1400 Pg C emitted under a business as usual sce-nario) would significantly accelerate climate change. Current field data sug-gests that the soil carbon release and enhanced plant growth are cancelling each other out for the moment, but it is unclear how long this balance will continue.

The second potentially unstable reserve is the biomass of tropical forests. Some climate change scenarios suggest that large areas of the eastern Amazon basin may become too dry to sustain rainforests, leading to a replacement of forest by woodland and savanna, and a potential release of about 100 Pg C of soil and plant carbon [Cox *et al.* (2004)]. In addition to global climate effects, such a change would have serious consequences for local climate and biodiversity. It is estimated that on a global scale such a positive feedback could enhance land surface temperatures by a further 2.5°C.

There is no certainty that either of these feedbacks will happen but, if they do, their consequences would be serious. There is a vital need to understand what the climatic and biophysical thresholds are that maintain the biosphere, and how the resilience of the biosphere to future atmospheric change can be enhanced. We are entering a century that is probably unprecedented in terms of global environmental change, but it will also be a century where our understanding of the earth's biogeochemical cycles will move from clumsy measurement and extrapolation to detailed and spatially explicit mechanistic understanding. Within half a century, I anticipate our current fumbling research into the biosphere-atmosphere system may seem as primitive, and as pioneering, as 19th century attempts to understand human physiology seem when compared to modern medical science.

Acknowledgements

Yadvinder Malhi gratefully acknowledges the support of a Royal Society University Research Fellowship. Work reported in this paper is supported from grants from the Natural Environment Research Council (NERC) and the European Union. The data for Fig. 4(b) were kindly supplied by Robert Clement.

References

Achard, F., Eva, H. D., Stibig, H. J., Mayaux, P., Gallego, J., Richards, T. & Malingreau, J. P. (2002) Determination of deforestation rates of the world's humid tropical forests. *Science* **297**(5583), 999–1002.

Baker, T. R., Phillips, O. L., Malhi, Y., Almeida, S., Arroyo, L., Di Fiore, A., Erwin, T., Higuchi, N., Killeen, T. J., Laurance, S. G., Laurance, W. F., Lewis, S. L., Monteagudo, A., Neill, D. A., Vargas, P. N., Pitman, N. C. A., Silva, J. N. M. & Martinez, R. V. (2004) Increasing biomass in Amazonian forest plots. *Philosophical Transactions of the Royal Society of London Series B-Biological Sciences* **359**(1443), 353–365.

Baldocchi, D., Falge, E., Gu, L. H., Olson, R., Hollinger, D., Running, S., Anthoni, P., Bernhofer, C., Davis, K., Evans, R., Fuentes, J., Goldstein, A., Katul, G., Law, B., Lee, X. H., Malhi, Y., Meyers, T., Munger, W., Oechel, W., Pilegaard, K., Schmid, H. P., Valentini, R., Verma, S., Vesala, T.,

Wilson, K. & Wofsy, S. (2001) Fluxnet: A new tool to study the temporal and spatial variability of ecosystem-scale carbon dioxide, water vapor, and energy flux densities. *Bulletin of the American Meteorological Society* **82**(11), 2415–2434.

Chave, J., Brown, S., Cairns, M. A., Chambers, J., Eamus, D., Folster, H., Fromard, F., Higuchi, N., Kira, T., Lescure, J.-P., Nelson, B. W., Ogawa, H., Puig, H., Riera, B. & Yamakura, T. (2005) Tree allometry and improved estimation of carbon stocks and balance in tropical forests." *Oecologia* DOI: 10.1007/s00442-005-0100-x.

Cox, P. M., Betts, R. A., Collins, M., Harris, P. P., Huntingford, C. & Jones, C. D. (2004) Amazonian forest dieback under climate-carbon cycle projections for the 21st century. *Theoretical and Applied Climatology* **78**(1–3), 137–156.

DeFries, R. S., Houghton, R. A., Hansen, M. C., Field, C. B., Skole, D. & Townshend, J. (2002) Carbon emissions from tropical deforestation and regrowth based on satellite observations for the 1980s and 1990s. *Proceedings of the National Academy of Sciences of the United States of America* **99**(22), 14256–14261.

Dixon, R. K., Brown, S., Houghton, R. A., Solomon, A. M., Trexler, M. C. & Wisniewski, J. (1994) Carbon pools and flux of global forest ecosystems. *Science* **263**(5144), 185–190.

Goodale, C. L., Apps, M. J., Birdsey, R. A., Field, C. B., Heath, L. S., Houghton, R. A., Jenkins, J. C., Kohlmaier, G. H., Kurz, W., Liu, S. R., Nabuurs, G. J., Nilsson, S. & Shvidenko, A. Z. (2002) Forest carbon sinks in the northern hemisphere. *Ecological Applications* **12**(3), 891–899.

Gurney, K. R., Law, R. M., Denning, A. S., Rayner, P. J., Baker, D., Bousquet, P., Bruhwiler, L., Chen, Y. H., Ciais, P., Fan, S., Fung, I. Y., Gloor, M., Heimann, M., Higuchi, K., John, J., Maki, T., Maksyutov, S., Masarie, K., Peylin, P., Prather, M., Pak, B. C., Randerson, J., Sarmiento, J., Taguchi, S., Takahashi, T. & Yuen, C. W. (2002) Towards robust regional estimates of CO_2 sources and sinks using atmospheric transport models. *Nature* **415**(6872), 626–630.

Houghton, R. A. (1996) Terrestrial sources and sinks of carbon inferred from terrestrial data. *Tellus Series B-Chemical and Physical Meteorology* **48**(4), 420–432.

Houghton, R. A. (2005) Aboveground forest biomass and the global carbon balance. *Global Change Biology* **11**(6), 945–958.

Houweling, S., Breon, F. M., Aben, I., Rodenbeck, C., Gloor, M., Heimann, M. & Ciais, P. (2004). Inverse modeling of CO_2 sources and sinks using satellite data: A synthetic inter-comparison of measurement techniques and their performance as a function of space and time. *Atmospheric Chemistry and Physics* **4**, 523–538.

IPCC (2002) *Climate Change 2001: The Scientific Basis.* Cambridge, Cambridge University Press.

Laurance, W. F., Oliveira, A. A., Laurance, S. G., Condit, R., Nascimento, H. E. M., Sanchez-Thorin, A. C., Lovejoy, T. E., Andrade, A., D'Angelo, S.,

Ribeiro, J. E. & Dick, C. W. (2004) Pervasive alteration of tree communities in undisturbed amazonian forests. *Nature* **428**(6979), 171–175.

Malhi, Y., Baker, T. R., Phillips, O. L., Almeida, S., Alvarez, E., Arroyo, L., Chave, J., Czimczik, C. I., Di Fiore, A., Higuchi, N., Killeen, T. J., Laurance, S. G., Laurance, W. F., Lewis, S. L., Montoya, L. M. M., Monteagudo, A., Neill, D. A., Vargas, P. N., Patino, S., Pitman, N. C. A., Quesada, C. A., Salomao, R., Silva, J. N. M., Lezama, A. T., Martinez, R. V., Terborgh, J., Vinceti, B. & Lloyd, J. (2004) The above-ground coarse wood productivity of 104 neotropical forest plots. *Global Change Biology* **10**(5), 563–591.

Malhi, Y., Baldocchi, D. D. & Jarvis, P. G. (1999) The carbon balance of tropical, temperate and boreal forests. *Plant Cell and Environment* **22**(6), 715–740.

Malhi, Y. & Grace, J. (2000) Tropical forests and atmospheric carbon dioxide. *Trends in Ecology & Evolution* **15**(8), 332–337.

Nepstad, D. C., Verissimo, A., Alencar, A., Nobre, C., Lima, E., Lefebvre, P., Schlesinger, P., Potter, C., Moutinho, P., Mendoza, E., Cochrane, M. & Brooks, V. (1999) Large-scale impoverishment of Amazonian forests by logging and fire. *Nature* **398**(6727), 505–508.

Norby, R. J. (1998). Nitrogen deposition: A component of global change analyses. *New Phytologist* **139**(1), 189–200.

Norby, R. J., Ogle, K., Curtis, P. S., Badeck, F. W., Huth, A., Hurtt, G. C., Kohyama, T. & Penuelas, J. (2001) Aboveground growth and competition in forest gap models: An analysis for studies of climatic change. *Climatic Change* **51**(3–4), 415–447.

Norby, R. J., Wullschleger, S. D., Gunderson, C. A., Johnson, D. W. & Ceulemans, R. (1999) Tree responses to rising CO_2 in field experiments: Implications for the future forest. *Plant Cell and Environment* **22**(6), 683–714.

Perlin, J. (1999). *A Forest Journey: The Role of Wood in the Development of Civilization.* Cambridge, MA, Harvard University Press.

Phillips, O. L., Malhi, Y., Higuchi, N., Laurance, W. F., Nunez, P. V., Vasquez, R. M., Laurance, S. G., Ferreira, L. V., Stern, M., Brown, S. & Grace, J. (1998) Changes in the carbon balance of tropical forests: Evidence from long-term plots. *Science* **282**(5388), 439–442.

Phillips, O. L., Martinez, R. V., Arroyo, L., Baker, T. R., Killeen, T., Lewis, S. L., Malhi, Y., Mendoza, A. M., Neill, D., Vargas, P. N., Alexiades, M., Ceron, C., Di Fiore, A., Erwin, T., Jardim, A., Palacios, W., Saldias, M. & Vinceti, B. (2002) Increasing dominance of large lianas in Amazonian forests *Nature* **418**(6899), 770–774.

Prentice, I. C. (2001) The carbon cycle and atmospheric carbon dioxide, *Climate Change 2001: The Scientific Basis.* IPCC, Cambridge, Cambridge University Press.

Rodenbeck, C., Houweling, S., Gloor, M. & Heimann, M. (2003) CO_2 flux history 1982–2001 inferred from atmospheric data using a global inversion of atmospheric transport. *Atmospheric Chemistry and Physics* **3**, 1919–1964.

Royal Society (2001). *The Role of Land Carbon Sinks in Mitigating Global Climate Change.* London, The Royal Society, **27**.

Stainforth, D. A., Aina, T., Christensen, C., Collins, M., Faull, N., Frame, D. J., Kettleborough, J. A., Knight, S., Martin, A., Murphy, J. M., Piani, C., Sexton, D., Smith, L. A., Spicer, R. A., Thorpe, A. J. & Allen, M. R. (2005) Uncertainty in predictions of the climate response to rising levels of greenhouse gases. *Nature* **433**(7024), 403–406.

Telles, E. D. C., de Camargo, P. B., Martinelli, L. A., Trumbore, S. E., da Costa, E. S., Santos, J., Higuchi, N. & Oliveira, R. C. (2003) Influence of soil texture on carbon dynamics and storage potential in tropical forest soils of Amazonia. *Global Biogeochemical Cycles* **17**(2), 1040.

Wright, S. J., Calderon, O., Hernandez, A. & Paton, S. (2004) Are lianas increasing in importance in tropical forests? A 17-year record from Panama. *Ecology* **85**(2), 484–489.

Dust in the Earth System: The Biogeochemical Linking of Land, Air, and Sea

Andy Ridgwell

School of Geographical Sciences
University of Bristol
University Rd, Bristol BS8 1SS, United Kingdom
andy@seao2.org

Karen E. Kohfeld

School of Resource and Environmental Management
Simon Fraser University
8888 University Drive, Burnaby
British Columbia V5A 1S6 Canada
office@kohfeld.com

Understanding the response of the Earth's climate system to anthropogenic perturbation is a pressing priority for society. To be successful in this enterprise we need to analyse climate change within an all-encompassing "Earth system" framework; the suite of interacting physical, chemical, biological, and human processes that, in transporting and transforming materials and energy jointly determine the conditions for life on the whole planet. To illustrate the integrative thinking that is required we review the diverse roles played by atmospheric transport of mineral 'dust', particularly in its capacity as a key pathway for the delivery of nutrients essential to plant growth, not only on land, but more importantly, in the ocean. Here, the global importance of dust arises because of the control it exerts on marine plant productivity and thus the uptake of CO_2 from the atmosphere. The complex way in which dust biogeochemically links land, air, and sea presents us with new challenges in understanding climate change and forces us to ask questions that transcend the traditional scientific disciplines.

1. Introduction

Winds can pick up soil particles that are smaller than a few tens of micrometres in diameter and carry them great distances through the atmosphere. Although these individual particles are often invisible to the naked eye, billions of tons of material are transported every year in this way. Some of these transport events are even visible from space, as shown in the accompanying satellite image (Fig. 1). This 'dust' can comprise viruses, pollen grains, and industrial emissions such as soot. Over the ocean, sea salt particles produced by breaking waves and subsequent evaporation of water droplets are a major constituent of atmospheric aerosols. In this review, however, we consider dust to be soil mineral fragments.

The entrainment of dust from the land surface into air depends on several factors, including the surface vegetation cover, wind speed, and the properties (texture and moisture content) of soil. Dust is primarily emitted from regions lacking dense vegetation, i.e. regions where less than approximately 15% of the ground is covered. It is not surprising then to discover that dust sources are predominantly restricted to arid and semi-arid regions

Fig. 1   Satellite (SeaWiFS) image taken on February 26th 2000 of a massive sandstorm blowing off northwest Africa and reaching over 1000 miles into the Atlantic. (The SeaWiFS image was provided by NASA DAAC/GSFC and is copyright of Orbital Imaging Corps and the NASA SeaWiFS project.)

with desert, grassland, or shrubland vegetation [Prospero *et al.* (2002)]. In these sparsely vegetated areas, the wind speed across the surface must be great enough to lift particles into the air. This critical wind speed is called the "threshold velocity" and depends on the surface properties of the soil. Silt-sized particles are easiest to lift and require the lowest surface wind speeds to become airborne, while larger particles are heavier and have higher threshold wind velocities. The smallest, clay-sized particles have a larger surface area-to-volume ratio and tend to adhere to each other. Thus higher wind speeds are required to overcome the cohesive forces holding these particles together. The ability to lift dust into the air also depends on the moisture content of the exposed soils, as moisture tends to increase cohesion between soil particles.

How much dust enters the atmosphere? The most recent studies estimate that between 1000 and 2500 Tg (10^{12} grams) of dust is emitted each year [Zender *et al.* (2004)]. This wide range of estimates results from two factors. First, it is very difficult to obtain observational datasets that are extensive and detailed enough to quantify dust emissions on a global scale. Second, global models used to estimate the mobilisation and transport of dust parameterise the controlling factors differently. For similar reasons, our knowledge of the atmospheric burden, or the amount of dust that remains in the atmosphere, is even less concise. Estimates of the atmospheric burden of dust vary by a factor of four, ranging from 8 to 36 Tg [Zender *et al.* (2004)].

While heavier particles rapidly settle out of the air and are deposited close to their source, finer particles remain suspended in air and can be transported great distances by the prevailing winds. Eventual deposition to the Earth's surface occurs either through 'dry' depositional processes such as gravitational sedimentation or turbulent transfer, or through 'wet' depositional processes such as entrainment into falling raindrops ('precipitation scavenging'). All of these factors, in conjunction with atmospheric circulation, combine to create the distribution of dust deposition shown in Fig. 2. Particularly high rates are observed immediately downwind of the Sahara and Sahel deserts of North Africa and extend across the Atlantic to the Caribbean and northeastern South America. High deposition rates are also found over the northwestern Pacific and northeast Indian Oceans, associated with the deserts of central Asia. Less extensive dust sources in Australia, southern Africa and Patagonia have more localised influences. In contrast, marine locations remote from any major dust sources, such as the Southern Ocean, are characterised by dust deposition rates that are

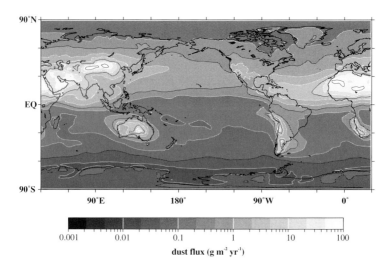

Fig. 2 **Model simulated distribution of the annual mean (1981–1997) rate of dust deposition to the Earth's surface** [Ginoux *et al.* (2002)].

100–1000 times lower than rates found immediately downwind of North Africa.

Dust affects the optical properties of the atmosphere by modifying incoming (ultraviolet and visible) and outgoing (infrared) radiation. According to climate models, dust aerosol can cause localised seasonal heating (over light-coloured surfaces) or cooling (over dark-coloured surfaces) by as much as $\pm 2°C$ [Miller and Tegen (1998)]. Another local heating effect can occur when dust is deposited on snow. Dust darkens the surface and decreases the fraction of sunlight that is reflected. This effect has been suggested as important in helping to melt the great ice sheets of the Northern Hemisphere at the end of the last ice age [Peltier and Marshall (1995)]. Finally, dust suspended in the atmosphere may affect climate by influencing cloud nucleation. It is when dust modifies the flow of carbon and nutrients within the Earth system ('global biogeochemical cycling'), however, that it arguably plays its most fascinating and intricate role.

2. Dust Deposition in the Terrestrial Realm

Wind-blown dust that settles on the land surface can accumulate to great thickness. For instance, over the past few million years, dust carried from Asian deserts to the Loess Plateau region of China has accumulated into soil sequences of up to 200 m thick. Dust influences soil structure even in places where deposition rates are considerably lower. For instance, aeolian quartz can become a major soil constituent when the underlying substrate is highly susceptible to weathering. On the basaltic bedrock of the Hawaiian Islands, much of the soil has literally come all the way from China and central Asian deserts [Kurtz *et al.* (2001)]. Dust exerts an important biogeochemical control upon ecosystem structure and plant productivity in these environments because its mineralogy and grain size strongly influence the water- and nutrient-holding properties of the soil.

Dust also plays a more direct biogeochemical role in terrestrial ecosystems. In parts of Amazonia, the soils are already highly weathered and nutrient-depleted, and the nutrient supply from rivers is likely not sufficient to maintain the nutrient balance of the rainforest on timescales of hundreds to thousands of years. In this region, aeolian deposition of nutrients such as phosphorous may be critical [Swap *et al.* (1992)]. Dust transported across the Atlantic from the deserts of the Sahara and Sahel (such as occurs during periodic dust storms — see Fig. 1) might then influence the maximum size of the ecosystem that can be supported. The highly weathered soils and phosphorous-limited ecosystems of some of the older Hawaiian Islands suggest an analogous situation, with losses due to leaching and immobilization of this vital nutrient exceeding local supply [Chadwick *et al.* (1999)]. Again, aeolian phosphorous transported across the ocean (this time the Pacific) is required to balance the nutrient budget of the ecosystem. Dust can therefore link the land surface of two physically remote landmasses. A change in one ecosystem, particularly a change to arid or semi-arid vegetation therefore has the potential to affect the productivity of the second ecosystem.

3. Dust Deposition in the Marine Realm

The major nutrients required by the primary producers of the open ocean (microscopic marine plants — 'phytoplankton') are phosphate (PO_4^{3-}) and nitrate (NO_3^-). Many species also require calcium or dissolved silica to construct their shells. These primary producers live and grow in the surface layers of the ocean where they receive sufficient sunlight for photosynthesis, and are kept for the most part from mixing into the deeper layers of the ocean by strong temperature and density gradients. As phytoplankton cells grow and divide, nutrients are removed from solution and transformed into cellular constituents. Most of this material is degraded by the action of

bacteria and zooplankton near the surface and returned into solution within
the mixed surface layer. However, a small (but important) fraction, in the
form of dead cells, zooplankton fecal pellets, and other particulate organic
debris sinks below the surface layer and is broken down much deeper down
in the ocean. Although the nutrients released even at depths of several
km will eventually be returned to the surface by upwelling and mixing, a
vertical gradient is created with lower nutrient concentrations at the surface
than in the deep ocean. This removal by the biota of dissolved constituents
from the surface waters and export of nutrients (in particulate form) to
depth is known as the 'biological pump'.

3.1. *Iron limitation in the ocean*

A long-standing puzzle in oceanography has been why phytoplankton do
not always fully utilise the nitrate that is supplied to them by the circulation of the ocean and thus why the biological pump does not always work
at its maximum possible rate. In certain areas of the world ocean and the
Southern Ocean in particular, high concentrations of NO_3^- remain in surface waters (Fig. 3). A similar situation prevails for PO_4^{3-} (not shown).
Despite the availability of NO_3^-, standing stocks of phytoplankton are relatively low, leading to the designation of such regions as 'High-Nutrient

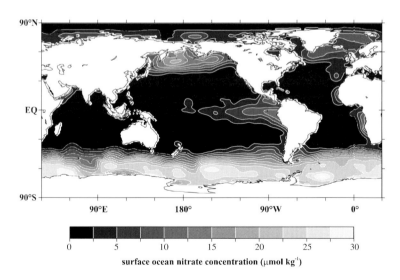

Fig. 3   Global distribution of near-surface (30 m depth) ocean nitrate (NO_3^-)
concentrations [Conkright *et al.* (1994)].

Low-Chlorophyll' (HNLC). Although physical conditions (such as low light levels) and the intensity of grazing by microscopic marine animals ('zooplankton') can help account for the HNLC condition they are not sufficient explanations on their own.

In the late 1980s came the idea that insufficient availability of iron might also restrict phytoplankton growth [Martin and Fitzwater (1998)]. Iron is essential for enzymatic activities associated with photosynthesis. Laboratory experiments demonstrated that the addition of Fe to HNLC water samples almost invariably stimulated phytoplankton growth and increased NO_3^- uptake. However, because the *in vitro* environment differs in a number of crucial respects from that of the ocean, the results of these small-scale experiments could not unambiguously tell us what was happening out in the open ocean.

A methodology for carrying out Fe fertilisation of the ocean was devised [Watson *et al.* (1991)], involving the dispersal of dissolved Fe from a ship whilst simultaneously marking the resulting 'patch' of enhanced Fe with an easily measurable label such as the inert tracer sulphur hexafluoride (SF_6). Following Fe release, the patch is crisscrossed, and observations made both within and outside the patch (as defined by the presence or absence of SF_6 in the water, respectively). The water outside acts as a 'control' on any changes measured in the Fe-enriched patch.

One such experiment was carried out in February of 1999 in the Southern Ocean — the 'Southern Ocean Iron RElease Experiment' (SOIREE) [Boyd *et al.* (2000)]. As hypothesised, the phytoplankton responded to the addition of Fe with a strong increase in the concentration of chlorophyll *a* within the fertilised patch but not outside it. (Chlorophyll *a* is a phytoplankton photosynthetic pigment whose concentration can be taken as a rough indicator of cell density.) In SOIREE, the impact of iron fertilisation was so striking that the results of the experiment were visible from space! Six weeks after the initial Fe release, gaps in the cloud cover allowed the remote sensing of surface ocean optical properties with the satellite image (Fig. 4) showing a 'bloom' of enhanced chlorophyll concentrations compared to the surrounding waters.

3.2. *Iron supply to the surface ocean*

Why should there be a deficit (relative to other nutrients) in the supply of iron to the biota, in some locations in the ocean but not others? Transport by rivers is the dominant route by which Fe is supplied to the ocean as

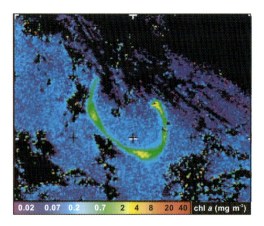

Fig. 4   Ocean colour satellite (SeaWiFS) image of surface ocean chlorophyll
a concentrations some 6 weeks after the deliberate release of iron in the
Southern Ocean. The fertilised ocean patch appears as a ribbon of high
chlorophyll a concentrations ∼100 km across. Cloud cover is indicated by
black regions. (SeaWiFS data provided by the NASA DAAC/GSFC and
copyright of Orbital Imaging Corps and the NASA SeaWiFS project, and
processed at CCMS-PML.)

a whole. But before it can reach the open ocean, rapid biological uptake
and sedimentation in highly productive estuaries and coastal zones tends to
remove much of the newly supplied Fe from the water. Rivers are therefore
not thought to be an important source of Fe to the open ocean. As with
NO_3^-, supply of Fe to the surface ocean occurs through upwelling and mixing
of ocean waters from below. However, dissolved Fe has a short lifetime
in the oxygenated seawater environment. Fe^{II}, the most soluble state, is
rapidly oxidized to Fe^{III} which is highly insoluble and tends to be removed
from solution by attaching to particulate matter settling through the water
column [Jickells et al. (2005)].

The result is that the concentration of Fe is low in upwelling water
relative to that of the highly soluble NO_3^-. Consequently, phytoplankton
in surface waters cannot fully utilise the abundant nitrate unless more Fe
is brought into the system. Dust becomes important in these iron-depleted
surface waters, because mineral aerosols contain about 4% iron by weight
(oxygen, silicon, and aluminum accounting for almost all the remainder).
Dissolution of this iron in surface waters has the potential to supply the
shortfall in upwelled Fe supply and enable phytoplankton to completely
utilise all available NO_3^-.

The distribution of dust deposition to the ocean (Fig. 2) shows that fluxes to the Southern Ocean are amongst the lowest anywhere on Earth. The aeolian supply is too small to compensate for the depleted Fe relative to NO_3^-. Consequently, phytoplankton cannot fully utilise the available NO_3^-. Similar reasoning also explains the high NO_3^- pool of the Eastern Equatorial Pacific (Fig. 3). Dust supply to the North Pacific appears moderately high, but fertilisation experiments in the Northwest Pacific [Tsuda *et al.* (2003)] suggest that this region is still iron-limited. Observations of natural dust-fertilisation events in the Northeast Pacific have shown that Asian dust is a significant source of iron today and can increase carbon biomass in the surface waters [Bishop *et al.* (2002)].

3.3. *Iron supply and the global carbon cycle*

The concentration of dissolved inorganic carbon (DIC) in the surface ocean exerts a fundamental control on air-sea CO_2 exchange along with other factors such as ambient temperature, pH, and wind speed. Processes that affect DIC will therefore influence the concentration of CO_2 in the atmosphere, and with it, climate (via the 'greenhouse effect'). One process that affects DIC concentrations is the biological removal of carbon from surface waters. Phytoplankton utilise carbon as well as nutrients at the ocean surface and incorporate it into cellular organic constituents. When biological activity reduces surface water DIC, the equilibrium concentration of gaseous CO_2 is depressed, driving a net transfer of CO_2 from the atmosphere into solution in the ocean. The concentration of CO_2 in the atmosphere will then exhibit an inverse relationship to the strength of the biological pump. Indeed, in the absence of any biological activity in the ocean, atmospheric CO_2 would be about 50% higher than it is today. Thus, changes in dust and iron supply to the ocean that modify the strength of the biological pump then have the potential to affect atmospheric CO_2 and climate.

4. Anthropogenic Modification of Dust Supply

The present-day supply of Fe to high-nitrate low-chlorophyll (HNLC) regions is not large enough for the ocean's biological pump to work at its maximum efficiency and fully utilise all supplied NO_3^-. In addition, other regions such as the central tropical Pacific and North Atlantic may be close to limitation or quasi-limited by Fe. Any reduction in dust supply will intensify limitation where it already exists and potentially induce limitation of productivity elsewhere. Either way, if aeolian Fe fluxes were lower, there should be a reduction in the rate of CO_2 uptake by the ocean, which has implications for atmospheric CO_2 concentrations and the rate and degree of future climate change. Under what circumstances might a reduction in dust supply to the ocean occur? Answering this question involves clarifying the two different types of anthropogenic contributions to atmospheric dust [Zender et al. (2004)]. Humans can influence the production of dust directly by altering the land surface. Alternatively, humans can indirectly affect the atmospheric dust burden through the cumulative impact of anthropogenic climate changes on the dust cycle.

The deliberate large-scale manipulation of terrestrial ecosystems has been proposed for the 'locking up' (sequestration) of carbon on land. These include, changes in soil management practices such as reducing tillage, enhancing the areal and seasonal extent of ground cover, and the 'setting-aside' of surplus agricultural land, in addition to the restoration of pre-viously degraded lands and forestation [Royal Society (2001)]. However, reduced disturbance, stabilisation of soils, and greater vegetation cover are also likely to reduce dust emissions. Since dust exerts an important control on the biological pump in the ocean, the effectiveness of carbon sequestra-tion on land may be diminished by a reduction in carbon uptake by the ocean.

Early models of dust transport and deposition suggested that a substan-tial (30–50%) component of the present-day global dust supply originated in disturbed soils [Tegen and Fung (1995)]. If these soils were stabilised in the future for sequestering carbon, a substantial decrease in global dust emis-sions would occur. Computer models suggest that a 30% reduction in dust flux to the global ocean would lower ocean productivity to such an extent that the weaker ocean carbon sink could potentially offset the benefit of sequestering carbon on land [Ridgwell et al. (2002)], leaving atmospheric CO_2 unchanged. However, subsequent satellite-based analyses suggest that the anthropogenic component is much smaller [Prospero et al. (2002)]. Fur-thermore, more recent attempts to match dust model simulations to the surface observations have suggested agricultural practices contribute less than 10% to the total global atmospheric dust burden [Tegen et al. (2004b)], although limited surface observations make this number difficult to ascer-tain exactly [Mahowald et al. (2004); Tegen et al. (2004a)]. These more recent results demonstrate that quantifying the possible reduction in dust production due to land-use changes remains an important challenge.

Socio-economic and political factors are likely to ultimately dictate any future large-scale alteration of the land surface, with changes in dust supply probably occurring on a regional scale rather than globally. For instance, a massive reforestation program is already under way in China with the specific intention of combating soil erosion and associated dust storms. Although several recent studies have demonstrated that climate factors play a strong role in determining the frequency of dust storms over China [Mukai *et al.* (2004); Zhao *et al.* (2004)], changes in reforestation could have (as yet unquantified) implications for marine ecosystems in the iron-sensitive equatorial and North Pacific.

The second means by which human activity could affect dust emissions is through anthropogenic climate change. A change in climate could drive an increase in vegetation cover in arid areas, reducing the supply of dust to the atmosphere [Harrison *et al.* (2001)]. The efficiency with which dust is transported through the atmosphere may also change, with any increase in global precipitation removing more dust before it reaches the open ocean. However, current model simulations have not reached a consensus regarding the impact of future climate on dust emissions. While some simulations suggest that dust emissions will decrease by as much as 60% by 2090 [Mahowald and Luo (2003)], other simulations suggest that dust emissions might even increase by as much as 10%. Thus, the response and even the direction of change of future dust emissions is highly dependent on the climate model used [Tegen *et al.* (2004b)].

Current computer carbon cycle models only give a relatively crude indication of the possible impacts. Such experiments do, however, serve to highlight the important link within the Earth system that is mediated by dust; a connection between carbon cycling and climate and human activities that was previously completely overlooked. The rather narrow and restricted land-atmosphere approach to carbon budgeting in the Kyoto Protocol that neglects important Earth system feedbacks involving the ocean is then too simplified.

5. The Demise of the Last Ice Age: A Role for Dust?

The Earth has experienced a series of intense ice ages over the course of the last million years or so. Each ice age ended rather suddenly, with a rapid warming transition ('termination') from cold glacial conditions into a (relatively brief) mild interglacial period (Fig. 5a). Many different theories have been advanced for how these cycles might be driven. These have typically focused on the physical climate system, particularly interactions between ice sheets and underlying bedrock with external forcing provided by orbitally-driven variations in the seasonal intensity of sunlight received

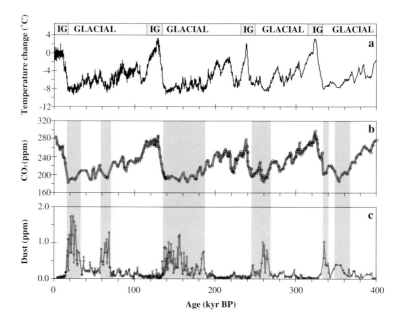

Fig. 5   **Key indicators of climatic state contained within the Vostok ice core
[Petit *et al.* (1999)]. (a) Isotopically-derived temperature change (relative
to the present) at the surface. Cold glacial and warmer interglacial ('IG')
intervals are indicated. (b) CO_2 concentration in air bubbles contained within
the ice. (c) Dust concentration in the ice. The correspondence between CO_2
minima and prominent dust peaks are highlighted.**

at the Earth's surface. However, such explanations fail to correctly predict
the amplitude and timing of the observed cyclically in global ice volume,
suggesting that additional factors might also be critical [Ridgwell *et al.*
(1999)].

Records of past atmospheric composition, in the form of microscopic
bubbles of ancient air trapped within the crystalline structure of ice sparked
a revolution in understanding of what drives these ice age cycles. Cores of
ice recovered from Antarctica and analysed for air bubble gas composition
revealed that atmospheric CO_2 varies cyclically between about ~280 ppm
during interglacials and ~190 ppm during the most intense glacial periods
(Fig. 5b).

What causes the observed variability in CO_2? A possible clue comes
from the changes in dust deposition, also recorded in the Vostok ice
(Fig. 5c). The concentration of dust contained within the ice exhibits a

series of rather striking peaks against a background of relatively low values; a much greater dynamic range than can be accounted for by dilution effects arising from changes in snow accumulation rate alone. The occurrence of these peaks correlates with periods of particularly low atmospheric CO_2 values. This is certainly consistent with increased dustiness during glacial times providing more iron to the surface and driving a more vigorous biological pump in the ocean [Martin (1990); Watson *et al.* (2000)]. However, investigations of the global carbon cycle using both numerical models [Archer *et al.* (2000); Bopp *et al.* (2003)] and observations [Bopp *et al.* (2003); Kohfeld *et al.* (2005)] suggest that an increase in the strength of the biological pump can only be part of the explanation for low glacial atmospheric CO_2 concentrations. Other mechanisms must be at work.

If dust is responsible for some of the observed glacial-interglacial variability in atmospheric CO_2, then we need a much better understanding of the factors that bring about changes in dust fluxes. Elevated glacial dust fluxes are not restricted to Antarctica. In fact, similar features are found in dust records from ice, marine, and terrestrial environments around the world. A colder, drier glacial climate, with a less vigorous hydrological cycle would result in decreased precipitation scavenging, more efficient transport of dust, and thus higher deposition rates. However, models of dust generation, transport, and deposition suggest that a reduction in the hydrological cycle alone is not sufficient to explain the increases in glacial dustiness. Greater source strengths of dust must also be invoked. The expansion of arid areas under cold, dry glacial climatic conditions, or even the exposure of continental shelves as the ice sheets grow and sea-level drops could result in new sources of dust. Furthermore, higher wind speeds during the glacial period could result in enhanced entrainment from existing source areas [e.g. Mahowald *et al.* (1999); Werner *et al.* (2003)]. Thus, changes in sea-level, aridity, and vegetation type and cover, as well as atmospheric circulation, and precipitation strength and patterns all combine to affect dust deposition, and with it, atmospheric CO_2 and climate [Ridgwell and Watson (2002)].

6. Conclusion and Perspectives

We are just embarking on a radical new integrated view of how the Earth system functions on a range of timescales [Schellnhuber (1999)]. This holistic view will be critical if we are to understand the complex and sometimes

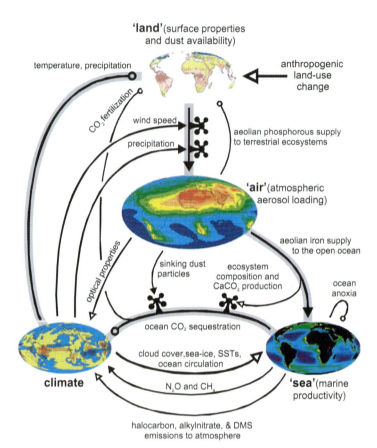

Fig. 6   Schematic view of the linking of land, air, and sea (and climate) by dust. [Adapted from Jickells *et al.* (2005)]. Highlighted are four critical components of the Earth system, (clockwise from top); state of the land surface and dust material availability ('land'), atmospheric aerosol loading and dust deposition ('air'), marine plankton productivity ('sea'), and climatic state (e.g. mean global surface temperature). The biogeochemical connections between them can have a positive correlation (e.g. increased atmospheric aerosol loading and dust deposition results in increased in marine productivity) indicated by a filled arrowhead, or a negative correlation (e.g. increased marine productivity leads to lower atmospheric CO_2 and a colder climate), indicated with an open circle. An open arrowhead indicates where the sign of the correlation is uncertain. The 'taps' represent where a mechanism affects the strength of a connection between two components rather than affecting a component directly. A change in global precipitation strength altering the efficiency with which entrained dust is transported out to the open ocean is a good example of this. Shown back-highlighted (grey lines) is the positive feedback; atmospheric aerosol loading → marine productivity → climatic state → dust availability → atmospheric aerosol [Ridgwell (2003); Ridgwell and Watson (2002)].

unexpected behaviour of the climate system that arises out of a high level of interaction and interconnectedness. Indeed, even individual sub-systems such as that involving dust (Fig. 6) can exhibit highly complex and nonlinear behaviour, because dust is much more than simply as a passive 'communicator' of events between components of the Earth system [Jickells *et al.* (2005)]. For instance, if changes in dust flux affect atmospheric CO_2 and climate, and dust fluxes are in turn responsive to global climate through changes in the land surface and the strength of the hydrological cycle, this raises the possibility of a 'feedback' [Ridgwell (2003); Ridgwell and Watson (2002)] (highlighted in Fig. 6). In such a system, any global cooling that occurs, such as the descent into a glacial state, could produce an increase in dust availability and transport efficiency. This could, in turn, drive a decrease in CO_2 through Southern Ocean iron fertilisation, causing a further cooling and thus further enhanced dust supply, etc. This 'positive' feedback will amplify the magnitude of the initial perturbation (climate cooling). This same dust feedback may also amplify any future climate change that is initially driven by CO_2 emissions from fossil fuel burning.

To fully understand the complicated role of feedbacks and overall behaviour of the operation of the climate system, we need fresh investigative tools — numerical models of the Earth system (e.g. 'genie' http://www.genie.ac.uk/). By coupling together representations of ocean and atmospheric circulation, cryosphere, and descriptions of the primary biogeochemical cycles that permeate the land, atmosphere, ocean, and sediments, the operation of the climate system on a range of time scales can be comprehensively explored. If these models are further extended by integration with socio-economic models, the interaction between the climate system and anthropogenic activities can be addressed. Climate-socio-economic 'Integrated Assessment Models' are currently being actively developed by institutions such as the UK Tyndall Centre (http://www.tyndall.ac.uk/), and will greatly aid us in deciding how we might mitigate and adapt to future climate change.

Acknowledgements

This work was supported by the Trusthouse Charitable Foundation. We would like to thank Ina Tegen for useful insights and discussions.

References

Archer, D., Winguth, A., Lea, D. & Mahowald, N. (2000) What caused the glacial/interglacial atmospheric pCO$_2$ cycles? *Rev. Geophys.* **38**, 159–189.

Bishop, J. K. B., Davis, R. E. & Sherman, J. T. (2002) Robotic observations of dust storm enhancement of carbon biomass in the North Pacific. *Science* **298**, 817–821.

Bopp, L., Kohfeld, K. E., Le Quéré, C. & Aumont, O. (2003) Dust impact on marine biota and atmospheric CO$_2$ during glacial periods. *Paleoceanogr.* **18**, 10.1029/2002PA000810.

Boyd, P. W., Watson, A. J., Law, C. S., Abraham, E. R., Trull, T. W., Murdoch, R., Bakker, D. C. E., Bowie, A. R., Buesseler, K. O., Chang, H., Charette, M., Croot, P., Downing, K., Frew, R. D., Gall, M., Hadfield, M., Hall, J., Harvey, M., Jameson, G., Laroche, J., Liddcoat, M., Ling, R., Maldonado, M. T., McKay, R. M., Nodder, S., Pickmere, S., Pridmore, R., Rintoul, S., Safi, K., Sutton, P., Strzepek, R., Tanneberger, K., Turner, S., Waite, A. & Zeldis, J. (2000) A mesoscale phytoplankton bloom in the polar Southern Ocean stimulated by iron fertilization. *Nature* **407**, 695–702.

Chadwick, O. A., Derry, L. A., Vitousek, P. M., Huebert, B. J. & Hedin, L. O. (1999) Changing sources of nutrients during four million years of ecosystem development. *Nature* **397**, 491–497.

Conkright, M. E., Levitus, S. & Boyer, T. P. (1994) World Ocean Atlas 1994 Volume 1. *Nutrients.* Washington, D.C., U.S. Department of Commerce.

Ginoux, P., Chin, M., Tegen, I., Prospero, J., Holben, B., Dubovik, O. & Lin, S.-J. (2001) Global simulation of dust in the troposphere: Model description and assessment. *J. Geophys. Res.* **106**(20), 20255–20273.

Harrison, S. P., Kohfeld, K. E., Roelandt, C. & Claquin, T. (2001) The role of dust in climate changes today, at the last glacial maximum and in the future. *Earth Sci. Rev.* **54**, 43–80.

Jickells, T. D., An, Z. S., Andersen, K. K., Baker, A. R., Bergametti, G., Brooks, N., Cao, J. J., Boyd, P. W., Duce, R. A., Hunter, K. A., Kawahata, H., Kubilay, N., laRoche, J., Liss, P. S., Mahowald, N., Prospero, J. M., Ridgwell, A. J., Tegen, I. & Torres, R. (2005) Global iron connections between desert dust, ocean biogeochemistry, and climate. *Science* **308**, 67–71.

Kohfeld, K. E., Le Quéré, C., Harrison, S. P. & Anderson, R. F. (2005) Role of marine biology in glacial-interglacial CO$_2$ cycles. *Science* **308**, 74–78.

Kurtz, A. C., Derry, L. A. & Chadwick, O. A. (2001) Accretion of Asian dust to Hawaiian soils: Isotopic, elemental, and mineral mass balances. *Geochim. Cosmochim. Acta* **65**, 1971–1983.

Mahowald, N., Kohfeld, K. E., Hansson, M., Balkanski, Y., Harrison, S. P., Prentice, I. C., Schulz, M. & Rodhe, H. (1999) Dust sources and deposition during the Last Glacial Maximum and current climate: A comparison of model results with palaeodata from ice cores and marine sediments. *J. Geophys. Res.* **104**, 15895–15916.

Mahowald, N. M. & Luo, C. (2003) A less dusty future? *Geophys. Res. Lett.* **30**, 1903, doi:10.1029/2003GL017880.

Mahowald, N. M., Rivera, G. D. R. & Luo, C. (2004) Comment on "Relative importance of climate and land use in determining present and future global soil dust emission" by I. Tegen *et al.*, *Geophys. Res. Lett.* **31**, doi:10.1029/2004GL021272.

Martin, J. (1990) Glacial-interglacial CO_2 change: The iron hypothesis. *Paleoceanogr.* **5**, 1–13.

Martin, J. H. & Fitzwater, S. E. (1998) Iron deficiency limits phytoplankton growth in the North-East Pacific subarctic. *Nature* **331**, 341–343.

Miller, R. L. & Tegen, I. (1998) Climate response to soil dust aerosols. *J. Clim.* **11**, 3247–3267.

Mukai, M., Nakajima, T. & Takemura, T. (2004) A study of long term trends in mineral dust aerosol distributions in Asia using a general circulation model. *J. Geophys. Res.* **109**, doi:10.1029/2003JD004270.

Peltier, W. R. & Marshall, S. (1995) Coupled energy-balance/ice-sheet model simulations of the glacial cycle: A possible connection between terminations and terrigenous dust. *J. Geophys. Res.* **100**, 14269–14289.

Petit, J. R., Jouzel, J., Raynaud, D., Barkov, N. I., Barnola, J. M., Basile, I., Bender, M., Chappellaz, J., Davis, M., Delaygue, G., Delmotte, M., Kotlyakov, V. M., Legrand, M., Lipenkov, V. Y., Lorius, C., Pepin, L., Ritz, C., Saltzman, E. & Stievenard, M. (1999) Climate and atmospheric history of the past 420,000 years from the Vostok ice core, Antarctica. *Nature* **399**, 439–436.

Prospero, J. M., Ginoux, P., Torres, O., Nicholson, S. E. & Gill, T. E. (2002) Environmental characterization of global sources of atmospheric soil dust identified with the Nimbus 7 Total Ozone Mapping Spectrometer (TOMS) absorbing aerosol product. *Rev. Geophys.* **40**, Art. No. 1002.

Ridgwell, A. J. (2003). Implications of the glacial CO_2 "iron hypothesis" for Quaternary climate change. *Geochem. Geophys. Geosyst.* **4**, 1076, doi:10.1029/2003GC000563.

Ridgwell, A. J., Maslin, M. A. & Watson, A. J. (2002) Reduced effectiveness of terrestrial carbon sequestration due to an antagonistic response of ocean productivity. *Geophys. Res. Lett.* **29**, doi:10.1029/2001GL014304.

Ridgwell, A. J. & Watson, A. J. (2002) Feedback between aeolian dust, climate, and atmospheric CO_2 in glacial time. *Paleoceanogr.* **17**, doi:10.1029/2001PA000729.

Ridgwell, A. J., Watson, A. J. & Raymo, M. E. (1999) Is the spectral signature of the 100 kyr glacial cycle consistent with a Milankovitch origin? *Paleoceanogr.* **14**, 437–440.

Royal Society (2001) The role of land carbon sinks in mitigating global climate change, pp. 10/01. Royal Society Document.

Schellnhuber, H. J. (1999) 'Earth system' analysis and the second Copernican revolution. *Nature* **402**, C19–C23.

Swap, R., Garstang, M., Greco, S., Talbot, R. & Kallberg, P. (1992) Saharan dust in the Amazon Basin. *Tellus, Series-B* **144B**, 133–149.

Tegen, I. & Fung, I. (1995) Contribution to the atmospheric mineral aerosol load from land surface modification. *J. Geophys. Res. Atmos.* **100**, 18707–18726.

Tegen, I., Werner, M., Harrison, S. P. & Kohfeld, K. E. (2004a) Reply to comment by N. M. Mahowald et al. on "Relative importance of climate and land use in determining present and future global soil dust emission". *Geophys. Res. Lett.* **31**, doi:10.1029/2004GL021560.

Tegen, I., Werner, M., Harrison, S. P. & Kohfeld, K. E. (2004b) Relative importance of climate and land use in determining present and future global soil dust emission. *Geophys. Res. Lett.* **31**, doi:10.1029/2003GL019216.

Tsuda, A., Takeda, S., Saito, H., Nishioka, J., Nojiri, Y., Kudo, I., Kiyosawa, H., Shiomoto, A., Imai, K., Ono, T., Shimamoto, A., Tsumune, D., Yoshimura, T., Aono, T., Hinuma, A., Kinugasa, M., Suzuki, K., Sohrin, Y., Noiri, Y., Tani, H., Deguchi, Y., Tsurushima, N., Ogawa, H., Fukami, K., Kuma, K. & Saino, T. (2003) A mesoscale iron enrichment in the western Subarctic Pacific induces a large centric diatom bloom. *Science* **300**, 958–961.

Watson, A., Liss, P. S. & Duce, R. (1991) Design of a small-scale *in situ* iron fertilization experiment. *Limnol. Oceanogr.* **36**, 1960–1965.

Watson, A. J., Bakker, D. C. E., Ridgwell, A. J., Boyd, P. W. & Law, C. S. (2000) Effect of iron supply on Souther Ocean CO_2 uptake and implications for glacial atmospheric CO_2. *Nature* **407**, 730–734.

Werner, M., Tegen, I., Harrison, S. P., Kohfeld, K. E., Prentice, I. C., Balkanski, Y., Rodhe, H. & Roelandt, C. (2003) Seasonal and interannual variability of the mineral dust cycle under present and glacial climate conditions. *J. Geophys. Res.* **108**, 10.1029/2002JD002365.

Zender, C. S., Miller, R. L. & Tegen, I. (2004) Quantifying mineral dust mass budgets: Terminology, constraints, and current estimates. *Eos* **85**, 509–512.

Zhao, C., Dabu, X. & Li, Y. (2004) Relationship between climatic factors and dust storm frequency in Inner Mongolia of China. *Geophys. Res. Lett.* **31**, doi:10.1029/2003GL018351.

The Late Permian Mass Extinction Event and Recovery: Biological Catastrophe in a Greenhouse World

Richard J. Twitchett

School of Earth, Ocean and Environmental Sciences
University of Plymouth
PL4 8AA, UK
richard.twitchett@plymouth.ac.uk

The extinction event at the close of the Permian period was the largest of the Phanerozoic. Understanding this event is crucial to understanding the history of life on Earth, and the past decade has witnessed a dramatic increase in the number of publications relating to this event. Four main areas of research and debate are considered to be the reason for the recent surge in scientific interest. These are (1) issues of dating and stratigraphy, (2) potential causes (specifically the debate of extra-terrestrial impact versus volcanically triggered global warming), (3) the patterns and rate of extinction, and (4) the nature of the post-extinction recovery. These key research areas are outlined below.

1. Stratigraphy and Dating

The base of the Triassic (i.e. the P/Tr boundary) is defined as the point at which the conodont taxon *Hindeodus parvus* first appears in the global stratotype section at Meishan, South China [Yin *et al.* (2001)] (Fig. 1). *H. parvus* is distributed worldwide and was chosen as the boundary-defining index fossil, after years of international debate and deliberation, in order to maximise the number of geological sections that can be directly correlated with each other. In the Meishan section, the first appearance datum (FAD) of *H. parvus* lies above the main extinction horizon [Jin *et al.* (2000)]. The extinction is thus a latest Permian (late Changhsingian) event. The FAD of *H. parvus* also occurs above a major (typically between 3 and 4%) negative

69

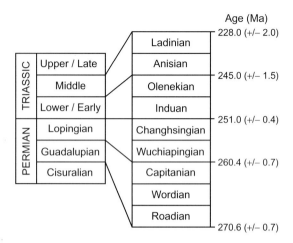

Fig. 1 Stratigraphy and dating of the Permian-Triassic interval. Ages from Gradstein *et al.* [2004]. The extinction event occurred in the late Changhsingian stage, and has been recently dated as 252.6 Ma by Mundil *et al.* [2004].

shift in δ^{13}C, which is occasionally used to identify "the P/Tr boundary" in sections where fossils are rare or absent [e.g. Baud *et al.* (1989)]. However, locally, this negative isotope shift may occur just below [Brookfield *et al.* (2003)], just above [Twitchett *et al.* (2001)], or be synchronous with the extinction horizon, and always begins within the late Changhsingian.

Volcanic ash beds at Meishan have also enabled absolute dating of the boundary interval through U/Pb dating of zircons. In 1991, zircons from an ash bed that was, at that time, inferred to be coincident with the P/Tr boundary (the bed actually lies below the FAD of *H. parvus*, and is now considered latest Changhsingian) was assigned the age of 251.2 ± 3.4 Ma [Claoué-Long *et al.* (1991)]. Subsequent reanalysis of the Meishan ash beds by Bowring *et al.* [1998] produced an age of 251.4 ± 0.3 Ma, confirming the earlier result, but with apparently far higher resolution, and cementing 251 Ma in the subsequent literature.

However, accurate dating is notoriously difficult. Mundil *et al.* [2001] confirmed that there were indeed problems with the earlier dates, due to diagenetic loss of lead from the zircons analysed and the presence of numerous zircons of disparate ages within a single bed; they reassigned the ash bed above the P-Tr boundary an (older) age of 252.7 ± 0.4 Ma. However, the ash bed below the P-Tr boundary, approximating to the extinction horizon, proved far more difficult to date: probably older than 254 Ma was the best estimate [Mundil *et al.* (2001)]. In the most recent development, improvements in the dating technique and analysis of ash beds from a section at Shangsi have resulted in a new age of 252.6±0.2 Ma for the extinction event, with the P/Tr boundary being slightly younger [Mundil *et al.* (2004)].

Do these problems with dating matter? For most P-Tr studies they are largely irrelevant. Correlation between sections relies on well-established methods of biostratigraphy and is unaffected by such geochronological revisions. The relative sequence of events can be deduced regardless. Problems may arise, however, when trying to link events in distant marine and terrestrial sections, where radiometric dating is the only means of correlation (see below), or when trying to determine absolute rates of faunal and/or environmental change.

2. The Question of Cause

2.1. *Extraterrestrial impact?*

The impact of large extraterrestrial bodies with the Earth is viewed by some as the major driver of global extinction, but it is clear that not every large impact event is associated with extinction. The publication by Alvarez *et al.* [1980] is widely credited with igniting the recent scientific interest in impact-related extinction, although a few other authors [e.g. McLaren (1970)] had suggested this possibility much earlier. Three pieces of evidence are considered crucial in unequivocally identifying an extraterrestrial impact event: a crater, an enrichment ('spike') of the otherwise rare element iridium, and a layer of impact debris containing shocked quartz and tektites. In order to conclude that bolide impact has caused a particular extinction event, such as the end-Permian event, three things are necessary: (1) unequivocal evidence of geologically instantaneous extinction; (2) unequivocal evidence of impact coincident with the extinction horizon; (3) absence of evidence for other potential extinction-causing physical mechanisms occurring at the same time.

Initial attempts to find evidence of a P-Tr impact were made in the early 1980s. Several Chinese studies reported significant iridium enrichment near the P/Tr boundary [e.g. Xu *et al.* (1985)] but, despite many attempts, none of these results have ever been replicated [Erwin (1993)] and these data are considered highly dubious. The small Ir peaks that have been recorded in uppermost Permian and lowermost Triassic marine strata are attributed to concentration under reducing conditions [e.g. Koeberl *et al.* (2004)]. No widespread impact layer has been observed at or near the P/Tr boundary.

Regardless, this debate has intensified in recent years. Becker *et al.* [2001] reported the presence of helium and argon trapped in fullerenes isolated from the P-Tr boundary beds in China and Japan. Isotopic profiles of the trapped helium indicated that the fullerenes had to have come from an extraterrestrial source [Becker *et al.* (2001)]. However, persistent problems remain with acceptance of these data, as other scientists have consistently failed to replicate the results, despite using samples from exactly the same sites and exactly the same laboratory procedures [Farley and Mukhopadhyay (2001)]. In addition, the Japanese samples came from well below the extinction level (Isozaki, 2001). Later that same year, Kaiho *et al.* [2001] described P-Tr sediment grains, from the GSSP at Meishan, that were supposedly formed by impact, as well as geochemical shifts that they interpreted as indicating a huge impact event. However, once more, these data were severely criticised by other geochemists [Koeberl *et al.* (2002)]. Basu *et al.* [2003] revived interest by claiming to have found forty tiny, (50–400 μm), unaltered fragments of meteorite in a sediment sample from a terrestrial P-Tr boundary section in Antarctica. Although the authors were at pains to dismiss contamination as a source of the fragments, other experts were immediately sceptical as meteoritic metals are highly reactive and in terrestrial settings oxidise extremely quickly [Kerr (2003)].

Most recently, announcement was made of the discovery of a purported impact crater situated off the coast of western Australia [Becker *et al.* (2004)]. The evidence included seismic profiles of the structure itself, a description of cores taken from the centre of the structure apparently showing an impact breccia, and ^{40}Ar/^{39}Ar dating of feldspars to provide the age constraint. However, every one of these lines of evidence has been subsequently criticised: The gravity anomaly map purporting to show a buried crater is significantly different to those of other confirmed impact structures; the impact breccia contains no unambiguous shocked minerals; and the age designation comes from a single sample of unknown stratigraphic horizon with a spread of inferred ages that do not define an objective "age plateau" [Renne *et al.* (2004)].

Doubtless, new data supporting a P-Tr extraterrestrial impact will continue to be published as this particular avenue of research continues to be explored. However, all previously published data are highly controversial and lack the key criteria of major impact that have been recorded time and again in K-T sections the world over. Independent replication of results is also crucial for scientific acceptance, especially when the data are unusual and/or controversial. So far, all the evidence proposed for a P-Tr boundary impact has failed this necessary test. On present evidence, the hypothesis that an extraterrestrial impact occurred near the P/Tr boundary is not supported.

2.2. *Volcanically triggered global warming?*

The leading alternative hypothesis is that rapid and severe climate change (specifically global warming) was responsible [e.g. Benton and Twitchett, (2003); Kidder and Worsley (2004)]. This model has evolved since the early 1990s, and incorporates a number of potential extinction mechanisms that were at one time [Erwin (1993)] considered to be separate, though not necessarily mutually exclusive, possibilities.

The present hypothesis is that flood basalt eruptions, represented by the Siberian Traps, vented large amounts of CO_2 into the atmosphere over a relatively short period of time. This resulted in rising global temperatures. Warming then led to the destabilisation and disassociation of shallow (marine and/or terrestrial) gas hydrate deposits, which vented large volumes of CH_4 into the oceans and atmosphere. This CH_4, although rapidly oxidised to CO_2, then caused more warming, which in turn would have melted further gas hydrate reservoirs. During this positive feedback loop, some sort of threshold was probably reached, beyond which the natural systems that normally reduce carbon dioxide levels could not operate and a 'runaway greenhouse' ensued (Fig. 2). Global warming would have had devastating effects on terrestrial ecosystems, and also in the marine realm, where it is believed to have caused a rise in sea level, stagnation, oceanic anoxia and a decrease in primary productivity.

What of the evidence? The basic premise of the model is that, climatically, the P-Tr interval was a time of global warming. Certainly, the Permian as a whole was a time of long-term global warming [Kidder and Worsley (2004)]. Evidence of extensive ice sheets are confined to the Early Permian and the last vestiges of glaciation in Australia and Siberia are now dated as Middle Permian [Erwin (1993)], although they had been thought by some to be late Permian in age [Stanley (1988)]. It has been argued that this long-term Permian warming trend is due to a global reduction in continental collisions and orogenic events, as the supercontinent Pangaea was fully assembled at this time [Kidder and Worsley (2004)]. Absence of mountain building would have resulted in a drop in global weathering rates and hence

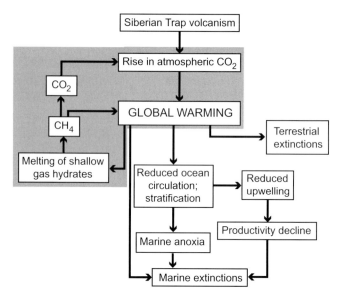

Fig. 2 **Schematic model of a volcanically triggered global warming scenario for the end-Permian mass extinction event (see text for detailed discussion on evidence for this model). Grey shading indicates positive feedback loop (the 'runaway greenhouse').**

a reduction in the draw down of CO_2, which would then slowly build-up in the atmosphere [Kidder and Worsley (2004)].

However, the extinction model requires that, at the culmination of this long-term temperature rise, there was an additional, rapid, warming episode resulting in an Early Triassic 'hothouse' world. A large decrease in the proportion of heavy oxygen isotopes ($\delta^{18}O$) in carbonates spanning the P-Tr boundary in the Gartnerkofel-1 core of southern Austria has been interpreted as indicating a 5–6°C increase in temperature [Holser et al. (1989)]. However, interpretation of the oxygen isotope record is problematic as the carbonate isotope values are very sensitive to alteration during burial and diagenesis. Certainly, the limestones of Gartnerkofel-1 have been heavily recrystallised, especially around the boundary interval, and thus the oxygen isotope data should be viewed with caution. Other potential archives of P-Tr seawater temperatures, such as the biogenic carbonates of brachiopods, have yet to be explored.

Currently, there is no unequivocal quantitative record of absolute temperature changes associated with the P-Tr extinction event [Kidder and Worsley (2004)]. The analysis of isotope changes in soil carbonates, which form in direct contact with the atmosphere, may provide this, provided they are diagenetically unaltered. Although such studies have yet to be undertaken, fossil soils (palaeosols) have provided qualitative evidence for global warming across the P/Tr boundary. Retallack [1999] has recorded fossil soils at high southerly palaeolatitudes (upto 85°S) with characteristics that, at the present day, typify low temperate latitudes, and which formed under warmer conditions than soils of the Late Permian from the same localities.

If the evidence for warming across the P-Tr event is accepted, then one might expect evidence for a rise in greenhouse gases in the atmosphere. However, such evidence remains equivocal. The best evidence provided for an increase in CO_2 derives from analysis of the stomatal density of plant cuticles by Retallack [2001], although the time resolution is somewhat coarse. Regarding methane venting, the most often-cited piece of 'evidence' for this cornerstone of the P-Tr global warming model is the large negative excursion in $\delta^{13}C$ recorded in shallow marine carbonates [e.g. Erwin (1993); Benton and Twitchett (2003); Kidder and Worsley (2004)]. The ca. 3–4% negative shift $\delta^{13}C$ (locally upto 8%) appears to be too large to be explained by any other mechanism, such as volcanic emissions, and methane venting from gas hydrate deposits is regarded as the only viable alternative [e.g. Erwin (1993)]. A negative shift is recorded in marine carbonates, terrestrial soil carbonates [e.g. Retallack (2001)], bulk organic matter [e.g. Twitchett *et al.* (2001)], and biomarker molecules [Grice *et al.* (2005)] and so appears to reflect a real atmospheric change.

However, some caution is required as large negative shifts could be caused by other mechanisms, such as productivity crash for marine carbonates [Kump (1991)]. The negative shift in bulk organic matter is likely due to a change in organic matter source, such as a reduction in the relative contribution of material from higher plants [Foster *et al.* (1997)]. The negative shift in individual biomarker molecules may likewise reflect an undetected change in source. Volcanically vented CO_2 is also enriched in ^{12}C, with the Siberian Traps and other volcanic centres, such as South China, providing potential sources. At the conclusion of a recent comprehensive study, Berner [2002] noted that is not possible to reject all of these other causes and the end-Permian negative shift in $\delta^{13}C$ was likely driven by methane release associated with mass mortality and volcanic degassing. Thus, methane venting (Fig. 2) is but one possible explanation for the observed negative shift in $\delta^{13}C$. The shift itself should not be regarded as unequivocal evidence of methane flux to the atmosphere. Until independent evidence for methane venting is found, the methane-induced runaway part of the 'runaway greenhouse' model (Fig. 2) will remain open to question. In addition, geochemical modelling suggests that oxidation of the released methane would not, in any case, have produced enough CO_2 to trigger catastrophic warming [Berner (2002)]. Finally, as noted above, in some localities the $\delta^{13}C$ shift occurs after the extinction crisis [Twitchett et al. (2001)].

What of the role of volcanism as trigger for the P-Tr changes? The largest outpouring of continental flood basalts in the Phanerozoic occurred in Siberia during the P-Tr interval. Some authors have suggested that this volcanism was initiated by decompression melting beneath a massive impact crater, which has since been obliterated by the subsequent igneous activity [e.g. Jones et al. (2002)]. However, more recent modelling has shown that even an extremely large impact, producing a final crater of 250–300 km, can "barely provoke an igneous event in normal lithosphere" [Ivanov and Melosh (2003)]. Including both the Siberian Platform basalts and newly discovered coeval deposits buried in the West Siberia Basin, the flood basalts covered an area of $1.6 \times 10^6\,km^2$ to maximum depths of 3.5 km [Reichow et al. (2003)]. If all other igneous rocks, such as pyroclastic flows, are included then this coverage increases to $3.9 \times 10^6\,km$. Dating the top and bottom of the lava pile shows that the eruptions occurred over a relatively short period of time, maybe just 600,000 years. Was this huge volcanic event a trigger of the P-Tr extinction crisis?

Radiometric dating is the only way to answer this question because there are no fossils interbedded with the basalts that provide adequate correlation with other regions. The most recent results, by different scientists using a variety of geochronological methods, date the bulk of the Siberian Trap deposits to 250 ± 1 Ma. In the 1990s, this was considered to be exactly the date required [Renne et al. (1995)] and the flood basalts were promoted to primary trigger for the catastrophic extinction [e.g. Benton (2003)]. However, the subsequent re-dating of the Meishan [Mundil et al. (2001)] and Shangsi [Mundil et al. (2004)] beds, if correct, imply that the extinction occurred between 252 and 253 Ma and the Siberian Traps might therefore be too young.

Excellent correlation between extinction episodes and flood basalt provinces through the Phanerozoic [Courtillot and Renne (2003)] means that it is difficult to accept that the Siberian Traps had no role to play in the end-Permian extinction event. Some of the apparent problems may be due to differences in the type of dating techniques used: Mundil et al. [2004] argue that because $^{40}Ar/^{39}Ar$ dating, used in some studies [e.g. Renne et al. (1995)], typically gives younger ages than U/Pb dating there is in fact no discrepancy. The oldest date for the onset of Siberian Trap volcanic activity (i.e. the emplacement of intrusive gabbros) is 253.4 ± 0.8 Ma [Reichow *et al.* (2003)].

Thus, while there is good qualitative evidence for a warming event around the P/Tr boundary and extinction level, there are still questions remaining concerning the role of the Siberian Traps as potential trigger. An alternative possibility, which has yet to be investigated, is that no trigger was required: The slow, long term warming that began in the Early Permian may simply have reached a critical threshold level in the late Changhsingian.

The P/Tr warming episode is supposed to have resulted in marine extinction by causing sea level rise, productivity crash, oceanic stagnation and anoxia [e.g. Benton and Twitchett (2003)] (Fig. 2). Temperature rise itself, if high enough, could be lethal, especially in shallow water at low palaeolatitudes [Kidder and Worsley (2004)]. Geological evidence for these environmental changes is good, but whether they can be linked directly to warming (and extinction) is not so easy to resolve. Data from modelling experiments suggest that warming may have triggered these environmental changes. For example, evidence from the most complete sections indicates that disappearance of the Permian taxa occurred during a time of global sea level rise [e.g. Wignall *et al.* (1996)]. Modelling results indicate that average whole-ocean temperature rise of 15°C could raise sea level by approximately 20 m through simple thermal expansion [Kidder and Worsley (2004)].

Regarding ocean stagnation and anoxia, a substantial body of data has accumulated to show that marine ecosystems of the Early Triassic, even those in very shallow water (storm wave base), were less well oxygenated than during Late Permian, pre-extinction times. Evidence derives from a variety of independent sources such as facies analysis, trace fossil studies, palaeoecology, geochemical data, isotopic analyses, and biomarker distributions [e.g. Wignall and Hallam (1992); Wignall and Twitchett (1996, 2002); Twitchett (1999); Grice *et al.* (2005)]. The deepest parts of the world's oceans were oxygen-restricted from the latest Changhsingian to the Middle Triassic: the 'Superanoxic Event' of Isozaki [1997]. During the Griesbachian, most shelf settings experienced episodic development of euxinic conditions, comparable to the present day Black Sea [e.g. Grice *et al.* (2005)]. These euxinic intervals alternated with intervals of slightly elevated, but still sub-normal, oxygen concentrations, allowing a limited, depauperate benthos to colonise [e.g. Twitchett (1999)]. Only the shallowest settings of Neotethys appeared to have escaped [e.g. Krystyn *et al.* (2003)].

Following this Griesbachian peak in oxygen-poor conditions, oxygenation of the marine shelf improved somewhat, and only the deeper basins remained oxygen-restricted [Wignall and Twitchett (2002)].

Computer modelling results show that global warming could have potentially caused oceanic anoxia [Hotinski *et al.* (2001)], although other possibilities exist [Erwin (1993)]. Significant global warming, and reduction of the pole-equator temperature gradient, would have severely curtailed the thermohaline conveyor that maintains oxygenation of the deep oceans, leading to a sluggish, near-stagnant ocean, dominated by warm saline bottom water (Kidder and Worsley (2004)]. Warmer water also holds less dissolved oxygen than cooler water. In addition, models of atmospheric oxygen concentrations through the Phanerozoic indicate that O_2 levels fell gradually through the entire Permian, reaching a minimum of just 15% (compared to the present day 21%) during the P/Tr transition [Berner (2001)]. Thus, any warming-related changes are likely to have been exacerbated in the low-oxygen Earth of the Late Permian.

Evidence for a crash in marine primary productivity is more circumstantial. Given that most shelf settings of the Early Triassic experienced low oxygen conditions, which should promote the preservation of organic matter, the total organic carbon (TOC) content of nearly all Lower Triassic shelf sediments is staggeringly low [Twitchett (2001)]. Only one, localised, Lower Triassic petroleum source rock is known [Grice *et al.* (2005)]. In most cases TOC content actually decreases from the oxygenated, well bio-turbated Changhsingian sediments to the overlying, unbioturbated Griesbachian sediments [Twitchett *et al.* (2001)], consistent with a decline in productivity levels. As surface productivity relies on efficient nutrient recycling, which itself depends on ocean circulation, the sluggish warm water oceans of the Early Triassic would be expected to support much lower levels of primary production [Wignall and Twitchett (1996); Kidder and Worsley (2004)]. Even minor warming produces dramatic productivity collapse: For example, ocean productivity declined by 50% between the last Ice Age and the present day [Herguera and Berger (1991)].

In summary, the volcanically triggered global warming model is better supported than the extraterrestrial impact hypothesis, but there are still issues to be resolved. While much of the geological evidence is consistent with the model, this is not necessarily proof of a cause-and-effect link. Key issues remain the lack of a quantitative high-resolution temperature record for the P/Tr interval, and the problems surrounding the absolute dating.

3. Patterns of Extinction

3.1. *Marine extinctions*

Estimating the severity of past extinction events is not easy, due to the vagaries of preservation and fossil recovery, and the difficulty in recognising true biological species from fossils. Measures of global extinction magnitude are derived from large databases of family diversity through time [e.g. Benton (1993)], which suggest that 49% of marine invertebrate families disappeared during the P-Tr interval.

Extinction at the species level is then estimated using the statistical technique of reverse rarefaction. From this method, figures of 95% or 96% loss of marine species are derived [Raup (1979); Erwin (1993)]. However, such a calculation involves a number of assumptions, including, for example, that species extinction was random, with no selectivity against certain groups. This assumption is clearly incorrect [Erwin (1993)] and so the true magnitude of species loss may be closer to 80% [McKinney (1995)]. At the local scale, observed extinction magnitude is often high, for example 94% at Meishan [Jin *et al.* (2000)]. However, such local studies depend heavily on accurate taxonomy and disappearance from a single section may indicate migration, rather than true extinction, and may also be influenced by facies and preservational biases. Statistical techniques may be employed to help counter this latter problem [Jin *et al.* (2000)], but do not reduce it entirely.

Despite these shortcomings, it is clear that several groups were severely decimated in the Late Permian, some to complete extinction, and that not all groups were affected equally [Erwin (1994)]. For example, the diverse and successful fusulinid foraminifera disappear suddenly during the Changhsingian with loss of some 18 families, whereas other benthic foraminifera suffered much lower levels of extinction [Erwin (1993)].

A pattern of gradual decline through the Permian, particularly during the latter half, followed by final disappearance of the last few remaining taxa in the Changhsingian is typical of many groups. Examples include the Palaeozoic corals (Rugosa and Tabulata), trilobites and goniatites, all of which became extinct, and the stenolaemate bryozoans and articulate brachiopods that were reduced to a handful of surviving taxa [Benton (1993); Erwin (1993, 1994)]. This pattern implies longer-term changes in the marine realm may have been largely to blame for the diversity loss in these cases. The obvious conclusion to draw is that these groups were responding to either the long-term Permian rise in global temperatures and/or the long-term Permian decrease in atmospheric oxygen levels discussed above.

A few groups also suffered catastrophic losses prior to the end-Permian event. For those echinoderm groups that suffered significant extinction, the major crisis interval appears to have been the late-Guadalupian, where crinoids experienced $> 90\%$ loss, and other groups of echinoderms, such as the Blastoidea, became extinct [Erwin (1994)]. Significant brachiopod extinctions also occur at this level [Erwin (1992)].

This end-Guadalupian peak in diversity loss has led some authors to discuss the Late Permian event in terms of two episodes of extinction [Stanley and Yang (1994)]. While there is some evidence of oceanographic change and local disappearances at this time, there are also good reasons why the end-Guadalupian event might not be a real global extinction event. The Middle Permian was a time of incredible biodiversity, but the majority of these fossil taxa were endemic to the comprehensively monographed, exquisitely preserved (typically silicified) faunas of the southern US, particularly west Texas. At the close of the Guadalupian, sea level fell across that region, and the overlying sediments are unfossiliferous evaporites. Disappearance of some proportion of these taxa is most likely the result of facies change, not real extinction.

At the global scale, Upper Permian fossiliferous marine rocks are also relatively poorly known and the quality (i.e. completeness) of the Late Permian fossil record is correspondingly low. Almost all of the documented, fossiliferous, shallow marine Upper Permian strata are located in China. Many taxa common in the Guadalupian have not been found in Upper Permian rocks, but must have been living (somewhere) at that time because they re-appear in the Triassic. The presence of many so-called Lazarus taxa [Flessa and Jablonski (1983)] indicates that the fossil record of this crucial time interval is appallingly incomplete [Twitchett (2001)].

The Simple Completeness Metric (SCM) of Benton [1987] is a measure of the quality of the fossil record. It is calculated as the proportion of taxa actually recorded by fossil specimens in a given time interval, compared to the total number of taxa that are known to have been present in that interval (i.e. those recorded as fossils plus the 'missing' Lazarus taxa). Assuming the taxonomy is correct, the SCM is actually a very optimistic measure, and provides a maximum estimate of completeness: If aspects like the phylogenetic relationships of the taxa are incorporated, the inevitable presence of 'ghost lineages' [Benton (1994)] will mean a reduction in apparent completeness. When applied to the P-Tr fossil record, the SCM records two dramatic declines in completeness: (1) at the end-Guadalupian, and (2) at the Permian/Triassic boundary [Twitchett (2001)] (Fig. 3). Typically,

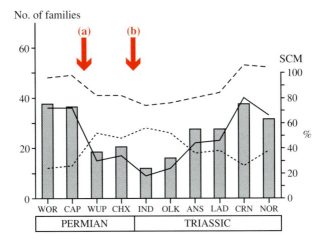

No. of families

Fig. 3   Permian-Triassic diversity of bivalve families showing the poor quality of the fossil record. Solid line = families represented by actual fossil specimens; dotted line = 'missing' Lazarus families; dashed line = total diversity. Arrows indicate extinction events at the end-Guadalupian (a) and end-Permian (b). At both (a) and (b) there is a decline in total diversity and the number of fossilised families, and an increase in the number of Lazarus families. SCM = Simple Completeness Metric [Benton (1987)], shown by the grey bars. Completeness declines at (a) and (b). See text for additional details.

the SCM for benthic invertebrate taxa during the Early Triassic is around 10–20%; in other words, 80–90% of Early Triassic taxa that we know must have been present somewhere on Earth are not recorded as fossils. And this is an optimistic measure of completeness! This change in the quality of the marine fossil record clearly affects our perceptions (including the perceived magnitude) of both the end-Guadalupian and end-Permian events.

3.2. *Terrestrial extinctions*

Extinction on land was just as severe as in the oceans, but this fact has been recognised only very recently [Benton (2003)]. In the Karoo Basin of South Africa, most vertebrate taxa disappeared gradually during the Changhsingian, with an additional peak in extinction rates approximately coincident with the negative shift in $\delta^{13}C$ values [Ward *et al.* (2005)]. When all fossil terrestrial organisms are considered together, the P-Tr event is the single largest extinction episode in the otherwise exponential rise of terrestrial diversity through the Phanerozoic.

A long-term ($> 20\,\mathrm{Ma}$) change in terrestrial vegetation (the Palaeophytic-Mesophytic transition occurred during the P-Tr interval and probably reflects gradual shifts in climate and palaeogeography. However, superimposed on this longer trend is a relatively catastrophic collapse of the dominant gymnosperm forests in the latest Permian, as evidenced by the disappearance of pollen taxa such as *Vittatina*, *Weylandites* and *Lueckisporites*. Studies in East Greenland have shown that this ecological crisis occurred simultaneously with the marine extinction event and took 10-100 kyr [Twitchett *et al.* (2001)]. It is also apparently coincident with the peak extinction of terrestrial vertebrates in the Karoo Basin.

Some workers [e.g. Eshet *et al.* (1995)] have suggested that a peak ('spike') in the abundance of *Reduviasporonites* (also called *Tympanicysta*) marks the sudden destruction of these forests. They interpreted this taxon as a saprophytic fungus, which thrived on the piles of dead and dying vegetation [Visscher *et al.* (1996)]. Unfortunately, this attractive scenario must now be rejected. Recent geochemical, structural and biomarker studies have shown that *Reduviasporonites* is definitely not a fungus but is (most probably) a photosynthetic alga [Foster *et al.* (2002)].

4. Post-Extinction Recovery

4.1. *Recovery of marine ecosystems*

The post-Permian marine recovery is the longest of any mass extinction, and is also proportionally longer than would be expected from the magnitude of the diversity loss [Erwin (1993)]. A literal reading of the fossil record shows that it took some 100 Myr (until the mid-Jurassic) for marine biodiversity at the family level to return to pre-extinction levels [Benton and Twitchett (2003)]. However, ecological recovery was somewhat quicker, with complex communities such as metazoan reefs becoming re-established in the Middle Triassic (within 10 Ma after the extinction). Thus, how one views the duration of the recovery interval, measured on a local or global scale, depends on how one defines 'recovery'.

Two main hypotheses have been advanced to explain the lengthy post-Permian recovery interval: (1) that the harsh palaeoenvironmental conditions associated with the extinction event continued well into the Early Triassic, and (2) that the delayed recovery is more apparent than real, being due to the biases of the fossil record. Evidence has been presented in support of both hypotheses, and is likely that a combination of factors were involved [Twitchett (1999)].

Fig. 4   The *Claraia* biofacies of the lower Induan, typified by laminated sediments (B, C) with occasional bedding plane assemblages of *Claraia* (A, D) indicating deposition under fluctuating low oxygen conditions. (A) and (B) from the lower Wordie Creek Formation of East Greenland; (C) and (D) core from the lower Kockatea Shale, Perth Basin, Australia.

Certainly, palaeoenvironmental conditions in most shelf settings in the first one or two million years after the extinction crisis were particularly severe, with widespread evidence for low oxygen conditions and low surface productivity (described above). The most commonly encountered facies of the Induan is the *Claraia* biofacies (Fig. 4). Sediments of this facies are typically laminated and were mostly deposited under anoxic or euxinic conditions [Wignall and Twitchett (1996, 2002)], with biomarker evidence indicating that at times euxinia extended from the seafloor to the photic zone [Grice *et al.* (2005)]. Intermittently, the ocean floor became weakly oxygenated for a few years, allowing a depauperate epifauna of small sized, thin-shelled taxa such as *Claraia* to colonise. Occasional horizons of mm-diameter *Planolites* burrows attest to temporary colonisation events by a scarce soft-bodied infauna of small, deposit feeders living just a few centimetres below the sediment surface [Twitchett (1999)]. These Early Induan benthic communities comprise low diversity assemblages of small-sized animals; e.g. the bivalves *Promyalina* and *Claraia*, the inarticulate brachiopod *Lingula* [Rodland and Bottjer (2001)] and rarer microgastropods. Bedding planes are typically dominated by a single taxon, which may occur in prodigious numbers. These taxa are considered by some authors, with varying degrees of evidence, to be pioneering, r-selected opportunists [Rodland and Bottjer (2001); Fraiser and Bottjer (2004)]. In some regions, stromatolites, and other evidence of microbial mats, are encountered [e.g. Schubert and Bottjer (1992)]. In the mid-high palaeolatitude regions of both the Boreal Ocean (Greenland, Spitsbergen) and NeoTethys (Madagascar, western Australia) a fairly diverse, but small-sized, nekton of fish and ammonoids is recorded.

One prediction of the hypothesis that harsh environmental conditions prevented rapid recovery is that in regions that were better oxygenated, such as the shallowest settings of Neotethys [Wignall and Twitchett (2002)], recovery should happen much faster. Support for this is provided by a mid-Griesbachian (early Induan) age post-extinction assemblage from Oman that was living in a shallow, well oxygenated, offshore (seamount) setting within wave base (Krystyn et al. (2003)). This fauna is the most diverse Induan fauna presently known and has a level of ecological complexity not recorded elsewhere, such as in the western USA [Schubert and Bottjer (1995)], until the late Olenekian [Twitchett et al. (2004)]. Thus, in the absence of low oxygen conditions post-Permian recovery was an order of magnitude faster, taking just a few hundred thousand years to reach the same level of ecological recovery that under oxygen-restriction took several million years.

Alternatively, the delayed recovery may be more apparent than real. As noted above, the Early Triassic fossil record is woefully incomplete (Fig. 3). Analysis of the gastropod record has demonstrated that between 24 genera [Wheeley and Twitchett (2005)] and 29 genera [Erwin (1996)] vanished in the Middle Permian, only to reappear in the Middle Triassic apparently unchanged. One possible reason for this lack of preservation is that these 24 Lazarus genera had aragonitic shells, which are easily dissolved postmortem and which are best recorded in fossil assemblages where the shells have been silicified. Such silicified faunas are almost completely absent from Lower Triassic rocks [Erwin (1996)], and those that have been recently documented do indeed contain some of these missing Lazarus taxa as predicted [Wheeley and Twitchett (2005)]. The apparent 'recovery' in the Middle Triassic coincides precisely with a return to a more complete fossil record (Fig. 3), suggesting, perhaps unsurprisingly, that the quality of the fossil record is affecting our perception of the timing of recovery.

There are at least two solutions to the problem of a poor quality Early Triassic fossil record. One is to assess extinction and recovery using a phylogenetic approach, rather than relying solely on first and last appearances in the fossil record. A study of K-T echinoids [Smith and Jeffery (1998)] demonstrated the power of this methodology for unravelling patterns of extinction and recovery. However, to date, none of the necessary work has been attempted for Permian-Triassic taxa. The other solution is to assess recovery by studying fossil taxa that do not suffer from problems of early dissolution.

Trace fossils are one possibility. Initial, empirical studies of the trace fossil record of northern Italy [Twitchett (1999)] suggested that changes in the types, size and depths of burrows present in Lower and Middle Triassic strata provides a measure of post-extinction recovery [Twitchett *et al.* (2004)]. With the disappearance of benthic oxygen restriction in the late Induan, burrows such as *Arenicolites*, *Skolithos* and *Diplocraterion*, (produced by deeper burrowing suspension feeders) reappear. Next to reappear are burrowing crustaceans (evidenced by trace fossils such as *Thalassinoides*). In the palaeotropics, small *Thalassinoides* first reappear (rarely) in the late Olenekian (Spathian), with pre-extinction sizes being recorded only in the Middle Triassic. In higher palaeolatitudes (East Greenland, Spitsbergen) small *Thalassinoides* reappear in the late Induan, suggesting faster recovery in the Boreal Realm [Twitchett and Barras (2004)].

4.2. *Recovery of terrestrial ecosystems*

Our understanding of the recovery of terrestrial ecosystems is presently less refined, with most data deriving from the Karoo Basin of South Africa [e.g. Ward *et al.* (2005)], although work is beginning on the Russian record too [Benton *et al.* (2004)]. How do the patterns and timing of terrestrial recovery compare to those of the marine realm? Like the marine survivors, terrestrial vertebrate survivors tend to be small: In the Karoo Basin five small carnivorous therocephalians, and one small anapsid (*Procolophon*) survive the P-Tr crisis. The other survivor, and the dominant terrestrial vertebrate for several million years, was the herbivore *Lystrosaurus*, which, like *Claraia*, was globally widespread.

Following collapse of the Late Permian gymnosperm forests, open herbaceous vegetation rapidly took over, with short lived blooms of pioneering, opportunistic lycopsids, ferns and bryophytes — stress tolerant forms that were subordinate members of the pre-crisis vegetation [Looy *et al.* (2001)]. Pollen from woody gymnosperms seems to indicate that a few surviving elements of the Permian forests lingered on for a while [Twitchett *et al.* (2001)], but equally these records could just represent reworking of the pollen into underlying, younger sediments. Certainly, by the earliest Triassic no tree-like gymnosperms remained and a stable, low diversity, open shrubland vegetation of cycads and lycopsids was established. Complex, diverse forest communities were absent until the latest Spathian and early Middle Triassic [Looy *et al.* (1999)], resulting in a significant stratigraphic gap in coal deposits during the Early Triassic [Retallack *et al.* (1996)]. Apparently, the return of ecological complexity on land closely mirrored that in benthic marine communities.

5. Conclusions

The greatest mass extinction event of the Phanerozoic is receiving an unprecedented level of scientific interest. Although the global coverage is still very patchy, new sections are being discovered and described, allowing hypotheses of causes and consequences to be tested, modified or rejected. Understandably, some sections and regions have received more study than others, but this imbalance needs to be addressed. Presently, an extraterrestrial cause can be rejected, but there are still questions concerning the current leading alternative of volcanic-triggered global warming. The uncertainty over the absolute dating, and the lack of high resolution, quantitative measures of P-Tr temperature change are crucial problems. Although there have been significant recent advances in stratigraphy and correlation, taxonomic advances have lagged somewhat. Given the problems of preservation, in particular, a thorough taxonomic revision of most P-Tr fossil groups is required, and up-to-date cladistic phylogenies need to be provided. Coupled with the new, high-resolution, stratigraphy this should lead to a greatly improved understanding of the patterns of global extinction and recovery.

References

Alvare, L. W., Alvarez, W., Asaro, F. & Michel, H. V. (1980) Extraterrestrial causes of the Cretaceous-Tertiary extinction. *Science* **208**, 1095–1108.
Baud, A., Magaritz, M. & Holser, W. T. (1989) *Geologische Rundschau* **78**, 649.
Basu, A. S., Pataev, M. I., Poreda, R. J., Jacobsen, S. B. & Becker, L. (2003) Chondritic meteorite fragments associated with the Permian-Triassic boundary in Antarctica. *Science* **302**, 1388–1392.
Becker, L., Poreda, R. J., Hunt, A. G., Bunch, T. E. & Rampino, M. (2001) Impact event at the Permian-Triassic boundary: Evidence from extraterrestrial noble gases in fullerenes. *Science* **291**, 1530–1533.

Becker, L., Poreda, R. J., Basu, A. R., Pope, K. O., Harrison, T. M., Nicholson, C. & Iasky, R. (2004) Bedout: A possible end-Permian impact crater offshore of northwestern Australia. *Science* **304**, 1469–1476.

Benton, M. J. (1987) Mass extinctions among families of non-marine tetrapods: the data. *Mém. Soc. Géol. France* **150**, 21–32.

Benton, M. J. (ed.) (1994) *The Fossil Record 2*. Chapman and Hall, 845 pp.

Benton, M. J. (1994) Paleontological data and identifying mass extinctions. *TREE* **9**, 181–185.

Benton, M. J. (2003) *When Life Nearly Died: The Greatest Mass Extinction of All Time*. Thames and Hudson, London.

Benton, M. J. & Twitchett, R. J. (2003) How to kill (almost) all life: The end-Permian extinction event. *TREE* **18**, 358–365.

Benton, M. J., Tverdokhlebov, V. P. & Surkov, M. V. (2004) Ecosystem remodelling among vertebrates at the Permian-Triassic boundary in Russia. *Nature* **432**, 97–100.

Berner, R. A. (2001) Modelling atmospheric O_2 over Phanerozoic time. *Geochim. Cosmochim. Acta* **65**, 685–694.

Berner, R. A. (2002) Examination of hypotheses for the Permo-Triassic boundary extinction by carbon cycle modelling. *Proc. Natl. Acad. Sci.* **99**, 4172–4177.

Bowring, S. A., Erwin, D. H., Jin, Y., Martin, M. W., Davidek, K. L. & Wei, W. (1998) U/Pb zircon geochronology and tempo of the end-Permian mass extinction. *Science* **280**, 1039–1045.

Brookfield, M. E., Twitchett, R. J. & Goodings, C. (2003) Palaeoenvironments of the Permian-Triassic transition sections in Kashmir, India. *Palaeogeog. Palaeoclimatol. Palaeoecol.* **198**, 353–371.

Claoue-Long, J. C., Zhang, Z., Ma, G. & Du, S. (1991) The age of the Permian-Triassic boundary. *EPSL* **105**, 182–190.

Courtillot, V. & Renne, P. R. (2003) On the ages of flood basalt events. *C.R. Geoscience* **335**, 113–140.

Erwin, D. H. (1993) *The Great Paleozoic Crisis: Life and Death in the Permian*. Columbia University Press, New York.

Erwin, D. H. (1994) The Permo-Triassic extinction. *Nature* **367**, 231–236.

Erwin, D. H. (1996) Understanding biotic recoveries: Extinction, survival and preservation during the end-Permian mass extinction. *Evolutionary Paleobiology*, Jablonski, D., Erwin, D. H. Lipps and J. H. (eds.), University of Chicago Press, Chicago, 398–418.

Eshet, Y., Rampino, M. R. & Visscher, H. (1995) Fungal event and paleontological record of ecological crisis and recovery across the Permian-Triassic boundary. *Geology* **23**, 967–970.

Farley, K. A. & Mukhopadhyay, S. (2003) An extraterrestrial impact at the Permian-Triassic boundary? Technical Comment. *Science* **293**, 2343–2344.

Flessa, K. W. & Jablonski, D. (1983) Extinction is here to stay. *Paleobiology* **9**, 315–321.

Foster, C. B., Logan, G. A., Summons, R. E., Gorter, J. D. & Edwards, D. S. (1997) Carbon isotopes, kerogen types and the Permian-Triassic boundary in Australia: Implications for exploration. *APPEA. J.* **37**, 472–489.

Foster, C. B., Stephenson, M. H., Marshall, C., Logan, G. A. & Greenwood, P. F. (2002) Revision of *Reduviasporonites* Wilson 1962: Description, illustration, comparison and biological affinities. *Palynology* **26**, 35–58.

Fraiser, M. L. & Bottjer, D. J. (2004) The non-actualistic Early Triassic gastropod fauna: A case study of the Lower Triassic Sinbad Limestone Member. *Palaios* **19**, 259–275.

Grice, K., Cao, C., Love, G. D., Böttcher, M. E., Twitchett, R. J., Grosjean, E., Summons, R. E., Turgeon, S. C., William Dunning, W. & Jin, Y. (2005) Photic zone euxinia during the Permian-Triassic Superanoxic Event. *Science* **307**, 706–709.

Herguera, J. C. & Berger, W. H. (1991) Paleoproductivity from benthic foraminiferal abundance — glacial to postglacial change in the western Equatorial Pacific. *Geology* **19**, 1173–1176.

Holser, W. P., Schönlaub, H. P., Attrep, M., Boekelmann, K., Klein, P., Magaritz, M. & Orth, C. J. (1989) A unique geochemical record at the Permian-Triassic boundary. *Nature* **337**, 39–44.

Hotinski, R. M., Bice, K. L., Kump, L. R., Najjar, R. G. & Arthur, M. A. (2001) Ocean stagnation and end-Permian anoxia. *Geology* **29**, 7–10.

Isozaki, Y. (1997) Permo-Triassic boundary superanoxia and stratified superocean: Records from lost deep sea. *Science* **276**, 235–238.

Isozaki, Y. (2001) An extraterrestrial impact at the Permian-Triassic boundary? Technical comment. *Science* **293**, 2344.

Ivanov, B. A. & Melosh, H. J. (2003) Impact do not initiate volcanic eruptions: Eruptions close to the crater. *Geology* **31**, 869–872.

Jin, Y. G., Wang, Y., Wang, W. & Erwin, D. H. (2000) Pattern of marine mass extinction near the Permian-Triassic boundary in South China. *Science* **289**, 432–436.

Jones, A. P., Price, G. D., Price, N. J., Decarli, P. S. & Clegg, R. A. (2002) Impact induced melting and the development of large igneous provinces. *EPSL* **202**, 551–561.

Kaiho, K., Kajiwara, Y., Nakano, T., Miura, Y., Kawahata, H., Tazaki, K., Ueshima, M., Chen, Z. & Shi, G. R. (2001) End-Permian catastrophe by bolide impact: Evidence of a gigantic release of sulfur from the mantle. *Geology* **29**, 815–818.

Kerr, R. A. (2003) Has an impact done it again? *Science* **302**, 1314–1316.

Kidder, D. L. & Worsley, T. R. (2004) Causes and consequences of extreme Permo-Triassic warming to globally equable climate and relations to the Permo-Triassic extinction and recovery. *Palaeogeog. Palaeoclimat. Palaeoecol.* **203**, 207–237.

Koeberl, C., Gilmour, I., Reimold, W. U., Claeys, P. & Ivanov, B. A. (2002) End-Permian catastrophe by bolide impact: Comment. *Geology* **30**, 855–856.

Koerberl, C., Farley, K. A., Puecker-Ehrenbrink, B. & Sephton, M. A. (2004) Geochemistry of the end-Permian extinction event in Austria and Italy: No evidence for an extraterrestrial component. *Geology* **32**, 1053–1056.

Krystyn, L., Baud, A., Richoz, S. & Twitchett, R. J. (2003) A unique Permian-Triassic boundary section from Oman. *Palaeogeog. Palaeoclimatol. Palaeoecol.* **191**, 329–344.

Kump, L. R. (1991) Interpreting carbon-isotope excursions: Strangelove oceans. *Geology* **19**, 299–302.

Krull, E. S. & Retallack, G. J. (2000) d^{13}C depth profiles from paleosols across the Permian-Triassic boundary: Evidence for methane release. *GSA Bulletin* **112**, 1459–1472.

Looy, C. V., Brugman, W. A., Dilcher, D. L. & Visscher, H. (1999) The delayed resurgence of equatorial forests after the Permian-Triassic ecologic crisis. *Proc. Nat. Acad. Sci. USA* **96**, 13857–13862.

Looy, C. V., Twitchett, R. J., Dilcher, D. L., Van Konijnenburg-Van Cittert, H. A. & Visscher, H. (2001) Life in the end-Permian dead zone. *Proc. Nat. Acad. Sci. USA* **98**, 7879–7882.

McKinney, M. L. (1995) Extinction selectivity among lower taxa — gradational patterns and rarefaction error in extinction estimates. *Paleobiology* **21**, 300–313.

McLaren, D. J. (1970) Presidential address: Time, life and boundaries. *J. Paleont.* **44**, 801–805.

Mundil, R., Metcalfe, I., Ludwig, K. R., Renne, P. R., Oberli, F. & Nicoll, R. S. (2001) Timing of the Permian-Triassic biotic crisis: Implication from new zircon U/Pb age data (and their limitations). *EPSL* **187**, 131–145.

Mundil, R., Ludwig, K. R., Metcalfe, I. & Renne, P. R. (2004) Age and timing of the Permian mass extinctions: U/Pb dating of closed-system zircons. *Science* **305**, 1760–1763.

Raup, D. M. (1979) Size of the Permo-Triassic bottleneck and its evolutionary implications. *Science* **206**, 217–219.

Reichow, M. K., Saunders, A. D., White, R. V., Pringle, M. S., Al'Mukhamedov, A. I., Medvedev, A. I. & Kirda, N. P. (2003) Ar^{40}/Ar^{39} dates from the West Siberian Basin: Siberian flood basalt province doubled. *Science* **296**, 1846–1849.

Renne, P. R., Zhang, Z., Richardson, M. A., Black, M. T. & Basu, A. R. (1995) Synchrony and causal relations between Permian-Triassic boundary crises and Siberian flood volcanism. *Science* **269**, 1413–1416.

Renne, P. R., Melosh, H. J., Farley, K. A., Reimold, W. U., Koeberl, C., Rampino, M. R., Kelly, S. P. & Ivanov, B. A. (2004) Is Bedout an impact structure? Take 2. *Science* **306**, 610–611.

Retallack, G. J. (1999) Postapocalyptic greenhouse paleoclimate revealed by earliest Triassic paleosols in the Sydney Basin, Australia. *GSA Bull.* **111**, 52–70.

Retallack, G. J. (2001) A 300 million year record of atmospheric carbon dioxide from fossil plant cuticles. *Nature* **411**, 287–290.

Retallack, G. J., Veevers, J. J. & Morante, R. (1996) Global coal gap between the Permian-Triassic extinction and Middle Triassic recovery of peat-forming plants. *GSA Bull.* **108**, 195–207.

Rodland, D. & Bottjer, D. J. (2001) Biotic recovery from the end-Permian mass extinction: Behavior of the inarticulate brachiopod *Lingula* as a disaster taxon. *Palaios* **16**, 95–101.

Schubert, J. K. & Bottjer, D. J. (1992) Early Triassic stromatolites as post-mass extinction disaster forms. *Geology* **20**, 883–886.

Schubert, J. K. & Bottjer, D. J. (1995) Aftermath of the Permian-Triassic mass extinction event: Palaeoecology of Lower Triassic carbonates in the western USA. *Palaeogeog. Palaeoclimatol. Palaeoecol.* **116**, 1–39.

Smith, A. B. & Jeffrey, C. H. (1998) Selectivity of extinction among seaurchins at the end of the Cretaceous period. *Nature* **392**, 69–71.

Stanley, S. M. (1998) Climatic cooling and mass extinction of Paleozoic reef communities. *Palaios* **3**, 228–232.

Stanley, S. M. & Yang, X. (1994) A double mass extinction at the end of the Paleozoic Era. *Science* **266**, 1340–1344.

Twitchett, R. J. (1999) Palaeoenvironments and faunal recovery after the end-Permian mass extinction. *Palaeogeog. Palaeoclimatol. Palaeoecol.* **154**, 27–37.

Twitchett, R. J. (2001) Incompleteness of the Permian-Triassic fossil record: A consequence of productivity decline? *Geol. J.* **36**, 341–353.

Twitchett, R. J. & Barras, C. G. (2004) Trace fossils in the aftermath of mass extinction events. *Geol. Soc. Lond. Spec. Publ.* **228**, 397–418.

Twitchett, R. J., Looy, C. V., Morante, R., Visscher, H. & Wignall, P. B. (2001) Rapid and synchronous collapse of marine and terrestrial ecosystems during the end-Permian mass extinction event. *Geology* **29**, 351–354.

Twitchett, R. J., Krystyn, L., Baud, A., Wheeley, J. R. & Richoz, S. (2004) Rapid marine recovery after the end-Permian mass extinction event in the absence of marine anoxia. *Geology* **32**, 805–808.

Ward, P. D., Botha, J., Buick, R., De Kock, M. O., Erwin, D. H., Garrison, G. H., Kirschvink, J. L. & Smith, R. (2005) Abrupt and gradual extinction among Late Permian land vertebrates in the Karoo Basin, South Africa. *Science* **307**, 709–714.

Wheeley, J. R. & Twitchett, R. J. (2005) Palaeoecological significance of a new Griesbachian (Early Triassic) gastropod assemblage from Oman. *Lethaia* **38**, 37–45.

Wignall, P. B. & Hallam, A. (1992) Anoxia as a cause of the Permian-Triassic mass extinction: Facies evidence from northern Italy and the western United States. *Palaeogeog. Palaeoclimat. Palaeoecol.* **93**, 21–46.

Wignall, P. B. & Twitchett, R. J. (1996) Oceanic anoxia and the end Permian mass extinction. *Science* **272**, 1155–1158.

Wignall, P. B. & Twitchett, R. J. (2002) Extent, duration and nature of the Permian-Triassic anoxic event. *GSA Spec. Paper* **356**, 395–413.

Wignall, P. B., Kozur, H. & Hallam, A. (1996) The timing of palaeoenvironmental changes at the Permian/Triassic (P/Tr) boundary using conodont biostratigraphy. *Hist. Biol.* **12**, 39–62.

Xu, D. Y., Ma, S. L., Chai, Z. F., Mao, X. Y., Sun, Y. Y., Zhang, Q. W. & Yang, Z. Z. (1985) Abundance variation of iridium and trace elements at the Permian-Triassic boundary at Shangsi in China. *Nature* **314**, 154–156.

Yin, H. F., Zhang, K. X., Tong, J. N., Yang, Z. Y. & Wu, S. B. (2001) The Global Stratotype Section and Point (GSSP) of the Permian-Triassic boundary. *Episodes* **24**, 102–114.

SECTION 2

DYNAMICS OF THE EARTH

Space-Plasma Imaging — Past, Present and Future

Cathryn N. Mitchell

University of Bath, UK

In 1986 Austen *et al.* proposed that a technique from medical imaging, tomography, could be used to image the Earth's ionosphere. Tomography was already very successful in creating images of the inside organs of human bodies by the mathematical manipulation of a series of X-ray measurements taken from multiple viewing angles around the body. The new idea was to use multiple satellite-to-ground radio signals to produce snapshots of the Earth's ionised environment. Tomography has now progressed into a technique for imaging the ionised plasma around the entire Earth. It is now feasible to create real-time movies of the ionised plasma, allowing us to watch the results of our planet's bombardment by the solar wind during events known as storms. The vision of the pioneering scientist Sir Edward V. Appleton for 'Ionospheric Weather' forecasting [1947], now a topical issue under the 'space weather' umbrella, is on the verge of being realised through new imaging-modelling approaches known as assimilation. It is becoming clear that ionospheric forecasting cannot be improved without storm warnings and consequently new research projects to link together models of the entire solar-terrestrial system, including the Sun, solar wind, magnetosphere, ionosphere and thermosphere, are now being proposed. The prospect is on the horizon of assimilating data from not just the ionosphere but the whole solar-terrestrial system to produce a real-time computer model and 'space weather' forecast. The application of tomographic imaging far beyond the ionosphere to include the whole near-Earth space plasma realm and possibly that of other planets is a likely possibility for the future.

1. Introduction

Medical imaging is familiar to most of us. The idea is to build up a picture of what is inside a patient in the least intrusive manner possible. The technique itself is called tomography and it has far-reaching applications in fields

93

beyond medicine. One such application is to build up pictures of our planet's upper atmosphere, the ionosphere.

The Earth's ionosphere exists in the shell of space around our planet extending from about 100 km above us into the regions of near-Earth space. The Earth's atmosphere decreases in density with increasing altitude and consists of different atoms and molecules at different heights. The ionosphere is formed by atoms or molecules absorbing energetic electromagnetic radiation (extreme ultra violet) from the Sun and becoming ionised releasing free electrons, which due to the low gas density remain apart from the positive ions, forming a plasma. The tenuous upper atmosphere contains mainly oxygen atoms up to a few hundred kilometres where it changes to mainly hydrogen. Charge exchange happens continuously and only a small percentage of these atoms remain ionised at one time. It is these ionised atoms and molecules that make up the ionosphere, embedded within a sea of neutral atoms.

If we brought a box full of the ionosphere down here to the surface of the Earth we might first think that it contains nothing at all — the upper atmosphere density is so low it would normally be considered to be a vacuum. In fact, if you were able to stand on a square metre of area at 100 km altitude the entire mass of the ionised atoms and molecules in the ionosphere above you would be far less than a milligram! It has been aptly described as a 'wispy thing.' Even so, the ionosphere is very important because the free electrons affect radio waves that propagate through it. The first time that this was observed was in 1901 when Marconi transmitted a radio signal from Cornwall, England to St. John's Newfoundland, using the ionosphere as a giant mirror to send his signal beyond the horizon and back to Earth on the other side of the Atlantic.

More recent scientific studies of the ionosphere have been pioneered by enormously powerful radars that provide measurements bringing insights into the complicated physics of this naturally occurring plasma. Figure 1 shows the European Incoherent Scatter (EISCAT) Radar UHF transmitter in Northern Norway. This remarkable instrument provides a community of ionospheric scientists with detailed observations of the Earth's ionosphere, sometimes during beautiful displays of the aurora borealis, allowing us to understand the complexity of interactions between our planet and the Sun. The study of these interactions is known as solar-terrestrial physics and the dynamical behaviour of the system is known as space weather — the weather in near-Earth space.

Fig. 1   The European Incoherent Scatter (EISCAT) radar UHF transmitter in Tromsø, Norway.

Tomographic imaging is important for observing the large-scale ionosphere in the context of space-weather events. It uses satellite-to-ground radio signals to build up a set of tomographic measurements just like the X-rays through the patient in medical imaging, only in this case the patient to be imaged is the Earth's ionosphere. Radio receivers are set up at suitable locations on the Earth and a series of measurements can be combined together and inverted to provide a picture of the free-electron concentration. Here, the technique is described in its current state and it is suggested that with more instrumentation the imaging technique could be extended to the whole solar-terrestrial environment.

2. Ionospheric Imaging

The mathematical problem of how to reconstruct a function from its projections was originally solved by Radon [1917], but the first practical application was not published until 1956, when the tomographic method was applied to radio astronomy [Bracewell (1956)]. Recent interest in tomographic imaging began with the invention of the X-ray computerized tomography scanner by Hounsfield in 1972. This original medical application, the CAT (computer aided tomography) scanner, took measurements of the

attenuation of X-rays passing through a human body from many different angles. By converting these measurements directly into digital impulses
and feeding them into a computer, a two-dimensional, cross-sectional image
of the body was obtained. More recent developments in the medical field
have seen the technique applied to nuclear medicine, magnetic resonance
imaging, ultrasound and microwave imaging.

Put into simple terms, tomography is a technique for finding unknown
numbers inside a grid. Imagine that you have a square grid divided into
four sections each containing an unknown number (Fig. 2) and you are
only allowed to know the sums of the numbers along certain paths. So take,
for example, the case where the sums are those four shown by the thick
grey lines in the figure. Now suppose that the sums of numbers along each
of the four paths labelled a to c is equal to ten and d is equal to 5. Then
you can easily set up four simultaneous equations and find a solution where
each of the original numbers was equal to five. Then to check it — five
plus five along each direction is equal to ten. This is the simplest case for
tomography — the measurements (sums) have no error and you have all
the measurements you need to find the solution.

In reality the problem is not quite so simple, although the underlying principles remain the same. The first difference is that you need to
account for the length of each path through each section (pixel). In the
case of Fig. 2, if each pixel has sides of unit length then the lengths of
each path a, b and c through each pixel would be one, but ray 'd' which
would have length root 2. So in reality the measurement 'd' would be root
2 times 5. The measurement 'a' would be two times one times five. The second difference is that sometimes the measurements cannot be taken from
all angles. Returning to Fig. 2, if you only have measurements a and b,
then you cannot discover unique values for all of the four unknown numbers. To overcome this problem you can try to obtain other information.

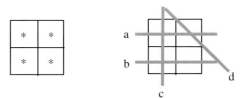

Fig. 2   Diagram showing a simple tomographic system with four unknown
numbers (*) to be found from four measurements a, b, c and d.

For example, if you know that all of the numbers are equal and that the sum along path 'a' is ten, then it is clear that all of the numbers must be equal to 5. In the case of ionospheric imaging this is a big problem; how do you compensate for the limited measurements? The satellite-to-ground geometry omits rays passing horizontally through the ionosphere. Fortunately, the missing information that is needed can be partly compensated for by bringing in some realistic assumptions about the distribution of electron density. Another problem in tomography is coping with systematic errors and noise on the measurements. In the ionospheric imaging case signals are often temporarily lost and regained causing discrete jumps in the measurement record. It is important to distinguish these jumps from real changes in the ionosphere so that actual ionospheric features can be imaged and not artificial features caused by these jumps in the phase records. A record of the changes in total electron content (TEC) in the ionosphere between a low-Earth-orbit satellite and ground-based receiver (the line integral of the electron density that we need to image) is shown in Fig. 3. Luckily this particular record had no phase jumps in it.

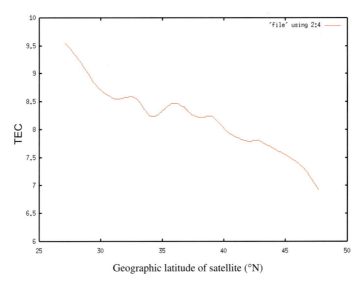

Fig. 3 **Total electron content ($\times 10^{16} \text{m}^{-2}$) record in January 1998 from a single receiver close to Rome in Italy. The undulations in the record are characteristic of ionospheric waves called travelling ionospheric disturbances.**

For ionospheric imaging, dual-frequency radio signals can be recorded by ground-based receivers to obtain relative phase shift and delay. The dispersive nature of the ionospheric component allows the ionospheric delay to be determined separately from effects caused by propagation through the non-ionised part of the atmosphere. This provides information that can be related directly to TEC (Fig. 3). The first experimental result showing a tomographic image of a slice of the ionosphere was published by Andreeva *et al.* [1990]. These authors, from the Moscow State University, used TEC data collected at three receivers located at Murmansk, Kem and Moscow. They made use of radio transmissions from Russian navigation satellites. For such preliminary tomographic results no other local measurements of the ionospheric electron density were available for comparison with the reconstruction and it was not until 1992 (Pryse and Kersley) that a tomographic image with independent verification was published. These images used data from the US Navy Navigation Satellite System (NNSS), the predecessor to the GPS (Global Positioning System). The verification for these early images over Scandinavia was provided by a scanning experiment of the European Incoherent Scatter (EISCAT) radar (Fig. 1). Subsequent co-ordinated studies between tomographic imaging and the EISCAT radar have contributed hugely to the general acceptance of the tomographic technique.

Many ionospheric features have been imaged using tomography. Beautiful images show snapshots of waves called travelling ionospheric disturbances (TIDs) [Pryse *et al.* (1995); Cook and Close (1995)]. These close relations of ocean waves are the manifestation in the ionosphere of internal atmospheric gravity waves. An example of an image showing a TID on Boxing Day of 1992 is shown in Fig. 4. Mitchell *et al.* [1995] have presented tomographic images of magnetic-field-aligned irregularities and E-region enhancements in the auroral ionosphere above northern Scandinavia. This region is particularly interesting to physicists because it is where particles from the Sun, having travelled though space in the solar wind, are able to enter the Earth's upper atmosphere. These high-speed particles whiz down the Earth's magnetic field lines like corkscrewing bullets, eventually colliding with atoms causing impact ionization and give up their energy in exchange for fantastic displays of the northern lights or aurora borealis.

Kersley *et al.* [1997] demonstrated that tomography could be used to make images of large-scale ionisation depletions known as troughs, generally found on the night-side auroral mid-latitude boundary. Results from the polar cap have revealed ionospheric signatures of processes occurring

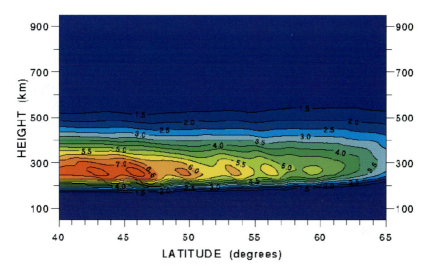

Fig. 4   Tomographic image of TIDs at 14:54 UT on 26 December, 1992. Contours show the electron density in units of $\times 10^{11} m^{-3}$.

further out in space, such as magnetic reconnection events [Walker *et al.* (1998)]. A novel idea by Bernhardt *et al.* [1996] proposed the inclusion of measurements taken from natural extreme ultraviolet emissions in the iono-sphere into tomographic inversions. These satellite-based observations can provide vertical profiles of ionized oxygen, which are essentially the same as the electron-density profiles at F-layer heights. More recently, Materassi *et al.* [2001] have applied tomographic techniques to measurements recorded from southern Italy and the Mediterranean to image and study the large-scale enhancement known as the equatorial anomaly. These great lumps of ionisation are produced by the 'fountain effect' where the plasma rises like a fountain over the geomagnetic equator and falls along the magnetic field lines forming distinct peaks on either side. An image of the northern peak of this peculiar structure is shown in Fig. 5.

3. Imaging Using GPS

New opportunities for ionospheric imaging have arisen with the introduc-tion of Global Positioning System (GPS) satellites and in particular because of the world-wide network of geodetic receivers that provide free ionospheric data. The problem with this data is the geometry. The ray-paths are in many different orientations and the GPS satellites, being in a much higher

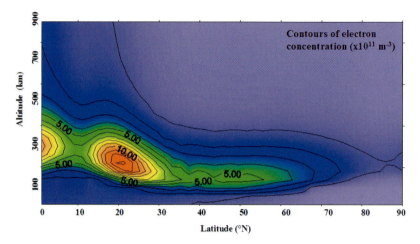

Fig. 5 Tomographic image of the northern crest of the equatorial anomaly in January 1998. Contours show the electron density in units of $\times 10^{11}\,\mathrm{m}^{-3}$.

orbit than the conventional NNSS satellites, move across the sky rather slowly in comparison to changes in the ionosphere. The problem is how to cope with many measurements taken over a large region, recorded at different times and through a changing ionosphere. It is like trying to image the entire human body while the patient keeps wriggling around!

The Global Positioning System consists of at least 24 satellites that transmit L-band radio signals at two frequencies, 1.575 and 1.228 GHz. GPS signals are already providing an important and inexpensive new tool for ionospheric measurement. Future possibilities of a European-lead navigation satellite system (Galileo) will further increase the density of such observations. This type of ionospheric research has the advantage of being very cost effective — no new satellites are required and the transmissions are at present free to everyone for scientific use. However, from any observation site the satellites appear at changing and oblique angles, making the observations of total electron content complicated to interpret. This network of TEC data drove forward a development of the mathematical ideas in tomography to produce a more generalised imaging system.

Another important data source for GPS imaging comes from the radio-occultation technique. This uses receivers located on Low-Earth-Orbit (L.E.O.) satellites to monitor the phase changes of GPS signals. Hajj *et al.* [1994] suggested using the satellite-to-satellite transmission of GPS to LEO satellite measurements in a tomographic framework to provide the so-called

(a)

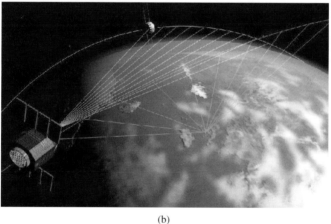

(b)

Fig. 6 **(a) Diagram illustrating the geometry for ionospheric tomography. Only one receiver site is shown in the diagram for clarity. (b) The additional rays provided by the radio-occultation satellite.**

'missing horizontal rays' and improve the vertical resolution. Figure 6 shows examples of the geometry involved first in ground-based-receiver tomography and second with the addition of the radio occultation rays. Importantly, L.E.O. satellites with GPS receivers on-board can provide measurements over the oceans and into the remote polar caps, thus enabling the iono-sphere to be studied on a truly global-scale. An exciting new prospect for ionospheric imaging is to combine observations from many different instruments such as occultation satellites to characterise the ionosphere globally [Spencer and Mitchell (2001)]. In future raw observations from vertical sounders (ionosondes) could be combined with multi-directional ground-based and radio-occultation observations of TEC but to accomplish these tasks properly, nonlinear mathematical and efficient computational techniques still need to be developed.

Figure 7 shows a sequence of four GPS ionospheric images over the northern hemisphere, six hours apart. The electron-density values in the images have been summed vertically to show the contours of vertical TEC, as if you are looking down on the ionosphere from above. A huge space-weather event, known as a storm, occurred during July of 2000 around the time of the peak in the 11-year solar cycle. Ionospheric storms are linked to flares and coronal mass ejections from the Sun, massive outpourings of electromagnetic radiation and particles that can collide with the Earth's magnetic field and cause dramatic disturbances to the Earth's ionosphere such as those seen here in the four images. The images show the uneven distribution of TEC globally with the build up of high TEC values (shown in red) over southern USA during the afternoon. The images were taken from a movie of the ionosphere lasting for the entire storm, which is not only useful for scientific understanding but also for showing why communication systems are disrupted in certain regions of the world and not in others.

Images such as these can be of interest to radio communication and radar surveillance planning. This was first noted in 1994 when Bust *et al.* investigated the application of ionospheric tomography to single site location range estimation; the determination of the location of an unknown transmitter by tracing the refractive path of a radar signal through a tomographic image of the ionosphere. They showed that better results could be obtained when using real images rather than empirical ionospheric models.

Another important requirement for ionospheric information comes from the navigation community. GPS receivers are very widespread in their use and the vast majority of receivers use only one of the two transmitted frequencies. These single frequency GPS receivers are less accurate than their dual frequency counterparts, but are also less costly. The main limitation to their accuracy is due to the unknown time delay of the signal as it traverses the ionosphere. Proposals have been made to use a network of dual-frequency receivers to create real-time ionospheric maps to provide

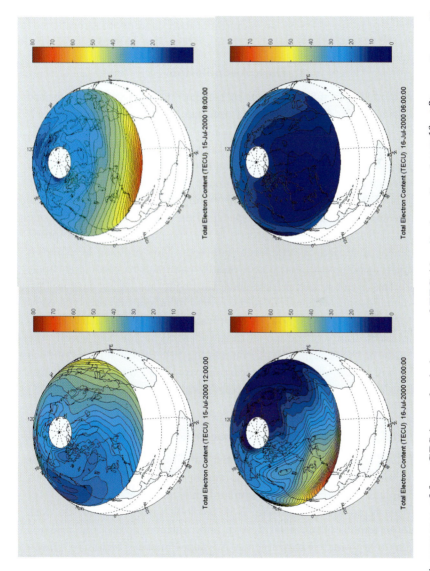

Fig. 7 A sequence of four GPS ionospheric images of TEC (density in units of $\times 10^{16} \mathrm{m}^{-2}$) over the northern hemisphere during a major storm.

position corrections to local GPS users. It is possible that extended tomographic approaches could assist in improving the accuracy achieved by the
proposed simple ionospheric mapping. Assessment and improvement of such
systems over the next few years may eventually lead to position accuracy
and reliability to use GPS to aid in automatically landing an aircraft.

4. Data Assimilation

Physical models driven by equations rather than just statistical records, are
now being tested in data assimilation programs. Essentially, this approach
combines physical laws governing the ionisation production and recombi-
nation processes with real-time observations. In regions of good coverage,
measurements can constrain the model and nonlinear inversions can be used
to solve for model parameters. One approach is to then to interpolate these
to regions of sparse measurements, for example over oceans. Ideally this
procedure should result in a global specification of the ionosphere, but one
present disadvantage of this approach is the computational load imposed
by nonlinear inversion of very large data sets. This problem is expected
to diminish as computational power rapidly continues to become cheaper,
although this is not a trivial issue since the requirements for data assimilation demand improvements of many orders of magnitude over the computational power available now. A significant advantage of such a system
is the inherent ability of the model to run forward in time and specify the
future state of the ionosphere. An interesting test for the success of such
a system will be the short-term forecasting capability. Whether or not this
promising approach will work in practice is still an open question.

5. Imaging Near-Earth Space and Other Planets

Back in 1947 the pioneering radio scientist Sir Edward V. Appleton gave his
Nobel lecture and talked of ionospheric-weather forecasting. He spoke about
the ionosphere as 'a region which human beings have not yet visited' — of
course this is no longer the case. Now we look further out to the dark
and remote regions of our solar system and continue most of our research
using remote automated instruments. Advances in space technology mean
that satellite sizes and costs can be greatly reduced. It is now possible to
build satellites that are smaller and lighter that ever before (microsatellites). Constellations consisting of many such satellites have already been
suggested as ways of solving a number of scientific problems requiring high

density of spatial sampling around the Earth. The majority of planets in the solar system have surfaces that are unsuitable for the remote operation of instrumentation or have no surface at all. Constellations of cheap, reliable microsatellites placed around such planets therefore have great potential for studies of their largely unexplored outer atmospheres.

The radio-occultation technique, mentioned earlier for improving the vertical resolution in studies of the Earth's atmosphere, was originally used to probe the atmosphere of Venus [see for example Luhmann (1991)]. In this previous application, a low-planet-orbit satellite transmitted signals through the planet's atmosphere at low elevations and these were received at the Earth. Subsequent analysis revealed structure in the planetary atmosphere. In a development of this idea, constellations of microsatellites could host radio beacons and receivers. Thus a network of satellite-to-satellite paths would sense the planet's atmosphere and ionosphere in a complex and dynamic frame. In addition, other instrumentation operated from each satellite such as Langmuir probes and ionosondes could provide a dense and varied network of measurements. Four-dimensional inversion algorithms, able to adapt simultaneously to the changing spatial and temporal environment of each measurement could provide the ideal analysis tool, making 'movies' of the atmospheres and ionospheres of other planets.

6. Solar-Terrestrial System Imaging and Data Assimilation

GPS signals recorded on the ground show the differential time-delay of the signals as they traverse the ionised regions. While the main contribution to the delay is from the ionosphere there is also some contribution from the plasmasphere (the even more tenuous plasma above the ionosphere), as the GPS satellites are at more than twenty-thousand kilometres altitude. This means that a GPS receiver on board a LEO satellite above the ionosphere could collect TEC observations for topside ionosphere and plasmasphere imaging. GPS receivers are already on-board many different satellites for accurate positioning. So far, very little scientific research has been done using this vast quantity of data. In the future all satellites could relay their GPS data to a common databank open to the scientific community. Even more would be possible if geostationary satellites (higher than GPS) carried GPS receivers and relayed the dual-frequency delays for scientific analysis. Instead of tomographic images limited to the region between GPS and Earth, it would then be possible to image tenuous plasma above GPS into the whole region of near-Earth space, as far out as beyond the magnetopause. There are already proposals for advanced 'topside' sounders operating at such altitudes [Cummer _et al._ (2001)] and such measurements are ideal for combination with GPS signals in a large inversion algorithm. With such strong prospects for many instruments to measure TEC or electron concentration out into near-Earth space the need for imaging techniques to unite multi-spatial and temporal observations seems to be secure.

The Sun streams out high-velocity charged particles forming the solar wind. This carries a magnetic field that interacts with Earth's magnetic field. The fast flowing plasma generates electric currents and fields that are transmitted to the ionosphere, sometimes dramatically showing stunning auroral displays. While many of the individual physical processes have been studied and modelled there are still many unanswered questions relating to coupling between the magnetosphere and ionosphere and the ionosphere-thermosphere interactions. It is apparent that the near-Earth space environment should not be studied as a set of separate regions but as an entire system. Such comprehensive study of the entire solar-terrestrial system during storm events will be a challenging research topic for the next few decades. In addition to the obvious scientific fascination, the success of this area has important practical applications. This is because processes originating at the Sun, such as the ejection of clouds of plasma associated with solar flares, can have dramatic effects on our communications by damaging satellites and causing disruption to electrical systems on Earth. If a solar flare is sufficiently energetic and occurs on the part of the Sun facing the Earth then some of the X-rays encounter the Earth's atmosphere and photo-ionise as low as the D-region, causing absorption of certain radio communication signals. Protons, produced within the flare site by the ionisation of hydrogen, could potentially cause damage to unshielded satellites and astronauts. Developing the capability to warn of these events, to gain knowledge of their occurrence statistics and to model their effects is of great importance. It is not yet clear if we have enough knowledge to link together all of the processes that run from coronal mass ejections all the way to the production of geomagnetic storms that affect the upper atmosphere. Understanding the limit of our ability to model these events and determining the predictability will be a huge challenge in this research area probably lasting over many years. New techniques, perhaps developments from tomography, will be needed to aide in the incorporation of many measurements into physical models. Discoveries about how to couple the models together could eventually lead to a super-model of the whole solar-terrestrial system in our computers of the future. Such a super-model could have practical use in a space-weather forecasting system and would deepen our understanding of planet's space environment and our complex relationship with the Sun.

Acknowledgements

The assistance of other members of the research group at the University of Bath, in particular Dr. P Spencer, is acknowledged. Figure 4 is adapted from a black and white graphic in the PhD thesis (University of Wales) of the author of this chapter. The author is grateful to the International GPS Service for the use of RINEX data and to the EPSRC for the Advanced Research Fellowship 'Effects of the ionized atmosphere on GNSS' (2004–2009).

References

Andreeva, E. S., Galinov, A. V., Kunitsyn, V. E., Mel'nichenko, Yu. A., Tereshchenko, E. D., Filimonov, M. A. & Chernykov, S. M. (1990) Radio tomographic reconstruction of ionisation dip in the plasma near the Earth. *J. Exp. Theor. Phys. Lett.* **52**, 145.

Appleton. (1947) *Nobel Lecture 'The Ionosphere'*.

Austen, J. R., Franke, S. J., Liu, C. H. & Yeh, K. C. (1986) Application of computerized tomography techniques to ionospheric research. *Proc. Int. Beacon Satellite Symp.* **25**, Oulu, Finland.

Bernhardt, P. A., Dymond, K. F. & Picone, J. M. (1996) Improved radio tomography of the ionosphere using EUV/optical instruments from satellites. *Proc. Ionospheric Effects Symp.*, Alexandria, USA.

Bracewell, R. N. (1956) Strip integration in radio astronomy. *Aust. J. Phys.* **9**, 198.

Bust, G. S., Cook, J. A., Kronschnabl, G. R., Vasicek, C. J. & Ward, S. B. (1994) Application of ionospheric tomography to single-site location range estimation. *Int. J. Imag. Syst. Technol.* **5**, 160.

Cook, J. A. & Close, S. (1995) An investigation of TID evolution observed in MACE '93 data. *Ann. Geophysicae* **13**, 1320.

Cummer, S. A., Reiner, M. J., Reinisch, B. W., Kaiser, M. L., Green, J. L., Benson, R. F., Manning, R. & Goetz, K. (2001) A test of magnetospheric radio tomographic imaging with image and wind. *Geophys. Res. Lett.* **28**(6), 1131.

Hajj, G. A., Ibañez-Meier, R., Kursinski, E. R. & Romans, L. J. (1994) Imaging the ionosphere with the Global Positioning System. *Int. J. Imag. Syst. Technol.* **5**, 174.

Hounsfield, G. N. (1972) A method of and apparatus for examination of a body by radiation such as X-ray or gamma radiation, *Patent Specification 1283915*, The Patent Office.

Mitchell, C. N., Jones, D. G., Kersley, L., Pryse, S. E. and Walker, I. K. (1995) Imaging of field-aligned structures in the auroral ionosphere. *Ann. Geophysicae* **13**, 1311.

Kersley, L., Pryse, S. E., Walker, I. K., Heaton, J. A. T., Mitchell, C. N., Williams, M. J. & Willson, C. A. (1997) Imaging of electron density troughs by tomographic techniques. *Radio Sci.* **32**(4), 1607.

Luhmann, J. G. (1991) Space plasma physics research progress 1987–1990 — Mars, Venus, and Mercury. *Rev. Geophys.*, Part 2, **29**, Supp. S, 965–975.

Pryse, S. E. & Kersley, L. (1992) A preliminary experimental test of ionospheric tomography. *J. Atmos. Terr. Phys.* **54**, 1007.

Pryse, S. E., Mitchell, C. N., Heaton, J. A. T. & Kersley, L. (1995) Travelling ionospheric disturbances imaged by tomographic techniques. *Ann. Geophysicae* **13**, 1325.

Radon, J. (1917) Uber die Bestimmung von Funktionen durch ihre Integralwerte längs gewisser mannigfaltigkeiten, Saechsische Berichte Akademie der Wissensxhaften **69**, 262.

Spencer, P. S. J. & Mitchell, C. N. (2001) Multi-instrument data analysis system. *Proc. Beacon Satellite Symp.*, Boston.

Walker, I. K., Moen, J., Mitchell, C. N., Kersley, L. & Sandholt, P. E. (1998) Ionospheric effects of magnetopause reconnection observed using ionospheric tomography. *Geophys. Res. Lett.* **25**(3), 293.

Fault Structure, Stress, Friction and Rupture Dynamics of Earthquakes

Eiichi Fukuyama

National Research Institute for
Earth Science and Disaster Prevention
3–1 Tennodai, Tsukuba
Ibaraki, 305-0006, Japan
fuku@bosai.go.jp

We are now able to simulate a dynamic rupture process of real earthquakes, once the fault geometry, stress field applied to the fault, and friction law on the fault surface have been provided. The next question will be what kind of information is now available and what are still required to reproduce more realistic rupture process of earthquakes. This procedure will provide us with physical insights of earthquake dynamics as well as clues to predicting the fault rupture of future earthquakes. In this paper, I review how to simulate earthquake dynamic rupture based on available information.

1. Introduction

An earthquake is a process of faulting where a rupture propagates along several fault segments and a dislocation is produced between two sides of the fault. To model this process, the medium outside the fault can be considered as elastic so that linear elastic theory can be applied [Aki and Richards (1980, 2002)]. Inside the fault, however, the material is no longer elastic and linear elastic theory cannot be applicable, but, instead, fracture theory can be employed [Freund (1990)]. Thus, an earthquake is a phenomenon in the framework of elastic theory combined with fracture theory [Scholz (1990, 2002)]. Our interest is then how to describe individual earthquakes under specific initial and boundary conditions [Fukuyama (2003a, b)]. Here the initial condition is the fault geometry and the stress field applied to the fault. The boundary condition is the constitutive relation on the fault.

Recent rapid progress in earthquake observations in the field as well as in rock friction/fracture experiments in the laboratory seems to be providing us with sufficient information on initial and boundary conditions for modelling earthquake rupture in a realistic way. In addition, recent considerable development in numerical computation techniques enable us to make a computer model of the whole process of earthquake rupture. In particular, the spatio-temporal evolution of dynamic rupture on the fault surface can be reproduced numerically.

A few decades ago, seismologists could only handle faults as a planar surface [Hartzell and Heaton (1983); Fukuyama and Irikura (1986)] because of the difficulties in handling its complexity, as well as the lack of detailed information on fault geometry. In contrast, geologists already recognized that faults have complicated structures, from the field observation [e.g. Chester et al. (1993)]. At that time there existed a huge gap between seismologists and geologists on the understanding of earthquake fault in spite of efforts to fill this gap by using a simplified model of fault complexity [Segall and Pollard (1980)]. Recently, however, a very dense seismographic network has been established [Fukuyama et al. (1996); Kinoshita (1998); Obara et al. (2005)], which enables us to see a very detailed, fine scale hypocenter distribution of earthquakes that images the underground fault structures [Fukuyama et al. (2003a)] similar to those observed on the surface by structural geologists [Chester et al. (1993)].

In addition, we are now able to measure *in-situ* stresses under the ground directly in a borehole by conducting hydraulic fracturing experiments, which gives us information on the total stress [Tsukahara et al. (1996)]. The total stress at depth cannot be obtained from seismic observations because seismic waves are described using linear elastic theory, which is intrinsically independent on absolute stress. Thus such direct measurements of *in-situ* stress are extremely important. Unfortunately, such experiments are limited at shallow depth in the crust (< 2 km) due to huge costs of drilling deep boreholes.

Constitutive properties during faulting have intensively been investigated in the laboratory under simulated conditions of stress, temperature and water saturation at seismogenic depth using fault zone material or its simulated one [e.g. Dieterich (1979); Ohnaka et al. (1987); Blanpied et al. (1995); Tsutsumi and Shimamoto (1997)]. Although the predominant scale is different between slip in the laboratory and that during earthquakes, laboratory-derived relations can be applicable to the simulation of earthquakes if an appropriate scaling relation has been derived.

There are two issues to be reminded. One is the spatially heterogeneous distribution of stress and constitutive parameters. Since the Earth is not composed of homogeneous materials without any structural heterogeneties, the simulation of earthquakes should be affected by these heterogenities. These heterogeneities are characterised by "asperities" and "barriers" [Madariaga (1983)]. An asperity is used to describe the region on the fault where a large stress drop occurred during the earthquake [Kanamori and Stewart (1978)]. Recently, in strong motion seismology, an asperity represents a large slip region on the fault [Somerville *et al.* (1999)]. In contrast, a barrier is a region where the strength is so high that there is no slip during the earthquake [Aki (1979)]. When considering a small scale fault system, a complicated fault geometry can be precisely described and the stresses can be measured based on the microscopic fault geometry. However, when a large scale fault system is considered, without taking into account its microscopic fault geometry, the macroscopic constitutive parameters might be different from microscopic ones. Thus the heterogeneity of earthquakes might be scale dependent.

The other issue is the scaling relation for earthquakes [Aki (1967); Kanamori and Anderson (1975)]. Earthquakes are self-similar and the average slip and slip region are scaled by the seismic moment. These relations suggest that the stress drop during an earthquake is independent of earthquake size [e.g. Abercrombie (1995); Ide and Beroza (2001)]. Actually, the material constants such as seismic wave velocities and densities cannot change by order of magnitudes, but the slip during the earthquake does vary by several orders of magnitudes. Thus even if stress is a scale dependent parameter, this dependence could be quite limited.

In order to focus on the basic understanding of earthquake rupture dynamics, I will not go into details concerning the heterogeneity and scaling of earthquakes in this section. Here I only point out that these issues are very important in order to understand the complexity of an earthquake as a natural phenomenon. Here I will review what kind of information is now available for numerical simulation of earthquake dynamic rupture by referring to two recent earthquakes. This consideration will be very useful in understanding the physics of earthquake dynamics, as well as the prediction of the generation and propagation of large earthquakes.

2. Fault Structure

From geological investigations of active fault traces an earthquake is considered to rupture several fault segments, which are connected by jogs, steps and branches with each other [Aydin and Du (1995)]. During an earthquake, the rupture propagates mainly along the preexisting fault segment and sometimes jumps to the neighbouring segment by creating a new fault segment.

Along the rupture zone of the 1992 Landers, California, earthquake (M_w7.3), a complicated fault system existed before the earthquake [Aydin and Du (1995)]. During the earthquake, the rupture propagated along the fault traces by selecting its route by itself. In Fig. 1, the preexisting fault

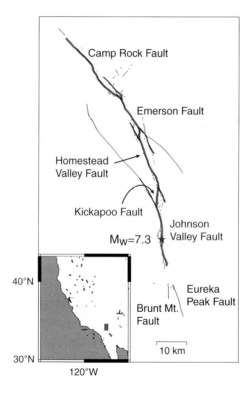

Fig. 1   Distribution of active fault traces around the source region of the 1992 Landers earthquake. Thick lines indicate the fault traces where the rupture propagated during the Landers earthquake. The star symbol stands for the hypocenter location of the Landers earthquake [Modified from Aochi and Fukuyama, (2002)].

traces and those followed by the rupture are shown. We found that how the rupture selected the fault trace at the branch basically depends on both the stress field applied to the fault branch and the rupture velocity [Aochi *et al.* (2000b)].

At seismogenic depths, we cannot see fault traces directly except for very ancient activity at depth, now exposed on the surface [e.g. Swanson (1988)]. We can see the currect fault structure at seismogenic depths in indirect ways. Along the fault trace, many microearthquakes occur. If these events are located very precisely, fault traces can be imaged. Very dense seismographic networks now enable us to locate very small earthquake with sufficient accuracy, which provides us with a detailed image of the fault system at seismogenic depth.

On October 6, 2000, a $M_w6.6$ earthquake with strike slip faulting occurred in western Tottori, southwest Japan. This was the first earthquake that occurred after the densely distributed seismographic network had been constructed. Following this earthquake, more than 10,000 aftershocks were recorded by the network and relocated [Fukuyama *et al.* (2003a)] by a very accurate technique called *Double Difference Method* [Waldhauser and Ellsworth (2000)]. The aftershock distribution shows a complicated image of the fault structure [Fukuyama *et al.* (2003a)] as shown in Fig. 2. The mainshock fault system consists of four fault segments (#1–#4 in Fig. 2(c)). Other fault segments (#5–#13) were created by the aftershock sequence, some of which were responsible for the post seismic deformation observed by GPS measurements [Sagiya *et al.* (2002)].

Focal mechanisms of aftershocks were calculated using the regional broadband seismic network [Fukuyama *et al.* (1998); Fukuyama *et al.* (2003a)], whose moment magnitudes are greater than 3.5. The fault strike directions of these aftershocks and the fault traces recognised from relocated hypocentral distribution are found to be consistent with each other, as shown in Fig. 2(b). Most aftershocks whose focal mechanisms were determined occurred along the pre-existing fault traces or parallel to them, and the lineaments inferred from the aftershock distribution are considered to be the fault structure at seismogenic depth.

We are therefore able to use the information on the geometry of the fault based on the active fault traces appearing on the surface, as well as seismic activities along the fault traces. But this information is sometimes insufficient, especially for an earthquake occurring in a seismically inactive region. This is sometimes called a blind fault. In order to overcome these situations, active seismic surveies such as a shallow reflection survey

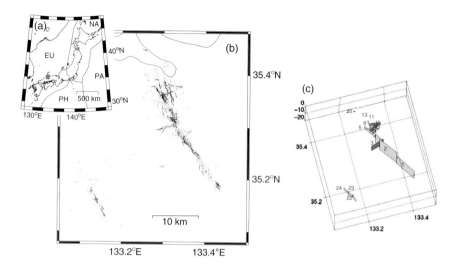

Fig. 2 (a) Location of the 2000 western Tottori earthquake plotted with plate configurations. PH, EU, NA and PA represent Philippine Sea, Eurasia, North American and Pacific plates, respectively. (b) Hypocentre distribution relocated by double difference method. Dots are hypocentres and straight lines are strike directions of the fault determined by the moment tensor inversion of regional broadband waveforms. Optimum fault direction for two possibilities for the focal mechanism are chosen based on the aftershock distribution. (c) Fault models based on the hypocenter distributions. Faults #1–#4 were created during the mainshock and other faults (#5–#13) are related to aftershock activity. Faults #22–#24 are caused by the largest aftershock two days later [Modified from Fukuyama *et al.* (2003a)].

would be useful [Sato *et al.* (2004)]. However, for strike slip faults, these experiments may not work properly because of the unclear vertical offsets of layers at depth.

3. Stress Field

For the simulation of earthquake rupture, information on absolute stress is crucial as an initial condition of the system. Absolute stress can be directly measured by *in-situ* experiments such as hydraulic fracturing experiments at the bottom of boreholes (Fig. 3) [Tsukahara *et al.* (1996, 2001); Ikeda *et al.* (2001)]. It can also be estimated by measuring the strain change of bore wall [Ishii *et al.* (2000)] and core samples taken from the borehole [Yamamoto and Yabe (2001)].

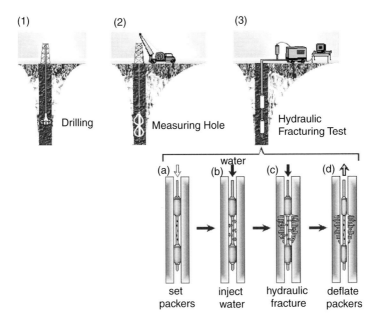

Fig. 3   Schematic illustration of hydraulic fracturing experiments. (1) A borehole is drilled. (2) Several measurements are done to check that there is no fracture on the bore wall where hydraulic fracturing experiments will be done. (3) Hydraulic experiments are done. (a) The region for the experiment is sealed by two packers. (b) Inject water to pressurise the region between packers. (c) Hydraulic fracture occurs. Then flow rate of the injected water is controlled to measure the shut in and shut out pressures of the fracture. (d) Deflate and pull up the packers.

Since these experiments are conducted at a single point around the fault, it is difficult to obtain a spatial variation of absolute stress. We only have local pin-point stresses. If we drilled at many sites around the fault we might be able to obtain the stress distribution. But, in reality, it is not possible because of the expensive drilling costs. Since what we need for the reconstruction of earthquake dynamic rupture is the distribution of absolute stress around the fault, a method is required to extrapolate the stress distribution from pin-point stresses.

To overcome this situation, earthquake focal mechanisms play an important role. Earthquake focal mechanisms are considered as strain changes at the focal area of each earthquake. Each focal mechanism does not indicate directly the stress field but an assembly of them does include the information on the stress field that caused the earthquake. If we assume that the stress is uniform inside the target area and that each earthquake slip occurs along the maximum stress direction, we are able to estimate the stress field from a group of focal mechanisms by using the variation in focal mechanism solutions in the dataset [Angelier (1979); Gephart and Forsyth (1984)]. The fault plane does not always direct to 45° to the maximum principal direction but is distributed around this direction depending on the frictional property of the fault. By processing many focal mechanism data statistically, we can estimate the stresses [Hardebeck and Hauksson (2001); Fukuyama et al. (2003a); Kubo and Fukuyama (2004)].

It should be noted that these estimated values are not sufficient to describe the total stress field; three principal stress directions and the stress ratio (R) are estimated by the stress tensor inversion. R is defined by $(\sigma_1 - \sigma_2)/(\sigma_1 - \sigma_3)$, where σ_1, σ_2 and σ_3 are maximum, intermediate and minimum principal stresses, respectively and compression is taken positive. An important advantage of this method is that when earthquake focal mechanisms are estimated in a region, we are able to estimate the stress field from the focal mechanisms. In order to calibrate the stress field esitmated by the focal mechanisms, one or two in-situ stress measurements are required.

During the 2000 western Tottori earthquake, about 75 focal mechanisms were estimated [Fukuyama et al. (2003a)] by the moment tensor inversion of broadband seismograms at regional distances [Fukuyama et al. (1998); Fukuyama et al. (2001); Kubo et al. (2002)] (Fig. 4a). The source region was divided into two: coseismic (#1–#4 in Fig. 2c) and post seismic slip (#5–#13) regions. By examining the focal mechanisms in Fig. 4(a), the predominant directions of the P-axes of the focal mechanisms appears different between coseismic and postseismic regions.

Figure 4(b) and (c) shows the results of the stress tensor inversion. Taking into account the 95% confidence region, the principal stress directions are well constrained by the data. Although the focal mechanisms are slightly differnet between northern postseismic region and southern coseismic region, stress field is considered to be similar [Fukuyama et al. (2003a)]. R value is estimated at 0.6, which is consistent with the fact that all the focal mechanisms are of strike slip type.

An alternative method is to measure the distribution of aseismic slips near the fault. The current stress field is considered to be the tectonically applied stress contaminated by the stress caused by aseismic slips. Since the materials around the fault are considered to be elastic, we can estimate the distribution of stress change around the fault due to the aseismic slip

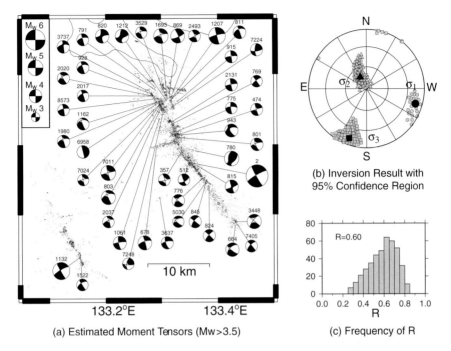

(a) Estimated Moment Tensors (Mw >3.5)

(b) Inversion Result with 95% Confidence Region

(c) Frequency of R

Fig. 4   Results of the stress tensor inversion using aftershock moment tensors. (a) Distribution of estimated moment tensor solutions whose moment magnitudes are greater than 3.5. The lower hemisphere projection is employed. (b) The result of the stress tensor inversion. Optimum solution for the principal stress directions are shown as solid big symbols. 95% confidence regions are shown for each stress direction. (c) Distribution of stress ratio R for the solutions within the 95% confidence region. The R value for the optimum solution was 0.60. Plot in (b) is in lower hemisphere projection [Modified from Fukuyama *et al.* (2003a)].

on the fault. Once we know the information on the tectonic stress applied to this region, the current stress field can be estimated by adding the stress change due to the aseismic slips on the fault. This aseismic slip distribution can be obtained by the analysis of strain distribution on the surface obtained by GPS (global positioning system) data [Hirahara *et al.* (2003)]. To calibrate the stress distribution estimated above, again, *in-situ* stress measurements around the fault become important.

This idea was applied to the estimation of the fault strength [Yamashita *et al.* (2004)], which could be equivalent to the shear stress value just before the earthquake. The *in-situ* stress measurements near the fault can only be done after the occurrence of earthquakes and the continuous monitoring of

stress is not now possible. The distribution of coseismic slip, however, can
be obtained by the waveform inversion analysis of mainshock seismograms,
and the coseismic stress change around the fault can be estimated. By
subtracting the stress change estimated from the coseismic slip from the
post seismic stress measured by the *in-situ* stress measurements, the pre-
shock stress can be estimated, which should be balanced by the strength of
the fault.

4. Constitutive Relation

A slip-weakening law was originally proposed for tension cracks [Barenblatt
(1959)], then extended to shear cracks [Ida (1972); Palmer and Rice (1973)]
based on the theoretical consideration of energy at the crack tip. After that,
based on constant slip-rate experiments with sudden velocity change in the
laboratory, rate and state dependent friction law [Dieterich (1979); Ruina
(1983); Perrin *et al.* (1995); Marone (1998)] has been proposed to describe
slip behavior on a fault. Since this relation is derived from very slow slip
friction experiments, where rate dependence is dominant, rate dependence
in this friction law was emphasised. The rate and state dependent friction
law is often applied to simulate the earthquake cycles where most slip is
quasi-static [Tse and Rice (1986); Rice (1993); Kato and Hirasawa (1999);
Lapusta *et al.* (2000)]. However, at high slip rate (e.g. during the earthquake rupture), the state effect becomes important [Okubo (1989); Bizzarri
and Cocco (2003)]. To handle this feature properly, a slip-weakening
friction law [Ohnaka *et al.* (1987); Matsu'ura *et al.* (1992)] was proposed to
describe the dynamic rupture of faulting. In this slip weakening constitutive
law, critical strength drop $(\Delta\tau_b)$ and slip-weakening distance (D_c) become
important parameters to characterise the rupture (Fig. 5).

The constitutive relation can be described by temporal variations of
both slip and stress on the fault. Once the spatio-temporal variation of slip
is obtained, the corresponding spatio-temporal variation of stress can be
uniquely obtained from the boundary integral equation below [Fukuyama
and Madariaga (1998)].

$$T_{ij}(\boldsymbol{x},t) = -\frac{\mu}{2\beta}\Delta\dot{u}_i(\boldsymbol{x},t)\nu_j + \int_S\int_0^t \mathcal{K}_{kl}^{ij}(\boldsymbol{\xi}-\boldsymbol{x},\tau-t)\Delta\dot{u}_k(\boldsymbol{\xi},\tau)\nu_l dSdt, \quad (1)$$

where T_{ij} is the traction on the fault. $\Delta\dot{u}$ is the slip velocity on the fault,
S is the fault surface. Positions $\boldsymbol{\xi}$ and $\boldsymbol{x}$ are at the source where slip occurs
and at the receiver where stress is measured, respectively. τ and t are corresponding times, respectively. $\boldsymbol{\nu}$ is the normal vector to the fault. μ and β

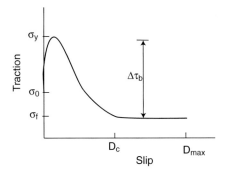

Fig. 5   Schematic illustration of slip-weakening friction law. σ_y, σ_0, and σ_f indicate yielding stress, initial stress and frictional stress, respectively. D_c and D_{max} correspond to the slip-weakening distance and final amount of slip, respectively critical strength drop ($\Delta\tau_b$) is defined as $\sigma_y - \sigma_f$.

are shear modules and shear wave velocity, respectively. $\mathcal{K}_{kl}^{ij}$ is the kernel that represents ij-component of stress at $(\boldsymbol{x}, t)$ when k-component of unit slip occurs at $(\boldsymbol{\xi}, \tau)$ on the fault. It should be noted that although Eq. (1) is derived by the boundary integral equation method, this relation is not limited to a particular numerical method but is a universal feature, because this relation is obtained based on the continuities and symmetries of stress and dislocation on the fault. Therefore, Eq. (1) explicitly shows that the spatio-temporal distribution of slip is uniquely related to the spatio-temporal distribution of stress change.

Using this idea, we are theoretically able to estimate the shape of slip weakening curves from the spatio-temporal evolution of slip velocity on the fault [Ide and Takeo (1997)] although we still cannot estimate the absolute stress by this method. But it should be noted that there is a resolution problem in this technique [Spudich and Guatteri (2004)]. Since the temporal resolution of slip velocity is generally not sufficient due to the band-limited nature in the waveforms which was caused by attenuation and scattering of the high frequency waves during the propagation. This limitation makes the estimation of small D_c values difficult [Guatteri and Spudich (2000)].

To overcome this situation, a new technique has been proposed [Fukuyama *et al.* (2003b); Mikumo *et al.* (2003)] in which temporal variation of stress is not required. In this method, stress is assumed to drop at frictional stress level (σ_f) at peak slip velocity time [Mikumo *et al.*, 2003] as shown in Fig. 6. This assumption is correct as long as the rupture propagates smoothly without strong barriers and asperities [Fukuyama *et al.* (2003b)].

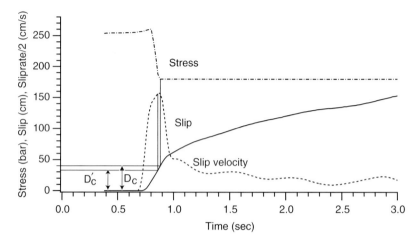

Fig. 6 Schematic illustration of a conventional method to estimate slip-weakening distance (D_c). D_c is defined as the amount of slip when the stress drops to the frictional level (σ_f). D'_c is an approximation of D_c which is defined as the amount of slip where slip velocity becomes maximum. This approximation is valid if the peak slip velocity time is close to the break down time [Modified from Mikumo *et al.* (2003)].

As long as the slip velocity function is obtained accurately enough we will be able to determine the range of the slip-weakening parameters. Even using the above technique, we cannot completely avoid the band-limited effect of the observation [Spudich and Guatteri (2004)]. Thus our discussion is still restricted to the order of these parameters and they are considered to be an upper bound of the real values.

5. Numerical Simulation

Many numerical techniques for simulating earthquake dynamic rupture have already been proposed. Since the end of 1970s, the boundary element method [Das and Aki (1977)], the finite difference method [Andrews (1976); Madariaga (1976); Mikumo and Miyatake (1978); Day (1982)] and the finite element method [Archuleta and Frazier (1978)] started to be developed. Recently, the boundary integral equation method [Andrews (1985); Koller *et al.* (1992); Cochard and Madariaga (1994); Fukuyama and Madariaga (1998); Aochi *et al.* (2000a)] was developed, which is a sophisticated version of the boundary element method.

In the finite difference method, since grids are located on evenly spaced points along the Cartesian coordinates, the fault model is constrained by this geometry [Harris *et al.* (1991); Mikumo and Miyatake (1993); Olsen *et al.* (1997); Kase and Kuge (1999)]. However, by introducing an inhomogeneous grid spacing and/or projection theory, the computation now becomes possible for a non-planar fault system [Inoue and Miyatake (1998); Cruz-Atienza and Virieux (2004)]. The finite element method is a very flexible technique to model a non-planar fault system. But the problem lies in the computational efficiency for temporal evolutions of the system. The equations to be solved at each time step are very complicated and solutions can only be found by solving the equations in a implicit way so that computations of dynamic rupture process become very heavy. But recent development of computer resources enables us to make it work [Oglesby *et al.* (1998, 2000)]. The distinct element method is also available [Mora and Place (1998); Dalguer *et al.* (2001)]. This method is more flexible than the finite elements in handling the fault geometry.

The above three methods are considered to be domain methods in which computations are done in a volume including the earthquake faults. However, boundary integral equation method is a boundary method so that we do not have to handle the volume but only consider the boundaries (fault surfaces). Thus the boundary integral equation method has a merit of applying boundary conditions (e.g. constitutive relation) more explicitly, which makes the computation more accurate [Fukuyama and Madariaga (1998); Aochi *et al.* (2000a)].

In the following section, we will show some examples of dynamic rupture propagation using the boundary integral equation method whose complete formulation of boundary integrals for non-planar fault in an unbounded elastic medium has already been derived [Tada *et al.* (2000); Tada (2005, 2006)].

6. Dynamic Rupture of Earthquake

Two results of simulations will be shown as examples of dynamic rupture simulation: The 2000 western Tottori, Japan (M_w 6.6) and the 1992 Landers, California (M_w 7.3) earthquakes. The western Tottori earthquake is a recent earthquake occurring inside a densely distributed seismographic network [Fukuyama *et al.* (2003a)]. There are several fault models for this earthquakes [Iwata and Sekiguchi (2001); Mikumo *et al.* (2003); Semmane *et al.* (2005)]. The Landers is a famous earthquake which occurred near the

San Andreas fault system in southern California for which many earthquake source analyses have been conducted [Cohee and Beroza (1994); Wald and Heaton (1994); Cotton and Campillo (1995); Olsen *et al.* (1997); Aochi and Fukuyama (2002)].

For the western Tottori earthquake, we have a very precise image of the aftershock distribution [Fukuyama *et al.* (2003a)] using the double difference method [Waldhauser and Ellsworth (2000)]. Based on this distribution, a fault model for the simulation was constructed, which consists of 4 fault planes [Fukuyama *et al.* (2003a)] (Fig. 2). Although, unfortunately, no *in-situ* stress measurements were conducted near the source region because of an unclear fault location on the surface, a stress tensor inversion of aftershock moment tensors was conducted [Fukuyama *et al.* (2003a); Kubo and Fukuyama (2004)] to obtain a relative stress field. By combining the relative stress information with the assumed absolute stress values, taking into account the lithostatic stress at seismogenic depth, a stress model for the simulation was constructed. The slip weakening distance for this earthquake has been estimated from the source time functions estimated by waveform inversions [Mikumo *et al.* (2003)].

Numerical simulations were conducted based on the above information of fault geometry, stress field and constitutive relation. The computational results are shown in Fig. 7. In this plot, three simulation results with

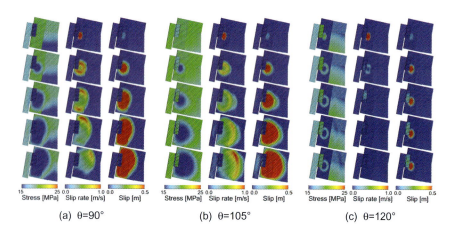

(a) θ=90°   (b) θ=105°   (c) θ=120°

Fig. 7 Result of dynamic rupture simulation of the 2000 western Tottori earthquake. Snapshots of stress (left column), slip velocity (center column) and slip (right column) are shown at a constant time step of 0.75 s. Scale of each column is shown as a color bar. Maximum principal stress directions of initial stress are (a) N90°E, (b) N105°E and (c) N120°E, respectively [Modified from Fukuyama (2003b)].

different maximum principal stress directions are shown. Other conditions such as slip weakening distances and magnitudes of principal stresses were kept the same. One can see that the rupture propagation is controlled by the stress field around the fault. This is convincing because the stress field applied to the fault controls the initial shear and normal stress on the fault. Shear stress corresponds to the coseismic slip ($D_{\max}$ in Fig. 5) allowed and normal stress applied to the fault controls the breakdown stress drop ($\Delta\tau_b$ in Fig. 5), which is also dependent on the fault geometry. Thus the fault geometry plays a very important role for the propagation of dynamic rupture on the fault.

In the simulation, although we did not assume any spatial heterogeneities on the fault, one can observe a non-uniform rupture along the fault, especially, whether the rupture extends to the northern small segment (#4 in Fig. 2), which is shifted from the main fault, depends on the stress field around the fault. Since all 4 fault segments were ruptured in the kinematic source model, the principal maximum stress direction of N105°E is the most probable, which is also consistent with the result of stress tensor inversion [Fukuyama *et al.* (2003a); Kubo and Fukuyama (2004)] (N107°E). The above simulations indicate that once the frictional property of the fault, stress field around the fault and the geometry of the fault are all obtained, we are able to estimate a scenario for earthquake rupture [Fukuyama (2003a, b)].

For the Landers earthquake, surface faults are very accurately traced [Hart *et al.* (1993)] (Fig. 1). The coseismic slip distribution was well estimated using near-field and teleseismic waveforms as well as using GPS data [Wald and Heaton (1994)]. Thus the fault model is constructed based on the surface fault traces. Since there is no information on the friction law, typical relations obtained in the laboratory were applied. For the stress field, principal stress directions were searched by trial and error.

Figure 8 shows the result of simulation of the Landers earthquake [Aochi (1999); Aochi and Fukuyama (2002)]. A uniform stress field cannot make the rupture propagate along several fault segments as obtained by the kinematic waveform inversion. In order to propagate through the Kickapoo fault (Fig. 1), the stress field should be different in the northern and southern regions. In this computation the most optimum solution was that with the stress rotated clockwise [Aochi and Fukuyama (2002)]. This is consistent with the fact that the northern and southern part of the faults belong to the different geological block, in which the stress field might be different [Unruh *et al.* (1994)].

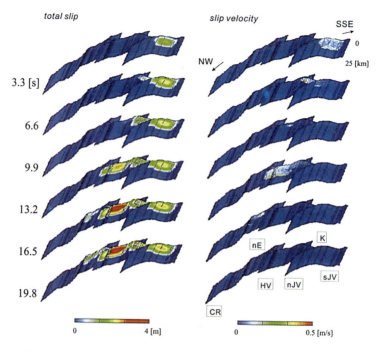

Fig. 8   Result of dynamic rupture simulation of the 1992 Landers earthquake. Snapshots of slip and slip velocity are shown in left and right
columns, respectively. nE, K, CR, HV, nJV, sJV stand for northern Emerson,
Kickapoo, Camp Rock, Homestead Valley, northern and southern Johnson
Valley faults, respectively [after Aochi (1999)].

In both cases, a complete set of initial and boundary conditions could
not be used to compute dynamic rupture propagation. Some of the parameters had to be assumed. However, this kind of situation is very common and
how to assume the missing information will be important if this is applied
to the prediction of future earthquake dynamic rupture.

7. Concluding Remarks

Through the numerical experiments of dynamic rupture simulation of earthquakes, we can see that the fault geometry plays an important role in the
propagation of earthquake rupture. Of course, the stress field applied to
the fault system and the frictional constitutive relation of the fault surface
control the rupture propagation, but these are simultaneously affected by
fault geometry.

The stress field around the fault system may not always be uniform, mainly due to contamination by creep on adjacent fault segments. Understandings of both the remotely applied tectonic stress and the local stress disturbance due to the deformation of the fault system become important. In addition, spatio-temporal variation of the constitutive parameters are not still understood well. The slip-weakening distance (D_c) can only be measured after the slip exceeds D_c. Thus in order to estimate the D_c values *a priori*, we need a physical model based on laboratory experiments including scaling properties with respect to the total slip [e.g. Ohnaka (2003)].

What we primarily need, and is feasible to obtain, is an accurate geometry of the fault system at various scales. Since the width of fault slip zone during an earthquake is known to within 1 cm [e.g. Swanson (1988); Sibson (2003); Di Toro (2004)], we need to know the microscopic fault geometry at a macroscopic scale to understand the essential feature of heterogeneous rupture.

Once the fault geometry, the applied stress field and the constitutive relation on the fault are obtained, we are ready to simulate dynamic rupture of the earthquake, which will help construct the scenario for future earthquakes. However, this is an ideal situation, and we need to make an effort to have physically reasonable assumptions when some of the parameters are unavailable.

Acknowledgements

Comments by an anonymous reviewer help improve the manuscript. This work was supported by the NIED project "Research on Mechanics of Earthquake Occurrence" and Grant-in-Aid SE(C) 15607020 by the MEXT, Japan.

References

Abercrombie, R. E. (1995) Earthquake source scaling relationships from 1 to 5 ML using seismograms recorded at 2.5-km depth. *J. Geophys. Res.* **100**(B12), 24015–24036.

Aki, K. (1967) Scaling law of seismic spectra. *J. Geophys. Res.* **72**, 1217–1231.

Aki, K. (1979) Characterization of barriers on an earthquake fault. *J. Geophys. Res.* **84**, 6140–6148.

Aki, K. & Richards, P. G. (1980) *Quantitative Seismology, Theory and Observation*, W. H. Freeman and Company.

Aki, K. & Richards, P. G. (2002) *Quantitative Seismology, Theory and Observation*, 2nd edn., University Science Books.

Andrews, D. J. (1976) Rupture velocity of plane strain shear crack. *J. Geophys. Res.* **81**, 5679–5687.

Andrews, D. J. (1985) Dynamic plane-strain shear rupture with a slip-weakening friction law calculated by a boundary integral method. *Bull. Seismol. Soc. Am.* **75**, 1–21.

Angelier, J. (1979) Determination of the mean principal directions of stresses for a given fault population. *Tectonophys.* **56**, T17–T26.

Aochi, H. (1999) Theoretical studies on dynamic rupture propagation along a 3D non-planar fault system. *PhD Thesis*, the University of Tokyo, 90pp.

Aochi, H., Fukuyama, E. & Matsu'ura, M. (2000a) Spontaneous rupture propagation on a non-planar fault in 3D elastic medium. *Pure Appl. Geophys.* **157**, 2003–2027.

Aochi, H., Fukuyama, E. & Matsu'ura, M. (2000b) Selectivity of spontaneous rupture propagation on a branched fault. *Geophys. Res. Lett.* **27**, 3635–3638.

Aochi, H. & Fukuyama, E. (2002) Three-dimensional nonplanar simulation of the 1992 Landers earthquake. *J. Geophys. Res.* **107**(B2), doi: 10.1029/2000JB000061.

Archuleta, R. J. & Frazier, G. A. (1978) Three-dimensional numerical simulations of dynamic faulting in a half-space. *Bull. Seismol. Soc. Am.* **68**, 541–572.

Aydin, A. & Du, Y. (1995) Surface rupture at a fault bend: The 28 June 1992 Landers, California, earthquake. *Bull. Seismol. Soc. Am.* **85**, 111–128.

Barenblatt, G. I. (1959) On equilibrium cracks formed in brittle fracture. General concepts and hypotheses. Axisymmetric cracks. *Prikl. Mat. Mek.* **23**(3), 434–444 (in Russian); *J. Appl. Math. Mech.* (PMM), **23**(3), 622–636 (English Translation).

Bizzarri, A. & Cocco, M. (2003) Slip-weakening behavior during the propagation of dynamic ruptures obeying rate- and state-dependent friction laws. *J. Geophys. Res.* **108**(B8), 2373, doi:10.1029/ 2002JB002198 (2003).

Blanpied, M. L., Lockner, D. A. & Byerlee, J. D. (1995) Frictional slip of granite at hydrothermal conditions. *J. Geophys. Res.* **100**, 13045–13064.

Chester, F. M., Evans, J. P. & Biegel, R. L. (1993) Internal structure and weakening mechanisms of the San Andreas fault. *J. Geophys. Res.* **98**(B1), 771–786.

Cochard. A. & Madariaga, R. (1994) Dynamic faulting under rate-dependent friction. *Pure Appl. Geophys.* **142**, 419–445.

Cohee, B. P. & Beroza, G. C. (1994) Slip distribution of the 1992 Landers earthquake and its implications for earthquake source mechanics. *Bull. Seismol. Soc. Am.* **84**, 692–712.

Cotton, F. & Campillo, M. (1995) Frequency domain inversion of strong motions: Application to the 1992 Landers earthquake. *J. Geophys. Res.* **100**, 3961–3975.

Cruz-Atienza, V. M. & Virieux, J. (2004) Dynamic rupture simulation of non-planar faults with a finite-difference approach. *Geophys. J. Int.* **158**, 939–954.

Dalguer, L. A., Irikura, K., Riera, J. D. & Chiu, H. C. (2001) Fault dynamic rupture simulation of the hypocenter area of the thrust fault of the 1999 Chi-Chi (Taiwan) earthquake. *Geophys. Res. Lett.* **28**(7), 1327–1330.

Das, S. & Aki, K. (1977) Fault plane with barriers: A versatile earthquake model. *J. Geophys. Res.* **82**, 5658–5670.

Day, S. M. (1982) Three-dimensional simulation of spontaneous rupture: The effect of nonuniform prestress. *Bull. Seismol. Soc. Am.* **72**, 1881–1902.

Dieterich, J. H. (1979) Modeling of roch friction 1. Exprimental results and constitutive equation. *J. Geophys. Res.* **84**, 2161–2168.

Di Toro, G., Goldsby, D. L. & Tullis, T. E. (2004) Friction falls towards zero in quartz rock as slip velocity approaches seismic rates. *Nature* **427**, 436–439.

Freund, L. B. (1990) *Dynamic Fracture Mechanics*, Cambridge University Press.

Fukuyama, E. (2003a) Numerical modeling of earthquake dynamic rupture: Requirements for realistic modeling. *Bull. Earthq. Res. Inst., Univ. Tokyo.* **78**, 167–174.

Fukuyama, E. (2003b) Earthquake dynamic rupture and stress field around the fault. *J. Geography* **112**(6), 850–856 (in Japanese with English abstract).

Fukuyama, E., Ellsworth, W. L., Waldhauser, F. & Kubo, A. (2003a) Detained fault structure of the 2000 western Tottori, Japan, earthquake sequence. *Bull. Seismol. Soc. Am.* **93**, 1468–1478.

Fukuyama, E. & Irikura, K. (1986) Rupture process of the 1983 Japan Sea (Akita-Oki) earthquake using a waveform inversion method. *Bull. Seismol. Soc. Am.* **76**, 1623–1649.

Fukuyama, E., Ishida, M., Hori, S., Sekiguchi, S. & Watada, S. (1996) Broadband seismic observation conducted under the FREESIA Project. *Rep. Nat'l Res. Inst. Earth Sci. Disas. Prev.* **57**, 23–31 (in Japanese with English abstract).

Fukuyama, E., Ishida, M., Dreger, D. S. & Kawai, H. (1998) Automated seismic moment tensor determination by using on-line broadband seismic waveforms. *Zisin (J. Seismol. Soc. Jpn.) Ser. 2* **51**, 149–156 (in Japanese with English abstract).

Fukuyama, E., Kubo, A., Kawai, H. & Nonomura, K. (2001) Seismic remote monitoring of stress field. *Earth Planets, Space* **53**, 1021–1026.

Fukuyama, E. & Madariaga, R. (1998) Rupture dynamics of a planar fault in a 3D elastic medium: Rate- and slip-weakening friction. *Bull. Seismol. Soc. Am.* **88**, 1–17.

Fukuyama, E., Mikumo, T. & Olsen, K. B. (2003b) Estimation of the critical slip-weakening distance: Theoretical background. *Bull. Seismol. Soc. Am.* **93**, 1835–1840.

Gephart, J. W. & Forsyth, D. W. (1984) An improved method for determining the regional stress tensor using earthquake focal mechanism data: Application to the San Fernando earthquake sequence. *J. Geophys. Res.* **89**, 9305–9320.

Guatteri, M. & Spudich, P. (2000) What can strong-motion data tell us about slip-weakening fault-friction laws? *Bull. Seismol. Soc. Am.* **90**, 98–116.

Hardebeck, J. L. & Hauksson, E. (2001) Crustal stress field in southern California and its implications for fault mechanics. *J. Geophys. Res.* **106**, 21859–21882.

Harris, R. A., Archuleta, R. J. & Day, S. M. (1991) Fault steps and the dynamic rupture process: 2D numerical simulations of a spontaneously propagating shear fracture. *Geophys. Res. Lett.* **18**(5), 893–896.

Hart, E. W., Bryant, W. A. & Treiman, J. A. (1993) Surface faulting associate with the June 1992 Landers earthquake. *Calif. Geol.* **46**, 10–16.

Hartzell, S. H. & Heaton, T. H. (1983) Inversion of strong ground motion and teleseismic waveform data for the fault rupture history of the 1979 Imperial Valley, California, earthquake. *Bull. Seismol. Soc. Am.* **73**, 1553–1583.

Hirahara, K., Ooi, Y., Ando, M., Hoso, Y., Wada, Y. & Ohkura, T. (2003) Dense GPS Array observations across the Atotsugawa fault, central Japan. *Geophys. Res. Lett.* **30**(6), 8012, doi:10.1029/2002GL015035.

Ida, Y. (1972) Cohesive force on longitudinal crack and Griffith's specific surface energy. *J. Geophys. Res.* **77**, 3796–3805.

Ide, S. & Beroza, G. C. (2001) Does apparent stress vary with earthquake size? *Geophys. Res. Lett.* **28**(17), 3349–3352.

Ide, S. & Takeo, M. (1997) Determination of constitutive relations of fault slip based on seismic wave analysis. *J. Geophys. Res.* **102**, 27379–27391.

Ikeda, R., Iio, Y. & Omura, K. (2001) *In situ* stress measurements in NIED boreholes in and around the fault zone near the 1995 Hyogo-ken Nanbu earthquake, Japan. *Island Arc* **10**, 252–260.

Inoue, T. & Miyatake, T. (1998) 3D simulation of near-field strong ground motion based on dynamic modeling. *Bull. Seismol. Soc. Am.* **88**, 1145–1456.

Ishii, H., Yamauchi, T., Matsumoto, S. & Ikeda, R. (2000) Initial stress measurements by means of intelligent type strainmeter by overcoring and strain observation in deep boreholes. *Proc. SEGJ Conf.* **102**, 109–113 (in Japanese).

Iwata, T. & Sekiguchi, H. (2002) Source process and near-source ground motion during the 2000 Tottori-ken Seibu earthquake. *Proc. 11th Japan Earthquake Engineering Symposium*, Earthquake Eng. Res. Liaison Comm., Sci. Counc. of Jpn., Tokyo.

Kame, N. & Yamashita, T. (1999) Simulation of the spontaneous growth of a dynamic crack without constraints on the crack tip path. *Geophys. J. Int.* **139**, 345–358.

Kanamori, H. & Anderson, D. L. (1975) Theoretical basis of some empirical relations in seismology. *Bull. Seismol. Soc. Am.* **65**, 1073–1095.

Kanamori, H. & Stewart, G. S. (1978) Seismological aspects of the Guatemala earthquake of February 4, 1976. *J. Geophys. Res.* **83**, 3427–3434.

Kase, Y. & Kuge, K. (1998) Numerical simulation of spontaneous rupture processes on twonon-coplanar faults: The effect of geometry on fault interaction. *Geophys. J. Int.* **135**, 911–922.

Kato, N. & Hirasawa, T. (1999) Nonuniform and unsteady sliding of a plate boundary in a great earthquake cycle: A numerical simulation using a laboratory-derived friction law. *Pure Appl. Geophys.* **155**, 93–118.

Kinoshita, S. (1998) Kyoshin Net (K-NET). *Seismol. Res. Lett.* **69**, 309–334.

Koller, M. G., Bonnet, M. & Madariaga, R. (1992) Modeling of dynamical crack propagation using time-domain boundary integral equations. *Wave Motion* **16**, 339–366.

Kubo, A., Fukuyama, E., Kawai, H. & Nonomura, K. (2002) NIED seismic moment tensor catalogue for regional earthquakes around Japan: Quality test and application. *Tectonophys.* **356**, 23–48.

Kubo, A. & Fukuyama, E. (2004) Stress fields and fault reactivation angles of the 2000 western Tottori aftershocks and the 2001 northern Hyogo swarm in southwest Japan. *Tectonophys.* **378**, 223–239.

Lapusta, N., Rice, J. R., Ben-Zion, Y. & Zheng, G. (2000) Elastodynamic analysis for slow tectonic loading with spontaneous rupture episodes on faults with rate- and state-dependent friction. *J. Geophys. Res.* **105**(B10), 23765–23790.

Madariaga, R. (1976) Dynamics of an expanding circular fault. *Bull. Seismol. Soc. Am.* **66**, 639–667.

Madariaga, R. (1983) High frequency radiation from dynamic earthquake fault models. *Ann. Geophys.* **1**, 17–23.

Marone, C. (1998) Laboratory-derived friction laws and their application to seismic faulting. *Ann. Rev. Earth Planet. Sci.* **26**, 643–696.

Matsu'ura, M., Kataoka, H. & Shibazaki, B. (1992) Slip-dependent friction law and nucleation processes in earthquake rupture. *Tectonophys.* **211**, 135–148.

Mikumo, T. & Miyatake, T. (1978) Dynamical rupture process on a three-dimensional fault with non-uniform frictions and near-field seismic waves. *Geophys. J. R. Astr. Soc.* **54**, 417–438.

Mikumo, T. & Miyatake, T. (1993) Dynamic rupture processes on a dipping fault, and estimates of stress drop and strength excess from the results of waveform inversion. *Geophys. J. Int.* **112**, 481–496.

Mikumo, T., Olsen, K. B., Fukuyama, E. & Yagi, Y. (2003) Stress-breakdown time and slip-weakening distance inferred from slip-velocity functions on earthquake faults. *Bull. Seismol. Soc. Am.* **93**, 264–282.

Mora, P. & Place, D. (1998) Numerical simulation of earthquake faults with gouge: Toward a comprehensive explanation for the heat flow paradox, *J. Geophys. Res.* **103**(B9), 21067–21090.

Obara, K., Kasahara, K., Hori, S. & Okada, Y. (2005) A densely distributed high-sensitivity seismograph network in Japan: Hi-net by National Research Institute for Earth Science and Disaster Prevention. *Rev. Sci. Instrum.* **76**, 021301.

Oglesby, D. D., Archuleta, R. J. & Nielsen, S. B. (1998) Earthquakes on dipping faults: The effects of broken symmetry. *Science* **280**, 1055–1059.

Oglesby, D. D., Archuleta, R. J. & Nielsen, S. B. (2000) The three-dimensional dynamics of dipping faults. *Bull. Seismol. Soc. Am.* **90**, 616–628.

Ohnaka, M., Kuwaahara, Y. & Yamamoto, K. (1987) Constitutive relations between dynamic physical parameters near a tip of the propagating slip zone during stick-slip shear failure. *Tectonophys.* **144**, 109–125.

Ohnaka, M. (2003) A constitutive scaling law and a unified comprehension for frictional slip failure, shear fracture of intact rock, and earthquake rupture. *J. Geophys. Res.* **108**(B2), 2080, doi:10.1029/2000JB000123.

Okubo, P. G. (1989) Dynamic rupture modelling with laboratory-derived constitutive relations. *J. Geophys. Res.* **94**, 12321–12335.

Olsen, K. B., Madariaga, R. & Archuleta, R. J. (1997) Three-dimensional dynamic simulation of the 1992 Landers earthquake. *Science* **278**, 834–838.

Palmer, A. C. & Rice, J. R. (1973) The growth of slip surfaces in the progressive failure of overconsolidated clay. *Proc. Roy. Soc. London* **332A**, 527–548.

Perrin, G., Rice, J. R. & Zheng, G. (1995) Self-healing slip pulse on a frictional surface. *J. Mech. Phys. Solids* **43**, 1461–1495.

Rice, J. R. (1993) Spatio-temporal complexity of slip on a fault. *J. Geophys. Res.* **98**, 9885–9907.

Ruina, A. (1983) Slip instability and state variable friction laws. *J. Geophys. Res.* **88**, 10359–10370.

Sagiya, T., Nishimura, T., Hatanaka, Y., Fukuyama, E. & Ellsworth, W. L. (2002) Crustal movements associated with the 2000 western Tottori earthquake and its fault models. *Zisin (J. Seismol. Soc. Jpn.) Ser. 2*, **54**, 523–534 (in Japanese with English abstract).

Sato, H., Iwasaki, T., Ikeda, Y., Takeda, T., Matsuta, N., Imai, T., Kurashimo, E., Hirata, N., Sakai, S., Elouai, D., Kawanaka, T., Kawasaki, S., Abe, S., Kozawa, T., Ikawa, T., Arai, Y. & Kato, N. (2004) Seismological and geological characterization of the crust in the southern part of northern Fossa Magna, central Japan. *Earth Planets Space* **56**, 1253–1259.

Scholz, C. H. (1990) *The Mechanics of Earthquakes and Faulting*, Cambridge University Press.

Scholz, C. H. (2002) *The Mechanics of Earthquakes and Faulting*, 2nd edn., Cambridge University Press.

Segall, P. & Pollard, D. D. (1980) Mechanics of discontinuous faulting. *J. Geophys. Res.* **85**, 4337–4350.

Semmane, F., Cotton, F. & Campillo, M. (2005) The 2000 Tottori earthquake: A shallow earthquake with no surface rupture and slip properties controlled by depth. *J. Geophys. Res.* **110**, B03306, doi:10.1029/2004JB003194.

Sibson, R. H. (1986) Rupture interaction with fault jogs. *Earthq. Source Mech., Geophys. Monogr. Ser.* **37**, edited by Das, S., Boatwright, J. and Scholz, C. H., Americal Geophysical Union, pp. 157–167.

Sibson, R. H. (2003) Thickness of the seismic slip zone. *Bull. Seismol. Soc. Am.* **93**, 1169–1178.

Somerville, P. G., Irikura, K., Graves, R., Sawada, S., Wald, D., Abrahamson, N., Iwasaki, Y., Kagawa, T., Smith, N. & Kowada, A. (1999) Characterizing earthquake slip models for the prediction of strong ground motion. *Seismol. Res. Lett.* **70**, 59–80.

Spudich, P. & Guatteri, M. (2004) The effect of bandwidth limitations on the inference of earthquake slip-weakening distance from seismograms. *Bull. Seismol. Soc. Am.* **94**, 2028–2036.

Swanson, M. T. (1988) Pseudotachylyte-bearing strike-slip duplex structures in the Fort Foster Brittle Zone, S. Maine. *J. Struct. Geol.* **10**, 813–828.

Tada, T., Fukuyama, E. & Madariaga, R. (2000) Non-hypersingular boundary integral equations for 3D non-planar crack dynamics. *Comput. Mechan.* **25**, 613–626.

Tada, T. (2005) Displacement and stress Green's function for a constant slip-rate on a quadrantal fault. *Geophys. J. Int.* **162**, 1007–1023.

Tada, T. (2006) Stress Green's functions for a constant slip rate on a triangular fault. *Geophys. J. Int.* **164**, 653–669.

Tse, S. T. & Rice, J. R. (1986) Crustal earthquake instability in relation to the depth variation of frictional slip properties. *J. Geophys. Res.* **91**, 9452–9472.

Tsukahara, H., Ikeda, R. & Omura, K. (1996) *In-situ* stress measurement in an earthquake focal area. *Tectonophys.* **262**, 281–290.

Tsukahara, H., Ikeda, R. & Yamamoto, K. (2001) *In situ* stress measurements in a borehole close to the Nojima fault. *Island Arc* **10**, 261–265.

Tsutsumi, A. & Shimamoto, T. (1997) High-velocity frictional properties of gabbro. *Geophys. Res. Lett.* **24**, 699–702.

Unruh, J. R., Lettis, W. R. & Sowers, J. M. (1994) Kinematic interpretation of the 1992 Landers earthquake. *Bull. Seismol. Soc. Am.* **84**, 537–546.

Wald, D. J. & Heaton, T. H. (1994) Spatial and temporal distribution of slip for the 1992 Landers, California, earthquake. *Bull. Seismol. Soc. Am.* **84**, 668–691.

Waldhauser, F. & Ellsworth, W. L. (2000) A double-difference earthquake location algorithm: Method and application to the northern Hayward fault, California. *Bull. Seismol. Soc. Am.* **90**, 1353–1368.

Yamamoto, K. & Yabe, Y. (2001) Stresses at sites close to the Nojima Fault measured from core samples. *Island Arc* **10**, 266–281.

Yamashita, F., Fukuyama, E. & Omura, K. (2004) Estimation of fault strength: Reconstruction of stress before the 1995 Kobe earthquake. *Science* **306**, 261–263.

Yoshida, S., Koketsu, K., Shibazaki, B., Sagiya, T., Kato, T. & Yoshida, Y. (1998) Joint inversion of near- and far-field waveforms and geodetic data for the rupture process of the 1995 Kobe earthquake. *J. Phys. Earth* **44**, 437–454.

Some Remarks on the Time Scales of Magmatic Processes Occurring Beneath Island Arc Volcanoes

Simon P. Turner

Department of Earth and Planetary Sciences
Macquarie University
Sydney, NSW 2109 Australia
sturner@els.mq.edu.au

Time scales and rates of change are fundamental to an understanding of natural processes in the Earth sciences. Short-lived U-series isotope studies are revolutionising this field by providing time information in the range 10^2–10^4 years. Here I review how their application has been used to constrain the time scales of magma formation, ascent and storage beneath island arc volcanoes. Different elements are distilled-off the subducting plate at different times and in different places. Contributions from subducted sediments to island arc lava sources appear to occur some 350 kyr to 4 Myr prior to eruption. Fluid release from the subducting oceanic crust into the mantle wedge may be multi-stage and occurs over a period ranging from a few 100 kyr, to < 1 kyr, prior to eruption. This implies that dehydration commences prior to the initiation of partial melting within the mantle wedge consistent with recent evidence that the onset of melting is controlled by an isotherm and thus the thermal structure within the wedge. Furthermore, time scales of only a few kyr require a rapid fluid transfer mechanism, such as hydrofracture. U-Pa disequilibria reflect the partial melting process, rather than fluid addition, and indicate that the matrix is moving through the melt region. The preservation of large ^{226}Ra disequilibria permit only a few kyr between fluid addition and eruption. This requires rapid melt segregation, magma ascent by channelled flow at 100–1000's m/yr and minimal residence time within the lithosphere. The evolution from basalt to basaltic-andesite probably occurs rapidly during ascent. Some magmas subsequently stall in more shallow crustal level magma chambers where they evolve to more differentiated compositions on time scales of a few 1000 yrs or less. Degassing typically occurs for a few decades prior to eruption but may not drive major compositional evolution of the magmas.

1. Introduction

What are the *physical processes* by which the Earths crust and atmosphere are made and recycled back into the mantle? The planet we live upon has evolved through repetition of geochemical cycles powered by its internal heat engine. Partial melting and melt migration are the principle mechanisms for transfer of solid and gaseous material to the Earths surface and atmosphere. Conversely, erosion and subduction return continental crust to the mantle during which rock-atmosphere equilibration influences climate change. For several decades Earth scientists have used a variety of long-lived isotopic and fossil chronometers to unravel the long-term evolution of the planet but a fuller understanding of the physical and chemical processes driving this evolution have remained elusive because these occur on time scales (10's–1000's years) which are simply not resolvable by conventional chronometers. Such information has only become available to Earth scientists relatively recently as analytical advances have enabled the measurement of the short-lived, U-series isotopes.

Island arcs are curved chains of volcanoes that occur where one of the Earths oceanic plates is being subducted beneath another plate (Fig. 1). They form one of the key geochemical cycles, being sites where melting and transfer of new material to the Earth crust occurs and also where

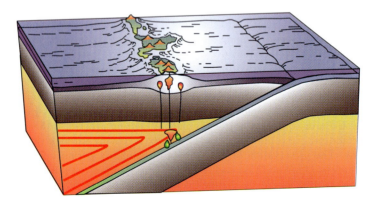

Fig. 1 Schematic cross-section of an island arc. Arrows indicate plate motion and the induced convection in the mantle wedge. Vertical arrow indicates magma ascent.

crustal materials are recycled back into the mantle. Here I review how U-series isotope studies are being used to constrain the time scales of magma formation, ascent and storage beneath island arc volcanoes [see Turner *et al.* (2003) for a more technical discussion]. Magmatism in this tectonic setting constitutes ~15% (0.4–0.6 km³/yr) of the total global output [Crisp (1984)] and the composition of the erupted magmas is, on average, similar to that of the continental crust [Taylor & McLennan (1981)]. In addition, many island arc volcanoes have been responsible for the most hazardous, historic volcanic eruptions (e.g. Mt Pelee, Tambora, Krakatau, Mt St. Helens).

2. U-Series Isotope Systematics

The decay series from ^{238}U and ^{235}U (Fig. 2) contains radioactive isotopes of many elements, and their varied geochemical properties cause them to be fractionated in distinctive ways in different geological environments. Provided that the decay chain remains unbroken, the parent and daughter nuclides will be in secular radioactive equilibrium. However, if the decay chain is broken by chemical fractionation, the system will be in isotope disequilibrium until equilibrium is restored by radioactive decay at a rate

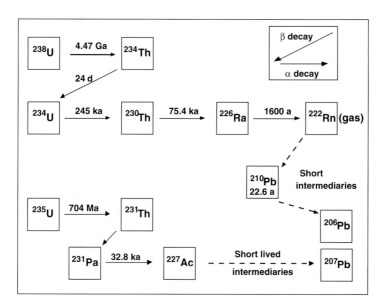

Fig. 2   The U-series decay chain showing the U-series nuclides of interest and their half-lives ^{230}Th (75 kyr), ^{231}Pa (32 kyr) and ^{226}Ra (1.6 kyr).

determined by the half-life of the daughter isotope involved. The isotopic ratios are usually reported as activity ratios (indicated by parentheses) and disequilibria (i.e. activity ratios $\neq 1$) are often referred to as excesses of the enriched nuclide, for example a $(^{230}\text{Th}/^{238}\text{U})$ ratio greater than 1 is often referred to as a ^{230}Th-excess. Of specific interest to the discussion here are ^{230}Th, ^{231}Pa, ^{226}Ra and ^{210}Pb which have half-lives of 75 000, 32 000, 1 600 and 22.5 years respectively (Fig. 1). A key aspect to the utility of U-series isotopes in the study of island arc lavas is that whereas Th and Pa behave as relatively immobile high field strength elements (HFSE), Ra and (under oxidising conditions) U behave like large ion lithophile elements (LILE) and form soluble complexes which are highly mobile in aqueous fluids (i.e. water containing dissolved solutes) [Brenan *et al.* (1995); Keppler (1996)].

3. Subducted Components and the Time Scales of their Transfer

The principle components of an island arc and the possible locations of various elemental fluxes in this tectonic environment are illustrated in Fig. 1. A distinctive geochemical signature of island arc lavas (Fig. 3) is the enrichment in LILE relative to HFSE inferred to reflect addition of fluids from the subducting plate [e.g. Gill (1981); Hawkesworth *et al.* (1997)]. Both the subducted altered oceanic crust and sediments are potential sources of this subduction component. Importantly, the highest LIL/HFSE ratios (e.g. Ba/Th) are found in those rocks with the lowest $^{87}\text{Sr}/^{86}\text{Sr}$ and $^{206}\text{Pb}/^{204}\text{Pb}$ ratios and from this it has been inferred that the fluid end-member was derived from the subducting altered oceanic crust rather than the overlying sediments [Miller *et al.* (1994); Turner *et al.* (1996); Turner & Hawkesworth (1997)]. Additionally, the presence of ^{10}Be and negative Ce anomalies in some island arc lavas can be taken as unambiguous evidence for a contribution from subducted sediments [e.g. Hole *et al.* (1984); Morris *et al.* (1990); George *et al.* (2003)]. Thus, most recent studies have argued for a three component model (Fig. 3) in which separate contributions from the mantle wedge, the slab fluid and the sediment have been identified [e.g. Kay (1980); Miller *et al.* (1994); Turner *et al.* (1996, 1997); Elliott *et al.* (1997); Hawkesworth *et al.* (1997)].

3.1. *Transfer of the sediment component*

Several studies have found evidence that the sediment component is characterised by fractionated U/Th, Nd/Ta and Th/Ce ratios and so it has been suggested that the sediment component is transferred as a partial

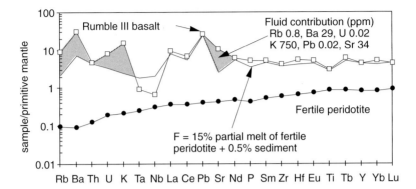

Fig. 3 Primitive mantle-normalised incompatible trace element diagram illustrating the three component mass balance model for a basalt ($SiO_2 = 51.7\%$) from the Rumble seamounts in the southern Kermadec island arc. The solid trace without symbols represents a 15% partial melt of a source composed of the fertile peridotite source shown (trace with circles) to which 0.5% sediment was added. This model composition provides a reasonable match to the Th, rare earth element, Zr, Hf and Ti contents of the arc basalt (squares). The over-estimate of the Ta and Nb concentrations in the model peridotite + sediment melt is one line of evidence that the sediment component is transferred as a partial melt (assumed to be formed in the presence of a residual phase which retains Ta and Nb). The excesses of Rb, Ba, U, K, Pb and Sr in the basalt are attributed to fluid addition (shaded area; composition indicated in ppm) to the source.

melt formed in the presence of a residual phase (mineral remaining in the source region after formation and extraction of a melt) which retains HFSE [Fig. 3; Elliott *et al.* (1997); Turner & Hawkesworth (1997); Johnson and Plank (1999)]. Because the sediment component requires a fractionated U/Th ratio yet appears to be in U-Th isotope equilibrium, Elliott *et al.* [1997] have argued for the Marianas arc that transfer of this component from the subducting plate must have occurred at least 350 kyr ago. In north Tonga, Turner & Hawkesworth [1997] used identification of the Louisville volcaniclastic sediment signature to estimate a sediment component transfer time of 2–4 Myr, whereas 0.5–1 Myr was inferred from [10]Be data in the Aleutian arc [George *et al.* (2003)].

3.2. *Transfer of the fluid component*

Fluids produced by dehydration reactions in the subducting, altered oceanic crust selectively add U (along with other fluid mobile elements) to the mantle wedge. So long as this is the principle cause of U/Th, fractionation (but see below), U-Th isotopes can be used to estimate the time elapsed since fluid release into the mantle wedge. The high U/Th ratios observed in the vast majority of arc lavas are generally accepted to reflect high U/Th ratios in the mantle source due to U addition by fluids from the subducting plate. In practice, information on the timing of fluid release can either be obtained from along-arc suites of lavas which form inclined arrays on U-Th equiline diagrams, or if the initial (^{230}Th/^{232}Th) ratio is constrained (Fig. 4a). For example, Elliott et $al.$ [1997] showed that lavas from the Marianas form an U-Th isotopes array which suggests that fluid was released from the subducting plate into the mantle wedge peridotite $\sim 30\,$kyr ago. Similarly, Turner & Hawkesworth [1997] showed that lavas from the Tonga-Kermadec arc scatter about a 50 kyr array (Fig. 4b). In total about 15 arcs have now been studied for U-Th disequilibria indicating that the time since U addition by fluids from the subducting oceanic crust appears to vary from 10 to 200 kyr prior to eruption. Excepting the Marianas, most of these arrays show variable degrees of scatter and any chronological interpretation should be viewed as the time-integrated effect of U addition rather than to imply that U addition occurred at a discrete and identical time along the length of an arc. An underlying assumption in these interpretations is that U addition by fluids is the only cause of U/Th, fractionation and if the same were true of U/Pa ratios then U addition should similarly produce U-Pa arrays which record a similar time to the U-Th arrays (Fig. 4c). So far the only island arc where this appears to be true is Tonga [Bourdon et $al.$ (1999)] because, in general, U/Pa are fractionated by an additional process such as partial melting [Pickett and Murrell (1997); Bourdon et $al.$ (1999)]. This is an important result because it supports the interpretation that the U-Th arrays have time significance but also places constraints on the melting process as discussed in further detail below.

^{226}Ra has a much shorter half life (1600 yr) than its parent ^{230}Th (75 kyr) and so provides the opportunity to look at very recent fractionations of Ra/Th. A recent global survey [Turner et $al.$ (2001)] has shown that most arc lavas preserve large ^{226}Ra-excesses and that these are generally well correlated with Ba/Th which is usually taken as a good index of fluid addition (Fig. 5). Like U/Th, Ba/Th is unlikely to be fractionated during crystal fractionation and crustal materials have low Ba/Th relative to most arc lavas. Moreover, the highest Ba/Th ratios occur in those arc rocks with the lowest SiO$_2$ and ^{87}Sr/^{86}Sr and so the observed ^{226}Ra-excesses are inferred to be a mantle signature resulting from fluid addition to the mantle wedge. At face value, the ^{226}Ra evidence for fluid addition

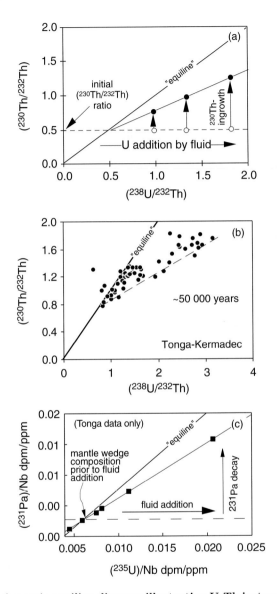

Fig. 4   (a) Schematic equiline diagram illustrating U-Th isotope systematics. Fluid addition, as shown by the horizontal arrow, moves data to the right after which isotope decay (^{230}Th-ingrowth) over time rotates the initially horizontal array towards the equiline. The slope of the array gives the time since fluid addition. (b) An example from the Tonga-Kermadec island arc from Turner & Hawkesworth (1997). (c) U-Pa isotopes normalised to Nb showing that lavas from Tonga record similar U addition times in both the U-Th and U-Pa systems [Modified from Bourdon *et al.* (1999)].

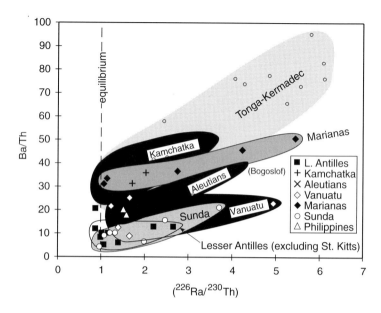

Fig. 5   The values of Ba/Th versus $(^{226}\text{Ra}/^{230}\text{Th})$ for the global arc data
[Turner *et al.* (2001)]. The Ba/Th ratio is sensitive to fluid additions (cf.
Fig. 3). The positive slopes shown here suggest that the $(^{226}\text{Ra}/^{230}\text{Th})$ ratios
greater than 1 result from Ra addition by fluids from the subducting plate.

in the last few 1000 years appears inconsistent with the interpretation that
U/Th disequilibria resulted from fluid addition 10–200 kyr ago. However,
unlike U, ^{226}Ra lost to the mantle wedge during initial dehydration continues to be replenished in the subducting altered oceanic crust by decay
from residual ^{230}Th (Fig. 6) on timescales of 10's kyr until all of the residual
^{230}Th has decayed away (350 kyr later). Thus, if dehydration reactions and
fluid addition occur step-wise or as a continuum [Schmidt & Poli (1998)],
the ^{226}Ra-excesses will reflect the last increments of fluid addition whereas
U-Th (and U-Pa in the case of Tonga) isotopes record the time elapsed
since the onset of fluid addition [Turner *et al.* (2000)].

4. The Mechanisms of Fluid Addition, Partial Melting and Magma Ascent

Island arc magmatism is widely regarded to reflect partial melting following
lowering of the peridotite solidus in the mantle wedge through addition of
fluids released by dehydration reactions in the subducted altered oceanic

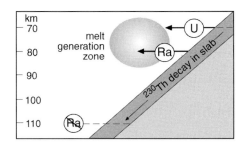

Fig. 6 **Illustration of the depth distribution of progressive distillation of fluid mobile elements from the subducting slab. Labelled arrows indicate the locations of U addition, Ra addition, and the point after which no further** ^{226}**Ra remains in the slab [after Turner** *et al.* **(2000)].**

crust [e.g. Tatsumi *et al.* (1986); Davies & Bickle (1991)] and U-series isotope constraints can be used to provide independent information on the mechanisms of fluid addition, partial melting and melt ascent in the mantle wedge above island arcs.

4.1. *Fluid addition*

Early work suggested that fluid transfer occurs horizontally across the wedge by a series of hydration-dehydration reactions [Davies & Stevenson (1992)]. In this model, fluid is added to the mantle where it forms amphibole which is dragged down by convective flow until it reaches an isotherm where it dehydrates releasing fluids which percolate upwards to a cooler zone where they again form amphibole and so the process repeats. Because of the inclined nature of convection in the mantle wedge (see Fig. 1) this results in horizontal translation in a zig-zag fashion across the wedge. However, this horizontal flow across the wedge can only occur at a rate controlled by the rate and angle of descent of the subducting plate which would predict fluid transfer time scales of several Myr [Turner & Hawkesworth (1997)]. Such a model could only be reconciled if the observed U- and Ra-excesses were generated by a final amphibole dehydration reaction in the mantle wedge prior to melting [Regelous *et al.* (1997)], however Ba is more compatible than Th in amphibole [La Tourette *et al.* (1995)] and so fluids produced in the presence of residual amphibole would be predicted to have low Ba/Th and $(^{226}Ra/^{230}Th) < 1$ (assuming that Ra behaves similarly to Ba). Instead, those lavas with the strongest fluid signature (e.g. largest $(^{238}U/^{230}Th)$ ratios) also have the highest Ba/Th ratios and the largest Ra-excesses (Fig. 4), and that implies that amphibole is not a residual phase but is consumed during partial melting. By implication, the major U/Th and Th/Ra fractionation is inferred to occur during fluid release from the subducting plate where redox conditions are the most strongly oxidising [Parkinson & Arculus (1999)]. Thus, the combined U-Th-Ra isotope data are in fact inconsistent with fluid transfer by a series of hydration-dehydration reactions and would seem to require that fluid transfer occurs via a much more rapid mechanism such as hydraulic fracturing [Davies (1999)].

4.2. *The thermal structure in the wedge*

The observation that island arc volcanoes occur at a relatively constant depth ($\sim$110 km) above the Benioff zone earthquakes which locate the top of the subducting plate has long been thought to indicate that partial melting occurs at the point of an isobaric dehydration reaction in the plate [Gill (1981); Tatsumi *et al.* (1986)]. However, recent analysis suggests that the depth to Benioff varies from sector to sector within island arcs depending on the thermal structure of the subducting plate [England *et al.* (2004)]. The implication from the different time scales obtained from the U-Th and Ra-Th data is that dehydration occurs in several stages and thus commences prior to the initiation of partial melting within the mantle wedge. This is consistent with the recent geophysical analysis which suggests that the onset of melting is controlled by an isotherm and thus is controlled by the thermal structure within the wedge [England *et al.* (2004)].

In contrast to the rapid fluid transfer time scales, contributions from subducted sediments appear to have longer transfer times and accordingly they are assumed not to be directly responsible for triggering partial melting and volcanism. Nevertheless, thermal models for the mantle wedge and the subducting plate are very sensitive to whether this sediment component is transferred by partial melts or by tectonic delamination (mechanical mixing). As outlined above, at present there appears to be growing evidence that the sediment component is transferred as a partial melt perhaps millions of years prior to magma eruption. In the simplest model, such long transfer times require the transfer of sediment into the mantle wedge at shallow levels and decoupling of convection in the wedge from the subducting plate in order to slow the rate of transfer of the sediment component to the site of partial melting [Turner & Hawkesworth (1997)]. In this regard it may be significant that there is increasing evidence that flow within the mantle wedge may often be oriented along the arc parallel to the trench rather than being directly coupled to the subducting plate [Turner & Hawkesworth (1998); Smith *et al.* (2001)]. Partial melting of sediments at relatively shallow levels requires temperatures of $\geq 800°C$ [Nichols *et al.* (1994); Johnson & Plank (1999); George *et al.* (2005)] and thus a thermal structure in the mantle wedge several hundred degrees hotter than that predicted by most current numerical thermal models [e.g. Davies & Stevenson (1992)]. However, higher wedge temperatures may help to reconcile the high eruption and equilibration temperatures inferred for some arc lavas [e.g. Sisson & Bronto (1998); Elkins Tanton *et al.* (2001)] and, in conjunction with the addition of volatiles, help to facilitate gravitational instabilities that lead to localised areas of upwelling (see below).

4.3. *A dynamic melt region*

As discussed above, addition of U by fluids will produce excesses of ^{235}U over ^{231}Pa, therefore an important observation is that the great majority of island arc lavas are characterised by the reverse sense of fractionation, or excesses of ^{231}Pa over ^{235}U [Pickett & Murrell (1997); Bourdon *et al.* (1999); Thomas *et al.* (2002); Dosseto *et al.* (2003)]. Bourdon *et al.* [1999] showed that only lavas from the Tonga-Kermadec arc preserve both ^{231}Pa excesses and deficits (Fig. 7). Their interpretation was that, in Tonga, fluid addition resulted in $(^{231}Pa/^{235}U) < 1$ and this is preserved because partial melting, in the absence of significant amounts of residual clinopyroxene, caused little subsequent fractionation of Pa/U. However, those from Kermadec, like most other arc lavas, have $(^{231}Pa/^{235}U)$ ratios > 1 and this is inferred to reflect the effects of the partial melting process [Pickett & Murrell (1997); Bourdon *et al.* (1999)]. This means that it is possible to distinguish elemental fractionation due to fluid addition from those of partial melting in the Tonga-Kermadec arc and $(^{231}Pa/^{235}U)$ ratios > 1 can be accounted for by ^{231}Pa-ingrowth during a dynamic melting process [e.g. McKenzie (1985)]. However that requires that there is relative movement between the melt and

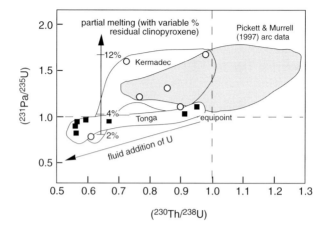

Fig. 7   Plot of U-Pa versus U-Th disequilibria for Tonga and Kermadec rocks [Bourdon *et al.* (1999)], along with the global data set from Pickett and Murrell (1997). The Tonga samples have $(^{231}\text{Pa}/^{235}\text{U})$ and $(^{230}\text{Th}/^{238}\text{U}) < 1$ as expected from U addition by subduction zone fluids. However, all the other samples have $(^{231}\text{Pa}/^{235}\text{U}) > 1$ suggesting that there was a subsequent increase in $(^{231}\text{Pa}/^{235}\text{U})$ due to the partial melting process.

the peridotite matrix. Thus, there is good evidence that the matrix undergoing melting within the mantle wedge beneath island arcs is not static but migrating through the melting zone at a few cm per year.

4.4. *Melt segregation and ascent rates*

Theoretical calculations based on compression of a porous matrix suggest that the segregation time scales for basaltic magmas are likely to be 1000's years or less [McKenzie (1985)]. One of the seemingly inescapable conclusions from the U-series disequilibrium data is that significantly less than a few half lives (i.e. 1600–3200 yrs) can have elapsed since the generation of the ^{226}Ra-excesses observed in the island arc lavas plotted on Fig. 5. Porous melt flow is likely to be unstable with this instability being resolved by a transition from porous to channelled magma flow [Aharonov *et al.* (1995)] and this may be swift if melting rates are high and the threshold porosity is quickly exceeded. The island arc Ra-Th disequilibria data indicates that the segregation of the melt from its matrix and channelled ascent does indeed occur on a very rapid time scale of a few kyr or less. If partial melting occurs at ~100 km depth beneath island arc volcanoes and the ascent time required to preserve the ^{226}Ra-excesses is 100–1000 years, then the required magma ascent rates are of the order of 100's to 1000's metres per year [Turner *et al.* (2001)]. These ascent rates rule out models of significant melt migration by porous flow or in diapers in this setting [Hall & Kincaid (2001)] and are so much faster than plate motions (cm/yr) that melt is unlikely to be deflected by convection within the mantle wedge. The corollary of near vertical melt ascent is that the principle site of melt production is likely to be defined by the surface distribution of volcanoes. Note that the ^{226}Ra-^{230}Th disequilibria do not preclude a partial melting origin for the ^{231}Pa-excesses but they do require that the residence time of Ra in the melting column was short enough ($<<$ 8000 years), and thus that the melt velocity was fast enough, to prevent ^{226}Ra from decaying back into secular equilibrium with ^{230}Th. These rising magmas initially traverse the inverted geothermal gradient in the lower half of the mantle wedge followed by the "right way up" geotherm prior to encountering the base of the lithospheric mantle and the density change at the Moho where they might be expected to slow or to occasionally stall and pond.

5. Magma Residence and Evolution Within the Crust

In principle, U-series isotope data can also be used to assess the residence times of lavas in shallow magma chambers beneath active volcanoes, either from variations in lavas shown to be derived from a common parental magma [e.g. George *et al.* (2004)] or from mineral isochrons whose ages exceed eruption ages [e.g. Volpe & Hammond (1991); Heath *et al.* (1998)]. Perhaps the simplest way forward is to look for systematic variations in the disequilibria observed in whole rocks (see below), so long as the half-life is appropriate to the magma residence times being investigated. One of the more robust observations is that the crustal residence time for magmas containing significant ^{226}Ra-excesses cannot have been greater than 8000 yrs, so long as those ^{226}Ra-excesses were produced in the mantle wedge (see Sec. 3 above). For example, in the Tonga-Kermadec and Aleutian arcs, Turner *et al.* [2000] and George *et al.* [2004] have found that Ra-Th disequilibria decrease with increasing SiO_2 suggesting that the time scale of differentiation was on the order of a few 1000 years. An encouraging point is that these time scales are similar to those obtained in recent numerical thermal modelling based around constraints on energy loss available from the thermal output at volcanoes [Hawkesworth *et al.* (2000)]. The main results are summarised in Fig. 9 which shows that $10\,km^3$ of basaltic magma losing heat at $100\,MW$ will undergo 20% crystallisation to reach a basaltic

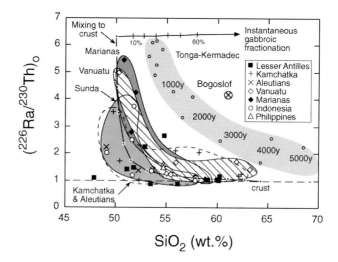

Fig. 8   Plot of $(^{226}\text{Ra}/^{230}\text{Th})_o$ versus SiO_2 showing that $(^{226}\text{Ra}/^{230}\text{Th})_o$ decreases with increasing SiO_2 which places constraints on the time scale of differentiation (time elapsed indicated in years along the light grey Tonga-Kermadec array). A model, instantaneous gabbroic fractionation vector shows that Ra/Th remains essentially constant during fractionation from basalt to dacite. Thus, if the observed decrease in $(^{226}\text{Ra}/^{230}\text{Th})_o$ is due to the time taken for differentiation, then this must have taken less than the time for ^{226}Ra-^{230}Th to return to equilibrium (8000 years).

andesitic composition in about 1000 years, 60% crystallisation to reach a dacitic composition in about 3000 years and closer to 5000–8000 years to reach a rhyolitic composition. These estimates can be directly compared with those derived from the Ra-Th disequilibria data on Fig. 8.

Studies of mineral ages have met with many complications [see Hawkesworth et al. (2004) for a recent review]. Detailed studies of mineral separates have revealed evidence that phenocryst populations may often have mixed ages and/or consist of old cores with young rims such that the U-Th and Ra-Th systems give differing ages [Cooper et al. (2003); Turner et al. (2003)]. Thus, there is growing evidence that phenocrysts within these lavas could be older than the estimated ages for the liquids and may reflect incorporation of older cumulate or wall rock materials into young magma batches [e.g. Pyle et al. (1988); Sparks et al. 1990; Heath et al. (1998)]. The corollary is that the observed phenocrysts were not the ones responsible for differentiation of their enclosing liquid and Sr isotope profiles in plagioclase phenocrysts provide independent evidence for complex crystal histories [Davidson & Tepley (1997)].

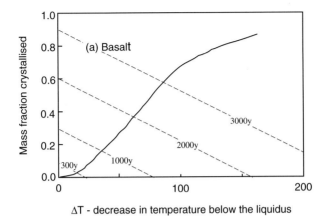

Fig. 9   **Results of a numerical power-output model for basaltic systems showing the time taken for crystallisation as a function of temperature decrease
below the solidus [after Hawkesworth *et al.* (2000, 2001)].**

These differentiation time scales imply that there is no direct link
between magma residence time and eruptive periodicity, since most
island volcano eruptions re-occur on the order of 10's to 100's years.
Rather, eruptive periodicity may be linked to degassing (Jaupart 1996)
and, for example, Tait *et al.* [1989] developed a model which predicted
eruptive periodicity on the scale of years to 10–100's years due to crystallisation induced increases in volatile over-pressure. Recently, Gauthier &
Condomines [1999] have exploited the fact that ^{210}Pb has a gaseous parent, ^{222}Rn (see Fig. 2), to constrain the time scales of magma degassing and
recharge at Stromboli and Merapi volcanoes. Applying their approach to
a global survey of ^{210}Pb systematics in arc lavas, Turner *et al.* [2004] estimated that most island arc magmas undergo degassing for several decades
prior to eruption. Moreover, in at least one example, from Sangeang Api volcano in the Sunda arc, there is evidence that this degassing occurred much
more recently than the bulk compositional differentiation of the magmas.
Therefore, the crystals formed by degassing did not separate from their liquid and cause bulk compositional changes. Finally, Berlo *et al.* [2004] found
that the ^{210}Pb systematics of lavas from Mount St. Helens varied with eruption style and through time following the cataclysmic May 1980 eruption.
Importantly, these studies have both recognised that ^{210}Pb-excesses may
indicate the presence of fresh degassing magma at depth. Since the build
up of gas pressure and the injection of hot mafic inputs into existing magma

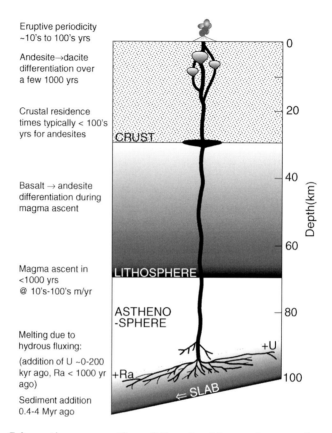

Fig. 10　Schematic cross section of the plumbing and reservoir system for an island arc volcano modified from Gill [1981] to include element transfer (+U and +Ra are intended to schematically illustrate the spatial and temporal separation of addition of U and Ra due to dehydration reactions in the subducting plate), magma transport and residence time scales discussed in text.

chambers [e.g. Sparks *et al.* (1977)] are likely triggers for eruption, such data may have important future application in eruptive hazard prediction.

6. Conclusions

The use of U-series disequilibria in unravelling the physical processes of fluid transfer, partial melting, melt migration and modification at convergent margins is still a new and rapidly expanding field of research. The presently available constraints on the time scales of magma formation, storage, ascent and degassing at island arcs are illustrated on Fig. 10. Advances in analytical techniques are allowing for more rapid and precise analysis and new data sets, particularly on fully characterised and well dated lavas, can only improve our understanding of convergent margin processes. However, these data will need to be combined with numerical models if their full significance is to be realised.

References

Aharonov, E., Whitehead, J. A., Kelemen, P. B. & Spiegelman, M. (1995) Channeling instability of upwelling melt in the mantle. *J. Geophys. Res.* **100**, 20433–20450.

Berlo, K., Blundy, J., Turner, S., Cashman, K., Hawkesworth, C. & Black, S. (2004) Geochemical precursors to volcanic activity at Mount St. Helens, USA. *Science*, **306**, 1167–1169.

Bourdon, B., Turner, S. & Allègre, C. (1999) Melting dynamics beneath the Tonga-Kermadec island arc inferred from ^{231}Pa-^{235}U systematics. *Science* **286**, 2491–2493.

Brenan, J. M., Shaw, H. F., Ryerson, F. J. & Phinney, D. L. (1995) Mineral-aqueous fluid partitioning of trace elements at 900°C and 2.0 GPa: Constraints on the trace element chemistry of mantle and deep crustal fluids. *Geochim. Cosmochim. Acta* **59**, 3331–3350.

Cooper, K. M. & Reid, M. R. (2003) Re-examination of crystal ages in recent Mount St. Helens lavas: Implications for magma reservoir processes. *Earth Planet. Sci. Lett.* **213**, 149–167.

Crisp, J. A. (1984) Rates of magma emplacement and volcanic output. *J. Volcanol. Geotherm. Res.* **20**, 177–211.

Davies, J. H. (1999) The role of hydraulic fractures and intermediate-depth earthquakes in generating subduction-zone magmatism. *Nature* **398**, 142–145.

Davies, J. H. & Bickle, M. J. (1991) A physical model for the volume and composition of melt produced by hydrous fluxing above subduction zones. *Phil. Trans. R. Soc. London* **335**, 355–364.

Davies, J. H. & Stevenson, D. J. (1992) Physical model of source region of subduction zone volcanics. *J. Geophys. Res.* **97**, 2037–2070.

Davidson, J. P. & Tepley, F. J. (1997) Recharge in volcanic systems: Evidence from isotope profiles of phenocrysts. *Science* **275**, 826–829.

Dosseto A., Bourdon B., Joron J. L. & Dupré B. (2003) U-Th-Pa-Ra study of the Kamchatka arc: New constraints on the genesis of arc lavas. *Geochim. Cosmochim. Acta* **67**, 2857–2877.

Elkins Tanton, L. T., Grove, T. L. & Donnelly-Nolan, J. (2001) Hot, shallow mantle melting under the Cascades volcanoc arc. *Geology* **29**, 631–634.

Elliott, T., Plank, T., Zindler, A., White, W. & Bourdon, B. (1997) Element transport from slab to volcanic front at the Mariana arc. *J. Geophys. Res.* **102**, 14991–15019.

England P., Engdahl R. & Thatcher W. (2004) Systematic variation in the depths of slabs beneath arc volcanoes. *Geophys. J. Int.* **156**, 377–408.

Gauthier, P.-J. & Condomines, M. (1999) ^{210}Pa-^{226}Ra radioactive disequilibria in recent lavas and radon degassing: Inferences on the magma chamber dynamics at Stromboli and Merapi volcanoes. *Earth Planet. Sci. Lett.* **172**, 111–126.

George, R., Turner, S., Hawkesworth, C., Morris, J., Nye, C., Ryan, J. & Zheng, S.-H. (2003). Melting processes and fluid and sediment transport rates along the Alaska-Aleutian arc from an integrated U-Th-Ra-Be isotope study. *J. Geophys. Res.*, **108**(B5), 2252, doi:10.1029/2002JB001916.

George, R., Turner, S., Hawkesworth, C., Nye, C., Bacon, C., Stelling, P. & Dreher, S. (2004) Chemical versus temporal controls on the evolution of tholeiitic and calc-alkaline magmas at two volcanoes in the Aleutian arc. *J. Petrol.* **45**, 203–219.

George, R., Turner, S., Morris, J., Plank, T., Hawkesworth, C. & Ryan, J. (2005) Pressure-temperature-time paths of sediment recycling beneath the Tonga-Kermadec arc. *Earth Planet. Sci. Lett.* **233**, 195–211.

Gill, J. B. (1981) *Orogenic Andesites and Plate Tectonics*, Springer-Verlag, New York, pp. 1–39.

Hall, P. S. & Kincaid, C. (2001) Diapiric flow at subduction zones: A recipe for rapid transport. *Science* **292**, 2472–2475.

Hawkesworth, C. J., Turner, S. P., McDermott, F., Peate, D. W. & van Calsteren, P. (1997) U-Th isotopes in arc magmas: Implications for element transfer from the subducted crust. *Science* **276**, 551–555.

Hawkesworth, C., Blake, S., Evans, P., Hughes, R., Macdonald, R., Thomas, L., Turner, S. & Zellmer, G. (2000) The time scales of crystal fractionation in magma chambers — integrating physical, isotopic and geochemical perspectives. *J. Petrol.* **41**, 991–1006.

Hawkesworth, C., George, R., Turner, S. & Zellmer, G. (2004) Timescales of magmatic processes. *Earth Planet. Sci. Lett.* **218**, 1–16.

Heath, E., Turner, S. P., Macdonald, R., Hawkesworth, C. J. & van Calsteren, P. (1997) Long magma residence times at an island arc volcano (Soufriere, St. Vincent) in the Lesser Antilles: Evidence from ^{238}U-^{230}Th isochron dating. *Earth Planet. Sci. Lett.* **160**, 49–63.

Hole, M. J., Saunders, A. D., Marriner, G. F. & Tarney, J. (1984) Subduction of pelagic sediments: Implications for the origin of Ce-anomalous basalts from the Mariana islands. *J. Geol. Soc. London* **141**, 453–472.

Jaupart, C. (1996) Physical models of volcanic eruptions. *Chem. Geol.* **128**, 217–227.

Johnson, M. C. & Plank, T. (1999) Dehydration and melting experiments constrain the fate of subducted sediments. *Geochem. Geophys. Geosys.* **1**, paper number 1999GC000014.

Kay, R. W. (1980) Volcanic arc magmas: Implications of a melting-mixing model for element recycling in the crust-upper mantle system. *J. Geol.* **88**, 497–522.

Keppler, H. (1996) Constraints from partitioning experiments on the composition of subduction-zone fluids. *Nature* **380**, 237–240.

La Tourette, T., Hervig, R. L. & Holloway, J. R. (1995) Trace element partitioning between amphibole, phlogopite and basanite melt. *Earth Planet. Sci. Lett.* **135**, 13–30.

McKenzie, D. (1985) The extraction of magma from the crust and mantle. *Earth Planet. Sci. Lett.* **74**, 81–91.

McKenzie, D. (1985) ^{230}Th-^{238}U disequilibrium and the melting process beneath ridge axes. *Earth Planet. Sci. Lett.* **72**, 149–157.

Miller, D. M., Goldstein, S. L. & Langmuir, C. H. (1994) Cerium/lead and lead isotope ratios in arc magmas and the enrichment of lead in the continents. *Nature* **368**, 514–520.

Morris, J. D., Leeman, B. W. & Tera, F. (1990) The subducted component in island arc lavas: Constraints from Be isotopes and B-Be systematics. *Nature* **344**, 31–36.

Nichols, G. T., Wyllie, P. J. & Stern, C. R. (1994) Subduction zone melting of pelagic sediments constrained by melting experiments. *Nature* **371**, 785–788.

Parkinson, I. J. & Arculus, R. J. (1999) The redox state of subduction zones: Insights from arc-peridotites. *Chem. Geol.* **160**, 409–423.

Pickett, D. A. & Murrell, M. T. (1997) Observations of ^{231}Pa/^{235}U disequilibrium in volcanic rocks. *Earth Planet. Sci. Lett.* **148**, 259–271.

Pyle, D. M. (1992) The volume and residence time of magma beneath active volcanoes determined by decay-series disequilibria methods. *Earth Planet. Sci. Lett.* **112**, 61–73.

Pyle, D. M., Ivanovich, M. & Sparks, R. S. J. (1988) Magma-cumulate mixing identified by U-Th disequilibrium dating. *Nature* **331**, 157–159.

Regelous, M., Collerson, K. D., Ewart, A. & Wendt, J. I. (1997) Trace element transport rates in subduction zones: Evidence from Th, Sr and Pb isotope data for Tonga-Kermadec arc lavas. *Earth Planet. Sci. Lett.* **150**, 291–302.

Schmidt, M. W. & Poli, S. (1998) Experimentally based water budgets for dehydrating slabs and consequences for arc magma generation. *Earth Planet. Sci. Lett.* **163**, 361–379.

Smith, G. P., Weins, D. A., Fischer, K. M., Dorman, L. M., Webb, S. C. & Hildebrand, J. A. (2001) A complex pattern of mantle flow in the Lau backarc. *Science* **292**, 713–716.

Sparks, R. S. J., Sigurdsson, H. & Wilson, L. (1977) Magma mixing: A mechanism of triggering acid explosive eruptions. *Nature* **267**, 315–318.

Sparks, R. S. J., Huppert, H. E. & Wilson, C. J. N. (1990) Comment on "Evidence for long residence times of rhyolitic magma in the Long Valley magmatic system: The isotope record in precaldera lavas of Glass Mountain" edited by Halliday, A. N., Mahood, G. A., Holden, P., Metz, J. M., Dempster, T. J. and Davidson, J. P. *Earth Planet. Sci. Lett.* **99**, 387–389.

Tait, S., Jaupart, C. & Vergniolle, S. (1989) Pressure, gas content and eruption periodicity of a shallow, crystallising magma chamber. *Earth Planet. Sci. Lett.* **92**, 107–123.

Tatsumi, Y., Hamilton, D. L. & Nesbitt, R. W. (1986) Chemical characteristics of fluid phase released from a subducted lithosphere and origin of arc magmas: Evidence from high-pressure experiments and natural rocks. *J. Volcan. Geotherm. Res.* **29**, 293–309.

Taylor, R. S. & McLennan, S. M. (1981) The composition and evolution of the continental crust: Rare earth element evidence from sedimentary rocks. *Phil. Trans. R. Soc. London* **301**, 381–399.

Thomas R. B., Hirschmann M. M., Cheng H., Reagan M. K. & Edwards R. L. (2002) $(^{231}Pa/^{235}U)$-$(^{230}Th/^{238}U)$ of young mafic volcanic rocks from Nicaragua and Costa Rica and the influence of flux melting on U-series systematics of arc lavas. *Geochim. Cosmochim. Acta* **66**, 4287–4309.

Turner, S. & Hawkesworth, C. (1997) Constraints on flux rates and mantle dynamics beneath island arcs from Tonga-Kermadec. *Nature* **389**, 568–573.

Turner, S., Hawkesworth, C., van Calsteren, P., Heath, E., Macdonald, R. & Black, S. (1996) U-series isotopes and destructive plate margin magma genesis in the Lesser Antilles. *Earth Planet. Sci. Lett.* **142**, 191–207.

Turner, S., Bourdon, B., Hawkesworth, C. & Evans, P. (2000) ^{226}Ra-^{230}Th evidence for multiple dehydration events, rapid melt ascent and the time scales of differentiation beneath the Tonga-Kermadec island arc. *Earth Planet. Sci. Lett.* **179**, 581–593.

Turner, S., Evans, P. & Hawkesworth, C. (2001) Ultra-fast source-to-surface movement of melt at island arcs from ^{226}Ra-^{230}Th systematics. *Science* **292**, 1363–1366.

Turner, S., Bourdon, B. & Gill, J. (2003) Insights into magma genesis at convergent margins from U-series isotopes. Bourdon, B., Henderson, G., Lundstrom, C., Turner, S. (eds.). *Uranium Series Geochemistry. Rev. Mineral. Geochem.* **52**, 255–315.

Turner, S., George, R., Jerram, D., Carpenter, N. & Hawkesworth, C. (2003) Case studies of plagioclase growth and residence times in island arc lavas from Tonga and the Lesser Antilles and Tonga, and a model to reconcile discordant age information. *Earth Planet. Sci. Lett.* **214**, 279–294.

Turner, S., Black, S. & Berlo, K. (2004) ^{210}Pb-^{226}Ra and ^{232}Th-^{228}Ra systematics in young arc lavas: Implications for magma degassing and ascent rates. *Earth Planet. Sci. Lett.* **227**, 1–16.

Volpe, A. M. & Hammond, P. E. (1991) ^{238}U-^{230}Th-^{226}Ra disequilibrium in young Mt. St. Helens rocks: Time constraint for magma formation and crystallization. *Earth Planet. Sci. Lett.* **107**, 475–486.

The Break-Up of Continents and the Generation of Ocean Basins

T. A. Minshull

National Oceanography Centre, European Way
Southampton SO14 3ZH, UK
tmin@soc.soton.ac.uk

Rifted continental margins are the product of stretching, thinning and, ultimately, breakup of a continental plate into smaller fragments. The rocks lying beneath them store a record of this rifting process. Earth scientists can read this record by direct sampling and with remote geophysical techniques. These experimental studies have been complemented by theoretical analyses of continental extension and associated melting of the mantle. Some rifted margins show evidence for extensive volcanic activity and uplift during rifting; at these margins, the record of the final stages of rifting is obscured by erosion and by the thick volcanic cover. Other margins have been underwater throughout their formation and have had rather little volcanic activity; here the ongoing deposition of sediment provides a clearer record. During the last decade, vast areas of exhumed mantle rocks have been discovered at such margins between continental and oceanic crust. This observation conflicts with well-established ideas that the mantle melts to produce new crust when brought close to the Earth's surface. In contrast to the steeply dipping faults commonly seen in zones of extension within continental interiors, faults with very shallow dips play a key role in the deformation immediately preceding continental breakup. Future progress in the study of continental breakup will depend on studies of pairs of margins that were once joined, and on the development of computer models which can handle rigorously the complex transition from distributed continental deformation to seafloor spreading focussed at a mid-ocean ridge.

1. Introduction

The Earth loses heat from its deep interior largely by slow convective motion of the mantle. The surface expression of this motion is the horizontal

movement of relatively rigid plates known as the 'lithosphere'. New lithosphere is formed at mid-ocean ridges, and eventually is destroyed by sinking
deep into the Earth at subduction zones. The global configuration of plate
boundaries is stable on human time-scales, but on time-scales of millions of
years, plates can break apart.

Evidence for the continuity of geological structures between widely separated continents provided evidence for so-called 'continental drift' long
before the development of the modern theory of plate tectonics. Geophysical
exploration and ocean drilling around the submerged edges of continents,
known as 'continental margins', led to the concept that continents stretch
and thin before finally breaking apart [Le Pichon and Sibuet (1981)]. Rifted
continental margins are the product of this stretching and thinning, and the
rocks lying beneath them store a record of the rifting process. Earth scientists have learned to read this record by drilling into the rocks at carefully
selected locations, by sampling the seabed in areas where crustal rocks are
not covered by thick sediments, and by probing the deeper structure with
a variety of geophysical techniques. Alongside these experimental studies,
theoretical analyses have related the style of deformation of the continental
crust to the properties of the rocks within and beneath it. Such approaches
have also related the subsequent formation of new oceanic crust following
break-up of the continent to seafloor spreading involving the upwelling and
melting of the underlying mantle.

The study of rifted margins is becoming of more than academic importance as they become frontier exploration provinces for the hydrocarbon
industry [White *et al.* (2003)] and as nations stake their claims to marine
territory with legal arguments based partly on geological criteria. Hydrocarbons are formed by burial and heating, over long periods, of organic
matter deposited with sediments. Hydrocarbon companies are therefore
interested in the burial and thermal history of these sediments, which determine the likely volume and composition of such hydrocarbons. Under the
United Nations Convention on the Law of the Sea, nations can lay claim
to parts of the ocean based on measurements of water depth, sediment
thickness and even crustal type.

At rifted margins, continental crust with a typical thickness of 30–40 km
is thinned by a factor of five or more before it finally ruptures. The associated lithospheric thinning results in subsidence of the surface and upwelling
of the mantle. During upwelling, mantle rocks cool, but their melting point
decreases more rapidly, so the rocks melt as they decompress. The oceanic
crust, typically 6–7 km thick, is formed from the frozen products of such

melting. On most margins, sediment supply is too slow to keep up with the subsidence, and the thinned crust now lies beneath several kilometres of water.

Detailed information on the deep structure of margins has come from seismic experiments. These experiments use acoustic waves generated by the release of compressed air from large 'airguns' towed from ships and recorded with hydrophones near the sea surface or on the seabed [Lonergan and White (1999)]. Two seismic techniques are widely used at rifted margins, often simultaneously. In seismic reflection experiments, seismic signals reflected from boundaries within the Earth are measured and processed to generate an image of the structure, to depths of typically a few tens of kilometres on continental margins. In seismic refraction, or 'wide-angle' seismic, experiments the travel times of the signals are used to infer variations of seismic velocity within the Earth, and hence to define large-scale structure and composition down to similar depths.

Modern experiments focused on restricted targets commonly extend these techniques to three dimensions, but the large scale (commonly several hundred kilometres) of rifted margins has to date precluded a three-dimensional approach. Seismic data from some rifted margins show evidence for extensive volcanic activity and uplift at the start of rifting. At these margins, the uplift results in erosion that partly removes the geological record of the rifting process. Rift-related structures are further obscured by the thick volcanic cover, which is difficult to penetrate with seismic techniques and difficult to drill through. Other margins, termed here 'magma poor', show evidence of continuous subsidence and had rather little volcanic activity; at these margins these key final stages are more readily documented. This chapter focuses mainly on such margins.

2. Magmatism During Continental Break-Up

During the last two decades, rifted continental margins around the North Atlantic have been extensively studied using seismic techniques [Eldholm and Grue (1994); Louden and Chian (1999); Fig. 1]. At many margins in the northern North Atlantic, these studies have identified several-kilometre-thick wedges of concave-downward seaward-dipping reflectors beneath the sediment column and above the thinned continental crust. These reflectors were interpreted initially as sedimentary deposits from deltas, but by drilling holes through these deposits, geologists learned that the reflections were generated by lava flows, with intercalated thin sediment layers,

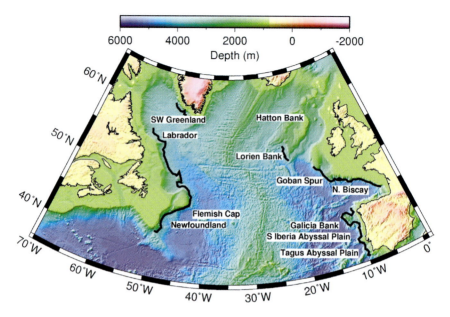

Fig. 1   Shaded bathymetry and topography of the North Atlantic region
[Smith and Sandwell (1997)]. Labelled thick lines mark the 3000 m contour,
which lies close to the base of the continental slope, at magma-poor rifted
margins. The Hatton Bank volcanic margin (Fig. 2a) is also labelled.

which acquired their observed shape due to an oceanward increase in subsidence beneath the weight of successive flows. Wide-angle seismic studies
of such margins also revealed up to 15–20 km thick regions at the base of
the stretched crust with seismic velocities of 7.2–7.6 km s^{-1}. These velocities are too high for continental crust and too low for the upper mantle,
and they are attributed to the presence of magnesium-rich igneous rocks.
Such rocks are formed by decompression melting of the mantle during continental break-up. The origin of the voluminous igneous material at such
'volcanic' margins remains controversial. The large volumes suggest that
either the mantle beneath these margins was hotter at the time of break-
up than the mantle beneath most of the ocean basins or an unusually
large volume of mantle moved up through the depth interval of melting.
There are a variety of views about the relative importance of these two
processes [White and McKenzie (1989); Holbrook and Keleman (1993)].
Other margins showed very little evidence for magmatic activity until new
oceanic crust was formed after the continents broke apart (Fig. 2). It was
inferred that magmatism at these margins was relatively straightforward,

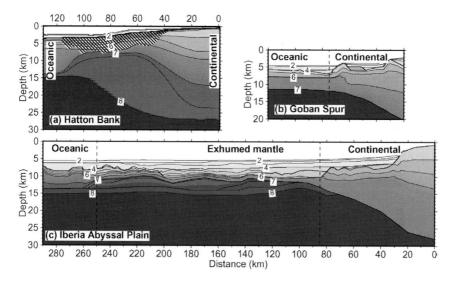

Fig. 2   Seismic velocity structures of some North Atlantic rifted margins, plotted at the same scale. In each panel, velocities are contoured at $0.5\,\mathrm{km\,s^{-1}}$ intervals up to $7.0\,\mathrm{km\,s^{-1}}$, and at $0.2\,\mathrm{km\,s^{-1}}$ between 7.0 and $8.0\,\mathrm{km\,s^{-1}}$. (a) Volcanic margin of Hatton Bank [Fowler *et al.* (1999)]. Diagonal shading marks regions of seaward-dipping reflections. The thick region of velocities between 7.2 and $8.0\,\mathrm{km\,s^{-1}}$ is interpreted as magmatic rocks accreted to the base of the crust during rifting. Transition from continental to oceanic crust occurs gradually between ca. 40 and 110 km. (b) Magma-poor Goban Spur margin [Horsefield *et al.* (1994)]. The dashed line marks inferred abrupt ocean-continent boundary; recent more detailed seismic work suggests that this picture is overly simplistic [Bullock and Minshull (2005)]. (c) West Iberia margin in the southern Iberia Abyssal Plain [Dean *et al.* (2000)]. The dashed lines mark the approximate edges of oceanic and of continental crust.

with substantial melting of the upwelling mantle only during the final stage of break-up or separation of the continental plate.

Typically a few times every million years, the Earth's magnetic field reverses its polarity. As new oceanic crust forms, magnetic minerals in the crust align themselves with the magnetic field, so the geologically frequent polarity reversals result in the formation of a series of 'stripes' of crust of alternating magnetic polarity. In the absence of more detailed information these magnetic stripes provide a straightforward means of identifying oceanic crust. A long-standing observation from many magma-poor margins in the North Atlantic is that magnetic anomaly stripes are often absent for tens or even hundreds of kilometres from the foot of the continental slope. This absence may be attributed in some places to continental break-up during periods when the Earth's magnetic field underwent no stable polarity reversals or when frequent reversals of a weak magnetic field resulted in weak to undetectable anomaly lineations. However, at margins where breakup occurred during periods of regular and well-spaced polarity reversals, an alternative explanation must be sought. Over the last decade, the exploration of magma-poor margins has extended out into these weakly magnetic abyssal plain regions. It is here that the process of transition from continental stretching to oceanic sea-floor spreading is recorded. Such regions have commonly been called 'transition zones', but this term may be misleading, since their structure is not transitional between oceanic and continental crust, as is observed at volcanic margins, but rather is distinct from both.

Intense surveying and sampling of one such weakly magnetic region, in the southern Iberia Abyssal Plain west of Portugal (Fig. 1), has led to the unexpected discovery that rocks from the mantle were exposed at the sea floor at the time of rifting in an area tens to hundreds of kilometres across [Whitmarsh *et al.* (2001)]. In this zone of exhumed mantle, seismic velocities rise steeply to ca. $7\,\mathrm{km\,s}^{-1}$ only $2\,\mathrm{km}$ beneath the basement (the top of the crystalline crust), and then increase gradually to normal mantle values of ca. $8\,\mathrm{km\,s}^{-1}$ ca. $6\,\mathrm{km}$ beneath the basement (Fig. 2c), without the abrupt discontinuity which normally marks the base of the crust (as in Fig. 2a; b). This pattern may be explained by a progressive reduction in the degree of alteration of mantle rocks with depth as chemical reactions with sea water become more limited. Drilling in the same region has found no evidence for magmatism at the time of rifting, and has reached altered mantle rocks beneath the sediments at several locations, confirming the conclusions drawn from geophysical data. A series of margin-parallel basement highs extend northward from the southern Iberia Abyssal Plain to a point where the basement outcrops at the seabed off Galicia Bank (Fig. 1). Here, sampling by submersible and by dredging has also recovered mantle rocks, suggesting that the zone of exhumed mantle extends several hundred kilometres along the margin. A full understanding of the formation of this region awaits similar detailed study of the once adjacent Newfoundland margin. Geophysical studies of other North Atlantic margins, such as the Labrador-southwest Greenland pair (Fig. 1), have also inferred the presence of such a zone [Louden and Chian (1999)], though in the absence of direct sampling, several interpretations are possible.

Computer models of the thermal structure of mid-ocean ridges, based on the separation of relatively rigid plates above a slowly convecting layer

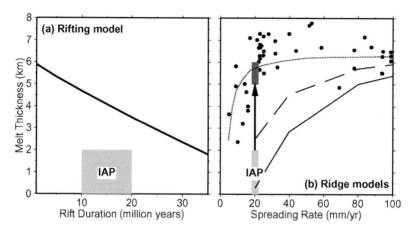

Fig. 3 Comparison between results from melting models and measured melt thicknesses [Minshull *et al.* (2001)] (a) The line marks the predicted melt thickness as a function of rift duration for lithospheric stretching by a factor of 50 (beyond which further stretching generates negligible melt volumes) over normal-temperature mantle for a model that takes account of heat loss by conduction. The shaded box marks the estimated melt thickness in the zone of exhumed mantle in the southern Iberia Abyssal Plain (IAP) and inferred rift duration from drilling and geophysical observations. (b) The lines mark the melt thickness as a function of spreading rate from models of melt generation at mid-ocean ridges: solid line, constant mantle viscosity; dashed line, temperature-dependent viscosity; dotted line, a lithosphere that thickens more steeply away from the ridge axis at slow spreading rates than would be predicted from temperature variations. The dots mark seismic measurements of oceanic crustal thickness as a function of spreading rate, and shaded regions mark the inferred melt thickness in the zone of exhumed mantle of Fig. 2(c), and the first-formed oceanic crust of Fig. 2(c), with the arrow indicating evolution over time.

called the 'asthenosphere', combined with laboratory-derived knowledge of the melting behaviour of mantle rocks, have successfully matched the volume and even the composition of oceanic crust formed at a variety of spreading rates [McKenzie and Bickle (1988)]. Such models also have been applied successfully to volcanic rifted margins, but they appear to fail when applied to the west Iberia margin (Fig. 3). In the zone of exhumed mantle, magnetic measurements and drilling indicate that the uppermost 2 km of the crystalline crust does not consist of melt products; at 2–6 km depth below basement the seismic velocities are too high for magmatic rocks to form more than ca. 50% of the volume, and beyond a depth of ca. 6 km the velocity is that of unaltered mantle. Therefore, the mean thickness of rocks formed by mantle melting at the time of rifting is unlikely to be more than ca. 2 km, compared to a predicted melt thickness of ca. 4 km (Fig. 3). The absence of voluminous magmatic rocks may be attributed to some special circumstance such as unusually low mantle temperatures at the time of rifting, but the apparent abundance of zones of exhumed mantle would then suggest that few margins form above 'normal' temperature mantle.

The volume of melt generated from the upwelling mantle is controlled primarily by the upwelling rate, which in turn is controlled by the plate-separation (spreading) rate. Faster upwelling leads to greater melt volumes, but these volumes are spread over a greater surface area created by plate separation, so if the upwelling is sufficiently rapid that little heat is lost, crustal thickness changes little with spreading rate. However, if upwelling is slower, significant heat is lost by conduction and the net effect is the formation of thinner crust. If the mantle viscosity depends only on temperature and the upwelling is a purely passive response to plate separation, computer models predict that crustal thickness should decrease steadily as spreading rate decreases (Fig. 3). Results from seismic experiments show that oceanic crustal thickness in fact changes little with spreading rate, except at very slow rates (less than ca. 20 mm yr^{-1}). These observations can be explained if at slow spreading rates the upwelling region is narrower and the upwelling rate therefore faster than that predicted by the models. Such a narrowing of the upwelling region may result from viscosity changes resulting from the actual melting process [Hirth and Kohlstedt (1996)]. Perhaps the lack of early magmatism suppresses this feedback between melting and upwelling at magma-poor margins.

3. Deformation During Continental Break-Up

At the very slow rates of geological processes, rocks are brittle if they are cold, but they can flow if they are warm in the same way that glass flows to the bottom of ancient window panes. Despite the development of sophisticated computer and analogue models [Hopper and Buck (1996); Brun and Beslier (1996)], the consequences of such flow for large-scale extension have remained controversial. Computer models have to make many idealizing assumptions, and analogue models, which can elegantly illustrate the consequences of particular flow laws, are limited by an inability to reproduce temperature-dependent changes in the flow behaviour during extension. A further complication is that the composition and, in particular, the water content of the lower crust and upper mantle beneath the continents are

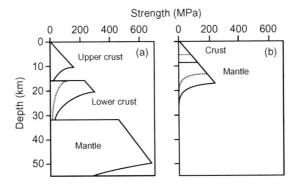

Fig. 4 Strength profiles for the continental lithosphere at different stages of extension [Perez-Gussinye and Reston (2001)]. Profiles assume that the crust is initially 32 km thick, the temperature at the base of the crust is 515°C, the upper crust deforms like wet quartz, and the mantle deforms like dry olivine. Solid and dashed profiles indicate a range of possible behaviours for the lower crust. Brittle strength increases linearly with depth, while in the lower crust strength decays rapidly with depth due to increasing temperatures — the 'jelly sandwich' model [Jackson (2002)]. (a) Initial conditions. (b) Strength profile when the whole crust is brittle, which occurs when the crust has stretched by a factor of 3.6 for the solid profile and 6.1 for the dashed profile.

poorly known. A key issue is the extent to which extension in the upper crust is decoupled from deeper deformation by a weak layer in the lower crust. Such a weak layer is expected on the basis of extrapolation of flow laws based on short-time-scale laboratory measurements to geological time-scales (Fig. 4). These laws predict that, for typical temperatures at the base of the crust of ca. 500°C, the upper crust and uppermost mantle are brittle but the lower crust deforms by flow, and that faults will dip steeply in the upper crust and flatten at depth. However, a fierce debate has been raging over the past few years regarding the validity of this picture [Jackson (2002)].

Whether or not a weak layer is present in the lower crust, flow laws predict that once the crust has been extended by a factor of 3–5, the entire crust becomes brittle (Fig. 4). Once this happens, the behaviour should be more predictable, since the strength of rocks varies little with composition when they are cool enough to be brittle. However, even in the absence of magmatism, a further complication arises in the last stages of continental break-up. If the mantle temperature is not unusually high and the rifting is not unusually rapid, by this time the crust lies beneath ca. 2 km of water and the temperature at the base of the crust has cooled significantly below

500°C. Sea water penetrates the entire crust through faults and fissures and comes into contact with mantle rocks. Olivine, the predominant mineral in the mantle, reacts with water at temperatures below 500°C to produce a weak mineral called serpentine. Laboratory studies have shown that once 10–15% serpentine is present, the strength of mantle rocks drops abruptly, so we might expect to see evidence for a weak layer at the top of the mantle under these conditions.

Seismic studies of highly extended crust at magma-poor rifted margins have imaged faults that flatten significantly with depth and merge with 'detachment' faults with very shallow dips [Reston *et al.* (1996)]. These detachment faults play a key role in the deformation immediately preceding continental break-up (Fig. 5). In the region of deep drilling west of Iberia, restoration of motion along such faults results in a crustal section only 7–10 km thick. Structures resulting from the extension which reduced the crustal thickness to 7–10 km from the initial thickness of ca. 30 km generally are not resolved [Whitmarsh *et al.* (2001)]. Seismic observations of such regions rarely resolve basement structures less than a few hundred metres across. Observations from fragments of rifted margins which have been lifted onshore by mountain-building processes, such as in the

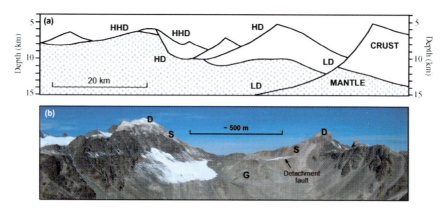

Fig. 5 (a) An interpretation of a seismic-reflection profile from the final stages of continental break-up west of Iberia [Whitmarsh *et al.* (2000); Manatschal *et al.* (2001)]. LD and HD are detachment faults which flatten at depth; HHD is a 'rolling-hinge' fault. (b) A similar detachment fault now exhumed above sea level in the eastern Swiss Alps. The Err detachment is highlighted by the snow cover and separates granite (G) below from schist and gneiss (S), overlain by dolomite (D), a sedimentary rock deposited before the main phase of rifting began [Manatschal and Nievergelt (1997)].

eastern Swiss Alps, where structures can be mapped on scales from a few centimetres to a few kilometres, fill in an important gap in horizontal scale between seismic and borehole observations. Similar styles of faulting have been observed within these fragments [Whitmarsh *et al.* (2001); Manatschal and Nievergelt (1997); Fig. 5), though unraveling the subsequent compressional deformation can be challenging.

A puzzling observation from drilling off west Iberia was that, where mantle rocks formed the basement, they were generally overlain by a layer of fractured rocks called breccia. Also, in the Swiss Alps, large near-horizontal expanses of exposed mantle rocks appear to represent fault surfaces, and this observation has led to the suggestion that regions of flat basement observed west of Iberia may also represent fault surfaces (Fig. 5). Computer models have shown that if faults weaken as motion along them proceeds and they form by extension of a relatively thin brittle layer, faults with an initial steep dip may rotate close to horizontal and acquire almost unlimited offset at these low angles [Lavier *et al.* (1999)]. These conditions are satisfied in the last stages of continental break-up, where a thin brittle layer overlies hot mantle rocks and faults are lubricated by serpentine minerals. Fault rotation allows large expanses of mantle rocks to be exposed with very low relief, giving a neat match with a variety of observations.

4. The Future

Advances in our understanding of continental break-up have been driven by a combination of computer and analogue modelling, laboratory measurements, geophysical studies at sea and drilling into rocks and sediments at continental margins. Future progress will depend on a similar combination of techniques. The transition from continental rifting to sea-floor spreading provides a particularly challenging problem for computer models. To date such models have often either considered continental deformation in isolation, without addressing the extraction of melt from the mantle, or focused on melt extraction in response to a predefined rift evolution. Models are only just beginning to consider the effect on the flow properties of the lithosphere of melt extraction and its subsequent addition to the crust [Nielson and Hopper (2004)]. Close to the time of break-up, the highly nonlinear effect of the penetration of water into the mantle must also be considered.

The low-angle faulting processes described above suggest that, at least to some degree, extension occurs in different places at different depths in the lithosphere. It may also lead to significant asymmetries in structure between 'conjugate' margin pairs that have rifted away from each other. Alternatively, some authors have suggested that such depth-dependent extension is a fundamental property of an essentially symmetric process of margin formation [Davis and Kusznir (2004)]. These alternatives can only be distinguished by future studies of conjugate margin pairs. Regions of present-day continental break-up are obvious targets: The Gulf of California, the Woodlark Basin east of Papua New Guinea, the western Gulf of Aden and the northern Red Sea. However, such margins may provide an incomplete picture of the breakup process because the process itself is incomplete, so studies of ancient conjugate margins also have an important role to play. The margin conjugate to west Iberia now lies offshore Newfoundland, and recently acquired geophysical and deep borehole data from that margin are still being evaluated. Key constraints on models of depth-dependent stretching may come from studies of thickly sedimented margins such as those of the Black Sea, where the subsidence history, which is controlled by the stretching history of the whole lithosphere, is better recorded than at many North Atlantic margins.

An important controlling parameter in continental break-up is the rate of plate separation. At the North Atlantic margins studied in detail to date, this rate is uniformly low: Typically ca. $20 \, \text{mm} \, \text{yr}^{-1}$. At higher extension rates, the melting model curves of Fig. 3(b) converge, so the transition to normal-thickness oceanic crust after continental break-up is expected to be much more rapid. Conductive cooling should be less significant, so lubrication of faults by serpentinisation may be less common and the resulting margin structures therefore more symmetrical. Studies of margins elsewhere in the world, formed at a variety of spreading rates, are needed to test these ideas. Finally, seismic reflection profiles collected parallel to margins sometimes show similar faulting patterns to those collected across them. The role of detachment faulting may remain obscure while we continue to rely on two-dimensional seismic images of an essentially three-dimensional process. The technology exists for fully three-dimensional imaging of these features [Lonergan and White (1999)], and an application of this technology to the large-scale tectonics of rifted margins, though expensive, is an obvious next step.

Acknowledgements

I thank R. B. Whitmarsh, G. Manatschal and N. White for their helpful discussions and the Natural Environment Research Council and the Royal Society for financial support.

References

Brun, J. P. & Beslier, M. O. (1996) Mantle exhumation at passive margins. *Earth Planet. Sci. Lett.* **142**, 161–173.

Bullock, A. D. & Minshull, T. A. (2005) From continental extension to sea-floor spreading: Crustal structure of the Goban Spur rifted margin, southwest of the UK. *Geophys. J. Int.* **163**, 527–546.

Davis, M. & Kusznir, N. J. (2004) Depth-dependent lithospheric stretching at rifted continental margins. *Proceedings of NSF Rifted Margins Theoretical Institute.* Karner, G. D., Taylor, B., Driscoll, N. W. and Kohlstedt, D. L. (eds.). pp. 92–137, Columbia University Press.

Dean, S. M., Minshull, T. A., Whitmarsh, R. B. & Louden, K. E. (2000) Deep structure of the ocean-continent transition in the southernIberia Abyssal Plain from seismic refraction profiles. II. The IAM-9 transect at 40°20′ N. *J. Geophys. Res.* **105**, 5859–5886.

Eldholm, O. & Grue, K. (1994) North Atlantic volcanic margins: Dimensions and production rates. *J. Geophys. Res.* **99**, 2955–2968.

Fowler, S. R., White, R. S., Spence, G. D. & Westbrook, G. K. (1989) The Hatton Bank continental margin. II. Deep structure from two-ship expending spread profiles. *Geophys. J. Int.* **96**, 295–309.

Hirth, G. & Kohlstedt, D. L. (1996) Water in the oceanic upper mantle: Implications for rheology, melt extraction and the evolution of the lithosphere. *Earth Planet. Sci. Lett.* **144**, 93–108.

Holbrook, W. S. & Keleman, P. B. (1993) Large igneous province on the US Atlantic margin and implications for magmatism during continental breakup. *Nature* **364**, 433–436.

Hopper, J. R. & Buck, W. R. (1996) The effect of lower crustal flow on continental extension and passive margin formation. *J. Geophys. Res.* **101**, 20175–20194.

Horsefield, S. J., Whitmarsh, R. B., White, R. S. & Sibuet, J.-C. (1994) Crustal structure of the Goban Spur rifted continental margin, NE Atlantic. *Geophys. J. Int.* **119**, 1–19.

Jackson, J. (2002) Strength of the continental lithosphere: Time to abandon the jelly sandwich? *GSA Today* **12**, 4–10.

Lavier, L. L., Buck, W. R. & Poliakov, A. N. B. (1999) Self-consistent rolling-hinge model for the evolution of large-offset low-angle normal faults. *Geology* **27**, 1127–1130.

Le Pichon, X. & Sibuet, J.-C. (1981) Passive margins: A model of formation. *J. Geophys. Res.* **86**, 3708–3720.

Lonergan, L. & White, N. (1999) Three-dimensional seismic imaging of a dynamic Earth. *Phil. Trans. R. Soc. London* **A357**, 3359–3375.

Louden, K. E. & Chian, D. (1999) The deep structure of non-volcanic rifted continental margins. *Phil. Trans. R. Soc. London* **A357**, 767–804.

Manatschal, G. and Nievergelt, P. (1997) A continent-ocean transition recorded in the Err and Platta nappes (eastern Switzerland). *Eclogae Geol. Helv.* **90**, 3–27.

Manatschal, G., Froitzheim, N., Rubenach, M. & Turrin, B. D. (2001) The role of detachment faulting in the formation of an ocean-continent transition: insights from the Iberia Abyssal Plain. *Non-Volcanic Rifting of Continental Margins: A Comparison of Evidence from Land and Sea.* Wilson, R. C., Whitmarsh, R. B., Taylor, B. and Froitzheim, N. **187**, 405–528, Geological Society London.

McKenzie, D. & Bickle, M. J. (1988) The volume and composition of melt generated by extension of the lithosphere. *J. Petrol.* **29**, 625–679.

Minshull, T. A., Dean, S. M., White, R. S. & Whitmarsh, R. B. (2001) Anomalous melt production after continental breakup in the southern Iberia Abyssal Plain. *Non-Volcanic Rifting of Continental Margins: A Comparison of Evidence from Land and Sea.* Wilson, R. C., Whitmarsh, R. B., Taylor, B. and Froitzheim, N. **187**, 537–550, Geological Society London.

Nielson, T. K. & Hopper, J. R. (2004) From rift to drift: Mantle melting during continental breakup. *Geochem. Geophys. Geosys.* **5**(art. 7), doi 10.1029/2003GC000662.

Perez-Gussinye, M. & Reston, T. J. (2001) Rheological evolution during extension at nonvolcanic rifted margins: Onset of serpentinization and development of detachments leading to continental breakup. *J. Geophys. Res.* **106**, 396–3975.

Reston, T. J., Krawczyk, C. M. & Klaeschen, D. (1996) *J. Geophys. Res.* **101**, 8075.

Smith, D. T. & Sandwell, W. H. F. (1997) Global seafloor topography from satellite altimetry and ship depth soundings. *Science* **277**, 1956–1962.

White, N., Thompson, M. & Barwise, T. (2003) Understanding the structural and thermal evolution of deep-water continental margins. *Nature* **426**, 334–343.

White, R. S. & McKenzie, D. (1989) Magmatism at rift zones: The generation of volcanic continental margins and flood basalts. *J. Geophys. Res.* **94**, 7685–7730.

Whitmarsh, R. B., Dean, S. M., Minshull, T. A. & Tomkins, M. (2000) Tectonic implications of exposure of lower continental crust beneath the Iberia Abyssal Plain, Northeast Atlantic Ocean: Geophysical evidence. *Tectonics* **19**, 919–942.

Whitmarsh, R. B., Manatschal, G. & Minshull, T. A. (2001) Evolution of magma-poor continental margins from final rifting to seafloor spreading. *Nature* **413**, 150–154.

Properties and Evolution of the Earth's Core and Geodynamo

F. Nimmo

Department Earth Sciences
University of California Santa Cruz
CA 95064, USA
fnimmo@es.ucsc.edu

D. Alfè

Department Earth Sciences
University College London
WC1E 6BT, UK

Department of Physics & Astronomy
University College London
WC1E 6BT, UK

INFM DEMOCRITOS
National Simulation Centre
via Beirut 2–4, 34125
Trieste, Italy
d.alfe@ucl.ac.uk

We review recent advances in the study of the Earth's iron core, focussing on three areas: The properties of the core-forming materials, the manner in which core motions generate the Earth's magnetic field (the dynamo), and the evolution of both the core and the dynamo. *Ab initio* computer simulations of the behaviour of iron alloys under core conditions suggest that the inner (solid) and outer (liquid) core contain 8% sulphur/silicon, and 8–10% sulphur/silicon plus 8–13% oxygen, respectively. The inner core boundary for these materials is at ~ 5500 K. Although computer simulations of the dynamo lack sufficient resolution to match likely terrestrial parameter values, such models can now reproduce the spatial and temporal behaviour of the observed magnetic field. The present-day dynamo occurs because the mantle is extracting heat from the core (at a rate of 9 ± 3 TW); the resulting inner core growth drives core convection and implies a young inner core age (< 1.5 Gyr). A long-lived dynamo requires rapid core cooling, which tends to result in an inner core larger than that observed. A possible solution to this paradox is that radioactive potassium may reside in the core. We also briefly review the current state of knowledge for cores and dynamos in other planetary bodies.

1. Introduction

The rocky exterior of the Earth conceals a Mars-sized iron body at its centre: The core. The core is of fundamental importance to the thermal and magnetic behaviour of the Earth as a whole. Recent advances in computational power and experimental techniques have galvanized the study of the core in at least three areas: The properties of the core-forming materials; the manner in which core motions generate the Earth's magnetic field (the geodynamo); and the evolution of both the core and the geodynamo through time. In this review we focus on these three issues.

We begin with a description of the Earth's interior structure, and how it is deduced. We also describe the main characteristics of the Earth's magnetic field, and how it has varied in time. These preliminaries completed, in Sec. 3 we examine how recent results have helped to pin down the composition and temperature structure of the core. In Sec. 4 we discuss the progress made in numerical models, which can now reproduce many aspects of the geodynamo. In Sec. 5 we discuss how the energy balance of the present-day core maintains the dynamo, and speculate how this energy balance may have changed over the course of Earth's history. We conclude with a look at the cores and dynamos of other Earth-like bodies in this solar system, and suggest likely directions of research for the next decade.

2. The Interior Structure of the Earth

With the exception of the Earth, our knowledge of the internal structure of silicate (rocky) planets is rather limited. Our knowledge of the Earth's structure is much more complete, thanks to the study of earthquakes (see below). However, even in the absence of this kind of information, there are several lines of evidence suggesting that silicate planets should possess a mostly iron core.

Firstly, the raw material of the solar system had a composition which was probably similar to that of the Sun and a common class of meteorites, the chondrites. The ratio of silicon to iron in these materials is roughly 1:1

by volume [Lodders and Fegley (1998)]; since neither element is particularly volatile, the Earth, and other similar bodies, should have retained their full complement. While some of the iron is likely to have been bound up in silicate minerals, in the absence of any other evidence one would expect the terrestrial planets to consist of silica-rich and iron-rich zones. The multiple collisions which gave rise to the terrestrial planets will have led to hot, and possibly molten, starting conditions [Tonks and Melosh (1993)]; under these conditions, the dense iron-rich material will have migrated to the centre, forming an iron core [Stevenson (1990)].

This rather theoretical argument is supported by observations. For instance, the near-surface of the Earth consists of a silicate mantle with a density of around $3300 \, \mathrm{kg \, m^{-3}}$. This density is considerably less than the mean density of the planet ($5500 \, \mathrm{kg \, m^{-3}}$). Even though the mantle density would increase with depth, due to the compressibility of rock, the likely increase is insufficient to account for the observed density. Thus, the presence of a subsurface, high density zone is required, and the compositional arguments outlined above imply that the dense material is primarily iron.

Further evidence comes from the moment of inertia of the Earth, which is accurately known from measurements of satellite orbits and the rate of precession of the rotation axis, and which constrains the distribution of mass within the Earth. A uniform sphere has a normalised moment of inertia of 0.4; the Earth's value of 0.3307 indicates that the mass is concentrated towards the centre of the planet. For a simple two-layer (core plus mantle) planet, the core radius can be determined if the densities of the two layers are known. Table 1 gives the densities and moments of inertia of several silicate bodies; these data and the arguments above strongly suggest that iron cores are a common planetary feature.

Perhaps surprisingly, we can also differentiate between a liquid and a solid core. The solid Earth deforms under the gravitational attraction of the Sun and the Moon, that is, it has tides. These tides are much smaller in amplitude (typically 0.2 m) than the tides associated with the oceans, but are still measurable. The deformation depends on the rigidity of the interior [Murray and Dermott (1999)]. Predicted tidal amplitudes, assuming a uniform (seismologically inferred) mantle rigidity, were too small; one solution to the problem was to postulate that the core had negligible rigidity, that is, it was fluid [Jeffreys (1929)].

Similar arguments can be applied to bodies other than the Earth. For instance, tidal deformation studies suggest that the cores of Mars [Yoder *et al.* (2003)], the Moon [Williams *et al.* (2001)], Venus [Konopliv and

Table 1　Geophysical parameters of silicate planetary bodies. Data are from Lodders and Fegley [1998], except as indicated otherwise. R_c is the core radius, C is the polar moment of inertia. m is the magnetic dipole moment of the body, measured in Tesla R^3, and indicates the strength of the magnetic field at the planet's surface [[a]Khan *et al.* (2004); [b]Schubert *et al.* (1988); [c]Konopliv and Yoder (1996); [d]Yoder *et al.* (2003); [e]Anderson *et al.* (1996)].

	Earth	Moon	Mercury	Venus	Mars	Ganymede
Radius R (km)	6371	1737	2438	6052	3390	2634
Mass M (10^{24} kg)	5.97	0.07	0.33	4.87	0.64	0.15
Bulk density (kg m^{-3})	5515	3344	5430	5243	3934	1940
Surface gravity g (m s^{-2})	9.8	1.6	3.7	8.9	3.7	1.4
C/MR^2	0.3307	0.394	—	—	0.366	0.31
R_c/R	0.55	0.2[a]	0.75[b]	0.5[c]	0.45–0.55[d]	0.15–0.5[e]
$m(\times 10^{-4} TR^3)$	0.61	—	0.003	—	—	0.008

Yoder (1996)] and Mercury [Margot *et al.* (2004)] are at least partially liquid. We will return to the cores of the other planets later in this article, but for now we will focus on that of the Earth, since we understand it in so much more detail.

Figure 1(a) shows a schematic cross-section of the Earth. The outer half of the planet consists of a silicate mantle. The near-surface is made up of rigid tectonic plates, roughly 100 km thick, which move laterally and are eventually recycled into the mantle at subduction zones. At a depth of 2890 km the mantle gives way to the liquid outer core; this interface is known as the core-mantle boundary (CMB). The liquid outer core in turn gives way to a solid inner core at a depth of 5150 km, the inner core boundary (ICB). The outer core is probably well-mixed and relatively homogeneous; however, both the inner core and the mantle are likely to be laterally heterogeneous. A particularly complex region is the base of the mantle, which forms a hot boundary layer from which convective plumes rise.

The remarkable detail in which the interior structure of the Earth is known is thanks almost entirely to seismology, the study of earthquakes [see Stein and Wysession (2000) for a comprehensive overview]. Earthquakes occur because of the relative motions of the Earth's different tectonic plates, and generate waves which propagate through the Earth. An earthquake caused by fault slip of about 1 m or more is easily detectable with modern seismometers on the other side of the globe; about 50 such earthquakes occur each year. Seismometers at different locations will detect

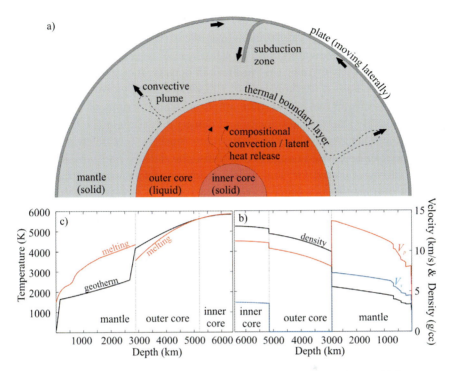

Fig. 1 (a) Schematic cross-section of the Earth. Boundary layer thicknesses are to scale. (b) Variation in P-wave (V_p) and S-wave (V_s) velocities and density with depth. From Dziewonski and Anderson [1981]. (c) Variation in temperature and melting curve (solidus) with depth. Mantle melting curve from Boehler [2000]; other curves from Nimmo *et al.* [2004].

seismic waves whose paths have sampled different parts of the Earth's interior; since the seismic velocity varies with depth, the relative wave arrival times at different seismometers can be used to infer the velocity structure of the Earth. The outermost core, having a slower velocity than the overlying mantle, refracts the waves so as to leave a "shadow zone" in which very few arrivals are observed. The existence of this shadow zone was the first seismological evidence for the core; later detection of arrivals within this zone confirmed the existence of an inner core.

Because the bulk of the interior of the Earth is solid, both transverse (S) and longitudinal (P) waves will propagate. The propagation velocities, which depend on the local density and elastic constants, are different. In particular, transverse waves do not propagate through liquids. This fact allowed identification of the outer core as a fluid: S waves cannot propagate directly through the outer core, and are therefore nearly absent at a characteristic range of angular distances from the earthquake source. Once both the P and S velocities are known as a function of radial distance, it is then possible to iteratively infer the variation in density and gravity with depth.

The very largest earthquakes set the whole Earth vibrating with periods of tens of minutes or less. The amplitudes and periods of the different oscillations depend on the velocity and density structure of the Earth, with shorter-wavelength oscillations being more sensitive to shallow structures and vice versa. Given enough oscillations, the density and velocity structures can be inverted for, and thus provide an independent check on the results obtained from seismic wave arrival times.

Results obtained by combining these two methods are shown in Fig. 1(b). The increase in velocity with depth is primarily due to the decrease in compressibility with increasing pressure. The S wave velocity drops to zero in the outer core because it is a fluid. The density increases with depth, as expected. It turns out that the density of the outer core is less than that expected for pure iron, suggesting the presence of a contaminant (see Sec. 3.4). The increase in density at the ICB, conversely, is larger than one would expect for a simple phase change, suggesting that the contaminant is not being incorporated into the solid core. This expulsion of a light constituent as the inner core solidifies provides a major source of energy to drive the dynamo (see Sec. 5.1).

Having established the radially-averaged structure of the Earth, it then becomes possible to look for lateral variations in seismic velocity, a process known as seismic tomography. Tomographic images now routinely detect both subducting plates and upwelling plumes [Montelli *et al.* (2004)]. Some of the velocity anomalies inferred, particularly those near the CMB, are too large to be caused simply by temperature changes, and may involve compositional or phase variations [Oganov and Ono (2004); Murakami *et al.* (2004); Tsuchiya *et al.* (2004)] or melting [Williams and Garnero (1996)]. Perhaps more surprisingly, the inner core also appears to contain structure: the outermost part is isotropic but of variable thickness, while the inner part has a fabric [Song (2003); Souriau and Poupinet (2003)]. This fabric is critical for observing inner core rotation (see Sec. 4).

2.1. *Thermal structure of the Earth*

As described above, the density structure of the Earth can be inferred directly from seismology. Equally important, however, is the temperature structure of the Earth. Within the rigid surface plates, the temperature increases roughly linearly with depth, and heat is transported by conduction. Somewhere within the range 1200–1600 K, mantle material stops behaving in a rigid fashion and starts to flow, so the base of the plate occurs within this temperature interval. The mantle below is sufficiently warm that it undergoes convection.

A packet of convecting mantle material moves sufficiently rapidly that it does not exchange appreciable heat with its surroundings. This so-called adiabatic situation means that, as the material rises and expands, it cools. Mantle material which rises all the way to the surface is sufficiently hot that it exceeds its melting temperature and generates a crust; the ~ 7 km thickness of the resulting oceanic crust implies a mantle surface (or potential) temperature of about 1600 K [McKenzie and Bickle (1988)].

The adiabatic effect depends on gravity, thermal expansivity and heat capacity, all of which are reasonably well known. The resulting adiabatic gradient is about 0.5 K/km (Fig. 1c); this gradient is shallower than the conductive gradient in the near-surface because heat is being transported by convection, not conduction. Following the adiabatic gradient (which decreases a little with depth), the temperature near the base of the mantle is about 2700 K. However, there will exist a boundary layer at the base of the mantle across which the temperature increases rapidly to that at the outer edge of the core.

The temperature at the ICB is (by definition) the core melting temperature at that pressure. Because the outer core is convecting, it will also have an adiabatic temperature gradient (of about 0.8 K/km). Thus, if the temperature at the ICB is known, the temperature everywhere in the outer core is also determined. This is why it is so important to determine the melting behaviour of iron: it provides a tie point from which temperatures elsewhere can be calculated.

Despite a great deal of experimental and theoretical effort in determining the melting temperature of iron at core conditions, only in the last few years have the uncertainties been reduced to even moderately acceptable levels. We will discuss recent progress in this field in Sec. 3 below; here, we will simply point out the example temperature structure shown in Fig. 1(c). It shows the conductive plate at the surface, the adiabatic gradient within the mantle, and the boundary layer at the base of the mantle. The location of the inner core is determined by the intersection of the adiabat with the melting curve. The adiabatic temperature drop across the liquid outer core is 1450 K, the ICB is at 5600 K, and the CMB is 4150 K. The uncertainties associated with some of these numbers are still large (see Sec. 3); however, the basic picture is certainly correct.

2.2. *Magnetic observations of the Earth*

The present-day magnetic field of the Earth is well characterised. As recognised by William Gilbert in 1600, it resembles that of a bar magnet, with north and south poles (dipolar). The dipole axis is currently inclined at about 10° to the rotation axis. At short wavelengths ($< 3000 \, km$) the surface magnetic field is dominated by crustal anomalies, but the longer wavelength features are due to processes occurring within the core. Short wavelength features are also undoubtedly generated within the core, but are not visible at the surface because they are strongly attenuated with radial distance. A further complication is that the so-called toroidal component of the core's magnetic field has field lines which are parallel to the surface of the core, and are thus unobservable at the Earth's surface. Thus, the field that we can measure at the surface is different in both frequency content and amplitude from the field within the core.

The behaviour of the Earth's magnetic field over time is of great interest [see reviews by Valet (2003) and Jacobs (1998)]. Modern measurements of the variation in intensity and orientation of the field date back only 150 years, to the time of Gauss. Less precise observations, mostly from naval expeditions, extend the historical record back to roughly 1500 A.D., and well-dated archaeological data cover the last $\sim 10 \, kyr$. Prior to this time, observations of field orientation and intensity rely on the natural (remanent) magnetisation of either volcanic rocks or sediments. The former are problematic because of the sporadic nature of volcanic eruptions. Dating the latter is usually more uncertain than for lavas; furthermore, the processes by which sediments acquire magnetic characteristics are not well understood, and may involve complicating effects such as changes in ocean chemistry.

Despite the difficulties, several time-dependent characteristics of the field are evident. Firstly, over the last four centuries, the tilt and the amplitude of the dipolar field have changed by tens of percent [Barton (1989)]. Over the same timescale, several features of the field appear to have drifted westwards with time, at a rate of about 0.5° per year. Secondly, over timescales $>$ a few thousand years the mean position of the magnetic axis coincides with the rotation axis [Valet (2003)]. Thirdly, the field appears to have remained predominantly dipolar over time [though see Bloxham (2000)], and has apparently persisted for at least 3500 Myrs [McElhinny and Senanayake (1980)], with the maximum field intensity having exceeded the present day value by no more than a factor of five [Valet (2003); Dunlop and Yu (2004)]. Fourthly, and much the most important, the polarity of the magnetic axis undergoes irregular reversals. Recent reversals have occurred roughly every 500 kyrs, and take place rapidly [about 7 kyr Clement (2004)]. However, there is a wide scatter in reversal frequency; for instance, there were no reversals at all in the period 125–85 Ma. The earliest reversal identified to date occurred at 3214 Ma [Layer *et al.* (1996)]. Incomplete reversals, or excursions, appear to take place more frequently. Finally, it is argued that the path swept out by the magnetic poles during reversals may be preferentially concentrated around the Pacific (Laj *et al.* (1991)], though this is controversial [Prevot and Camps (1993); Love (2000)]. These observations provide constraints on the models for magnetic field generation, to be discussed below.

3. Core Properties

As discussed above, inferring the thermal structure of the core requires a knowledge of the melting temperatures of iron and iron alloys at core pressures. Similarly, understanding how light impurities partition between the solid and liquid core phases is necessary to infer the core composition. The traditional way of answering these questions is by experiments, in which the high pressures may be either static (e.g. diamond anvil cells) or transient (shock waves). Carrying out such experiments is exceedingly challenging, and typically results in melting temperature uncertainties of ± 500 K. More recently, computational methods based on quantum mechanics have been used to predict the behaviour of iron and iron alloys at core conditions. Here we will focus on the computational approach, and in particular that of the group based at UCL. As discussed below, several other groups have obtained similar results using slightly different approaches [Laio *et al.* (2000); Belonoshko *et al.* (2000)].

The basic approach of the computational methods is to calculate the chemical potential of the material at the conditions of interest. This approach relies on the fact that the minimum of the chemical potential defines the stability. Thus, for example in the context of melting, at a temperature above the melting temperature the chemical potential of the liquid will be lower than that of the solid, and conversely at a temperature below the melting temperature the chemical potential of the solid is lower.

It follows that at the equilibrium between two phases (e.g. solid and liquid at the melting temperature) the chemical potentials of the two phases are equal. These chemical potentials can be calculated, and therefore the melting curve can be determined.

Analogously, in a mixture of elements A and X, equilibrium between solid and liquid implies the continuity of the chemical potentials of *both* A and X across the phase boundary. The equality of the chemical potentials of A and X in the solid and in the liquid determines the partitioning of A and X between solid and liquid. This information can be used to infer the composition of the Earth's core, as explained below.

In what follows we will first briefly describe the theoretical framework on which the calculations were based, and then present the results for the melting curve of iron, and the partitioning of light elements in the core.

3.1. *First principles calculations*

With *first principles simulations* one usually means calculations in which no adjustable parameter and no experimental input is allowed (apart from some fundamental constants, such as the charge on the electron and the Planck constant). In the context of simulating the properties of matter, this means solving the Schrödinger equation $\mathcal{H}\Psi_N = \mathcal{E}\Psi_N$, where $\mathcal{H} = \mathcal{T} + \mathcal{V}$ is the *Hamiltonian* of the system which contains N particles (both electrons and nuclei), with $\mathcal{T}$ the kinetic energy and $\mathcal{V}$ the potential energy, $\mathcal{E}$ the energy of the system, and Ψ_N the many-body *wavefunction*, which is a complicated function of the positions of the N particles in the system. Since the electrons are at least three order of magnitudes lighter than the nuclei, it is customary to introduce the so called *adiabatic approximation*, in which the motion of the electrons is decoupled from the motion of the nuclei. This in practice means solving a new Schrödinger equation $\mathcal{H}\{R\}\Psi_n\{R\} = \mathcal{E}\{R\}\Psi_n\{R\}$, where now the Hamiltonian is only a function of the n electronic degrees of freedom, and depends parametrically on the nuclear degrees of freedom $\{R\}$. The energy $\mathcal{E}\{R\}$ is interpreted as a potential energy for the motion of the nuclei. From this, one can calculate the forces on the nuclei, which, for example, can be used to integrate the Newton's equation of motion and perform *molecular dynamics* simulations.

Even with this simplification though, the problem of solving the modified Schrödinger equation remained very difficult, at least until the introduction of *Density Functional Theory* (DFT) in the mid 60s. DFT was a breakthrough in the state of the art of quantum mechanics [Hohenberg and Kohn (1964); Kohn and Sham (1965)]. In this section we review only main points of the theory; for a rigorous description the reader should consult the original papers or, for example, the excellent books of Parr and Wang [1989] or Dreizler and Gross [1990].

The central idea of DFT is that the complicated many-body wavefunction Ψ_n is not needed, and the important physical quantity is the electronic charge $\rho(\mathbf{r})$, which is only a function of the three-dimensional variable $\mathbf{r}$. The energy of the system is a *functional* of the density, and can be written as $\mathcal{E}[\rho] = \mathcal{F}[\rho] + \mathcal{V}[\rho]$, where $\mathcal{V}[\rho]$ is the potential energy of the electronic charge density ρ in the external potential $\mathcal{V}$ (e.g. the potential due to the nuclei), and $\mathcal{F}[\rho]$ a *universal functional of* ρ. The ground state energy of the system is given by the minimum of $\mathcal{E}[\rho]$ with respect to ρ, and the electronic charge density which minimises the total energy is the ground state electronic charge density.

Of course, it rarely happens that a simple reformulation of a problem solves all the difficulties, and indeed this is not the case for DFT: The functional $\mathcal{F}[\rho]$ is unknown. However, Kohn and Sham [1965] proposed a simple approximation called the *local density approximation* (LDA), which made it possible to define an approximated $\mathcal{F}[\rho]$. Although the LDA was constructed to work for homogeneous systems, this approximation turned out to also work amazingly well for highly inhomogeneous systems like molecules and surfaces, and it is probably fair to say that DFT owes its tremendous success in the past forty years to this approximation. Recently, more sophisticated approximations have been proposed, like the so called *generalised gradient corrections* (GGA) [e.g. Wang and Perdew (1991); Perdew *et al.* (1996)], but the LDA is still playing a major role in the DFT community.

An additional approximation which contributed to the great success of DFT was the so called *pseudo-potential* approximation [e.g. Bachelet *et al.* (1982)]. The main point here is the recognition that only the outermost electrons of the atoms are involved in bonding, the so called *valence* electrons. This means that the *core* electrons which are tightly bound to the nuclei can be treated as frozen in their atomic configurations, and included only implicitly in the calculations. This is done by replacing the potential generated by the bare nuclei, with a *pseudo-potential* generated by the ionic cores, which are formed by the nuclei surrounded by the frozen core electrons. The quality of this approximation can be easily tested by explicitly including more electrons in *valence*.

The increasing popularity of DFT in the physical, chemical, and more recently geological and biological community, is due to its exceptional reliability in reproducing experimental results, giving DFT-based methods unparalleled predictive power. The success of these kinds of first principles calculations is also due to the increasingly widespread availability of large computational resources, as well as to more and more efficient computer codes.

In the next section we briefly describe the main points relevant to the calculations of chemical potentials from first principles, and we report the results for the melting curve of iron and the partitioning of light elements in the core in the following two sections.

3.2. *Free energies*

Chemical potentials are closely related to *free energies*. In particular, in a one component system the chemical potential μ is given by the Gibbs free energy per molecule, $\mu = G/N = F/N + pV/N$, where F is the Helmholtz free energy of the system containing N particles, $p = -\partial F/\partial V|_T$ is the pressure and V is the total volume.

To calculate F at a given state (V, T), it is possible to use the technique known as *thermodynamic integration* [see e.g. Frenkel and Smit (1996)]. This is a general scheme to compute the free energy difference $F - F_0$ between two systems, whose potential energies are U and U_0 respectively. The idea is that F is the "difficult" free energy of the quantum mechanics system, and F_0 the free energy of a system where the interactions between the atoms are approximated by some simple relation.

The free energy difference $F - F_0$ is the reversible work done when the potential energy function U_0 is continuously and reversibly switched to U. This switching does not correspond to a physical process, but it is a well defined mathematical procedure which can be carried out in a computer. The computational effort is proportional to the "distance" between the reference and the quantum mechanical systems. Therefore, the crucial point here is that the reference system should be chosen to be as close as possible to the quantum mechanics system, because this minimises the number of quantum mechanics calculations needed. To calculate the free energy of the reference system F_0 one can use the same procedure, and evaluate $F_0 - F^*$, where F^* is the free energy of some other simple reference system whose free energy is known. Here, since the calculations do not involve heavy quantum mechanics calculations, one can afford "large distances" between these two systems, and possibly F^* could even be the free energy of a perfect gas (i.e. a system with no interactions between the particles).

To compute the melting temperature of iron, Alfè *et al.* [1999, 2001, 2002b, 2003, 2004] performed calculations at a number of thermodynamic states spanning the conditions of the Earth's core, and fitted the calculated Helmholtz free energies $F(V, T)$ to polynomials in volume and temperature. From this it was possible to obtain all the relevant thermodynamical properties by appropriate differentiation of F.

Similar techniques can be used to evaluate the chemical potentials of the elements of a mixture. For example, the chemical potential μ_X of an element X in a solution A/X (we may identify element A with the solvent and X with the solute, but the description is completely general) is equal to the change of Helmholtz free energy of the system as one atom of the element X is added to the system at constant volume and constant temperature. This change of free energy can be evaluated using the techniques of thermodynamic integration described above. A detailed explanation of how this is done can be found in Alfè *et al.* [2002a]. We shall see below how the ability of calculating the chemical potentials of the elements in a solution can be used in conjunction with seismological data to put constraints on the composition of the core.

3.3. *The melting curve of iron*

To determine the melting curve of iron, Alfè *et al.* [1999, 2001, 2002b, 2003, 2004] calculated the chemical potential μ of pure iron as a function of pressure and temperature for both solid and liquid. In a one component system this is the same as the Gibbs free energy per atom G/N. In fact, Alfè *et al.* [1999, 2001, 2002b, 2003, 2004] calculated the Helmholtz free energy F as a function of volume and temperature and then obtained G from its usual relation $G = F + pV$. As mentioned above, for any fixed pressure the continuity of G with respect to temperature defines the melting transition, which is found by the point where the Gibbs free energies of liquid and solid become equal, $G_l(p, T_m) = G_s(p, T_m)$.

Figure 2 shows the melting curve of iron from pressures of 50 to 400 GPa. The solid black line is the result obtained by combining the calculated free energies of solid and liquid. The predicted melting temperature at the ICB is 6350 ± 300 K, where the error quoted is the result of the combined statistical errors in the free energies of solid and liquid. Systematic errors due to the approximations of DFT are more difficult to estimate, and a

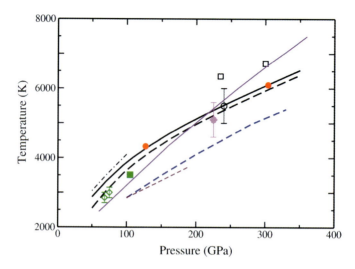

Fig. 2   Comparison of the melting curve of Fe from *first principles* calculations and experiments. Heavy solid and dashed curves: results from Alfè
et al. [2002b] without and with pressure correction (see text); filled red circles: corrected coexistence results (see text) from Alfè *et al.* [2002c, 2004];
blue dashed line: results of Laio *et al.* [2000]; solid purple line: results of
Belonoshko *et al.* [2000]; black chained and dashed maroon curves: DAC
measurements of Williams *et al.* [1987] and Boehler [1993] respectively; open
green diamonds: DAC measurements of Shen *et al.* [1998]; green filled square:
DAC measurements of Ma *et al.* [2004]; black open squares, black open circle and pink filled diamond: shock experiments of Yoo *et al.* [1993], Brown
and McQueen [1986], and Nguyen and Holmes [2004]. Error bars are those
quoted in original references.

definite value will only be obtained after the problem is explored with a
more accurate implementation of quantum mechanics. We hope that quantum Monte Carlo techniques may serve this cause in the near future. However, it is possible to gauge *probable* errors by analysing the performance
of the current techniques in describing known physical properties of iron.
For example, a comparison of the dependence of the pressure of pure iron
as a function of volume (a form of the *equation of state*) between the calculations and experiments shows a slight underestimation of the pressure
(by 2–7% in the high-low pressure region respectively). An intuitive way to
see how this error propagates in the melting curve is to realise that on the
melting line the actual pressure p is higher than the calculated one, and
therefore the right melting curve is "shifted" to the right. This "corrected"

melting curve is also shown in Fig. 2, as a dashed black line, and provides a prediction of the temperature at the ICB of about $6200 \pm 300 \, \text{K}$.

We also report on the same figure the theoretical results of Laio *et al.* [2000] (dashed blue line) and Belonoshko *et al.* [2000] (solid purple line). These two groups used first principles in a different way. They used DFT calculations to fit some cleverly constructed model potential, and then used the model potential to calculate the melting curve using the technique of the "coexistence of phases". In this technique, a large system containing solid and liquid is simulated using molecular dynamics. This "coexistence of phases" technique can be implemented in a number of different ways. For example, by constraining the pressure and the temperature to some chosen values (NPT ensemble), the system always evolve towards a single phase, either solid or liquid, depending on whether the temperature is below or above the melting temperature for the chosen pressure. In this way it is possible to bracket the melting temperature. Other approaches include simulating the system in the NVE ensemble, where the volume and the total energy of the system are kept constant, or the NVT, NPH ensembles, where volume and temperature or pressure and enthalpy are kept constant respectively.

The method of the coexistence of phases and the one which relies on the explicit calculation of free energies are of course equivalent, if applied consistently. This was demonstrated recently for the calculation of the melting curve of aluminium, in which both methods were used in the context of first principles calculations and delivered the same results [Vočadlo and Alfè (2002); Alfè (2003)]. For the melting curve of iron however, the two methods cannot be directly compared, because the coexistence method was not employed directly in the context of DFT. Instead, a model potential was used for the coexistence simulations. Most likely, this is the reason for the discrepancy between the melting curves of Laio *et al.* [2000], Belonoshko *et al.* [2000], and that of Alfè *et al.* [1999, 2002b]. However, once the differences between DFT and the model potential are taken into account, it is possible to devise corrections which deliver exactly the same results [Alfè *et al.* (2002c)]. To show this, in Fig. 2 we also report two melting points obtained by applying these corrections to the model potential of Belonoshko *et al.* [2000], which agree closely with the melting curve obtained from the free energy calculations of Alfè *et al.* [2002b].

Experimental results are also displayed on the figure for comparison. In the low pressure region we report the diamond anvil cell (DAC) experiments of Williams *et al.* [1987], Boehler [1993], Shen *et al.* [1998] and the recent experiments of Ma *et al.* [2004]. The "corrected" melting curve of

Alfè *et al.* [2002b] is in good agreement with the latter two more recent experiments. In the high pressure region only shock waves experiments are available. The measurements of Yoo *et al.* [1993] fall about 1000 K above the calculated melting curve of Alfè *et al.* [2002b], which is in good agreement with the results of Brown and McQueen [1986] and the recent experiments of Nguyen and Holmes [2004].

3.4. *Constraints on the composition of the Earth's core*

As mentioned above, the Earth's core is mainly composed of iron, but the seismologically inferred density means that it must also contain some light impurities. The most popular candidates are sulphur, silicon, oxygen, carbon and hydrogen. The presence of these impurities modifies the melting temperature of the mixture with respect to that of pure iron. Moreover, the crystallisation of the inner core gives rise to compositional convection in the outer liquid core, and this convection helps to drive the geodynamo (Sec. 4). It is therefore also important to investigate what the exact composition of the core is. Evidence from seismology indicates that at the ICB the density contrast between the solid and the liquid is between 4.5% [Dziewonski and Anderson (1981); Masters and Shearer (1990); Shearer and Masters (1990)] and 6% [Masters and Gubbins (2003)]. This is much larger than that expected if the core were pure iron, and indicates a significant partitioning of light impurities between the solid and the liquid. This partitioning has been investigated by calculating the chemical potential of some of these light impurities in solid and liquid iron. At the ICB equilibrium between solid and liquid implies continuity of the chemical potential of both iron and a chosen light impurity: By imposing this continuity, it is possible to extract the equilibrium concentration of the chosen impurity in solid and liquid, and from this work out the density contrast. This strategy has been applied to sulphur, silicon and oxygen [Alfè *et al.* (2000, 2002d, 2002a, 2003)]. The results showed that sulphur and silicon demonstrate very little partitioning between solid and liquid iron, mainly because their size is very similar to the size of the iron atoms under ICB conditions. Conversely, oxygen partitioning is almost complete, with very little of it going into the solid inner core. This is intuitively explained by the fact that oxygen is significantly smaller than iron, and therefore would fit rather loosely in the solid. This waste of space results in an increase of its chemical potential, which tends to push it out into the liquid, where it can be accomodated much more efficiently.

Putting all this information together Alfè *et al.* [2000, 2002a, 2002d] suggested a composition for the inner core near the ICB of about 8% of sulphur/silicon and no oxygen (sulphur and silicon cannot be distinguished at this stage), and an outer core which contains about 8–10% of sulphur/silicon plus 8–13% of oxygen (the exact values depend on the exact value of the density contrast at the ICB, for which new estimates are still being published [Masters and Gubbins (2003)]). This large partitioning of oxygen between the inner and the outer core is responsible to a large extent for the generation of the Earth's magnetic field (see below). The melting temperature of the mixture is reduced by $\sim 800\,\mathrm{K}$ with respect to that of pure iron due to this large partitioning of light impurities, so that the Alfè *et al.* (2002d) best estimate for the temperature of the ICB is $\sim 5500\,\mathrm{K}$.

4. Dynamo Models

The time-variable behaviour of the Earth's magnetic field, discussed above, shows that its source cannot be a permanent magnet; furthermore, the high interior temperatures of the Earth would prevent minerals from retaining any permanent magnetism. Instead, the Earth's magnetic field is maintained by a gigantic dynamo — the outer core. Motion of a fluid conductor in a magnetic field induces an electric current, and consequently a secondary magnetic field. Under the right circumstances, this field can reinforce the original magnetic field and lead to a "self-exciting" dynamo. The complicated motions generated by a rotating, convecting fluid such as the outer core are well-suited to generating a dynamo. Furthermore, these fluid motions allow for both a slow drift, and a complete reversal, of the poles. However, although the basic theory has been understood for at least 80 years, the actual calculations are exceedingly challenging. As we discuss below, considerable progress has been made in modelling the behaviour of the dynamo in the last ten years; excellent summaries may be found in Busse [2000], Kono and Roberts [2002] and Glatzmaier [2002], while a useful summary of the underlying theory is in Hollerbach [1996].

There are four reasons why numerical simulations of the core dynamo are more challenging than simulating the convecting mantle. Firstly, there are more governing equations to deal with. Modelling mantle convection requires solving two coupled differential equations: One describing the change in temperature due to fluid motion and diffusion of heat; and one describing the change in fluid velocity as a result of viscous and buoyancy forces. To model the dynamo, the effect of electromagnetic forces has to be added to the fluid motion, and an additional equation which links the change in magnetic field to fluid motion and magnetic diffusion is also required. Another way of viewing this same problem is mantle convection problem has only one characteristic timescale — that of thermal diffusion — but the core dynamo problem has three, wildly different, ones. These are the rotational period (1 day), the magnetic diffusion timescale ($\sim 10^4$ years), and the viscous diffusion timescale (> 1 Gyr).

Secondly, the spatial resolution required is orders of magnitude higher for dynamo simulations than mantle convection models. The resolution required is set by the thickness of the fluid boundary layer, which for the core is determined mainly by the effective viscosity of the turbulently convecting material. Although this turbulent viscosity is much larger than the molecular viscosity of liquid iron (which is comparable to water), the likely boundary layer thickness is of order 0.1 km [Glatzmaier (2002)]. For comparison, typical boundary layer thicknesses for the mantle are of order 100 km. In practice, the kind of resolution required is not currently attainable, an issue we discuss further below.

Thirdly, the computational timestep is determined by the smaller of two transit times, those for the convecting material, and for propagating magnetic (Alfven) waves. Mantle convection velocities are millimetres to centimetres per year, while those of core convection are centimetres per second, and are similar to the Alfven speed. As a result of the high core convective velocity, rotational effects (Coriolis forces) are important in the core while they can be neglected in the mantle. Because of the short transit times, timesteps in numerical dynamo calculations have to be much shorter than for mantle convection simulations. Typical dynamo simulations rarely last more than about 1 Myr (about 100 magnetic diffusion times), while 2D (though not 3D) mantle simulations can be run for the 4600 Myr age of the Earth [e.g. Nakagawa and Tackley (2004)].

Finally, in mantle convection models, higher spatial resolution (or longer model durations) can be attained by using 2D models, without producing grossly different results to 3D models. This option is not available for geo-dynamo models, because a purely axisymmetric self-sustaining geodynamo is not possible. This result, known as Cowling's theorem [see Hollerbach (1996)], means that 3D models are required to simulate the geodynamo.

The limitations imposed on dynamo models by current computer technology are severe. For instance, even with variable grid sizes, current models would need 10 times higher radial resolution to capture the turbulent boundary layer [Glatzmaier (2002)]. The corresponding increase in computer power required is unlikely to occur for at least a decade. As a result, the parameter space attainable with numerical models is a long way from that occupied by the real Earth. Current models have to either assume a core viscosity which is 10^4 times too large, or a rotation timescale which is 10^4 times too long [see Glatzmaier (2002)]. Perhaps surprisingly, despite these issues, several models have recently started to produce results which resemble the behaviour of the Earth's magnetic field.

As discussed above, the main characteristics of the Earth's magnetic field are its predominantly axial dipolar nature, its slow westwards drift, and its tendency to show excursions and complete reversals on ~ 0.1 Myr timescales. The first two characteristics were relatively easy to obtain in numerical models. Figure 3 shows a comparison between the present-day Earth's magnetic field and a numerical simulation. The agreement is generally very good in terms of field intensity and geometry; models which are run for long enough also tend to generate a mean field aligned with the rotation axis [Kono and Roberts (2002)]. Figure 3 also makes the point that the short-wavelength structure in the core field is not visible at the surface, due to attenuation. Westwards drift has been obtained in many, though not all, models [Kono *et al.* (2000); Glatzmaier *et al.* (1999); Christensen and Olson (2003)].

The last characteristic, polarity reversal, was much harder to obtain in numerical models, partly because of the very long computational times required, and partly because of worries that the results might be artefacts of the numerical procedures adopted, or initial transients [Ochi *et al.* (1999); Coe *et al.* (2000)]. Nonetheless, the first reversing dipoles were achieved in the mid 1990s [Glatzmaier and Roberts (1995); Kuang and Bloxham (1997)] and can now be produced and studied routinely. Figure 4 compares a typical numerical result with a real record. Figure 4(a) is the numerical result: The bold line is the total field intensity, and the black and white boxes denote episodes of normal and reversed polarities. The total model time elapsed is just over 1 Myr (though an unrealistically slow Earth rotation is assumed). Figure 4(b) shows a time sequence (2.8 Myr duration) of the observed dipolar field intensity and polarity reversals. Although Fig. 4(a) shows more reversals, and appears to have more power at high frequencies, overall the two plots are at least qualitatively similar.

Evidently, now that Earth-like dynamo behaviour is being routinely modelled, it is possible to discriminate between different models using quantitative (statistical) approaches [e.g. Dormy *et al.* (2000); McMillan *et al.*

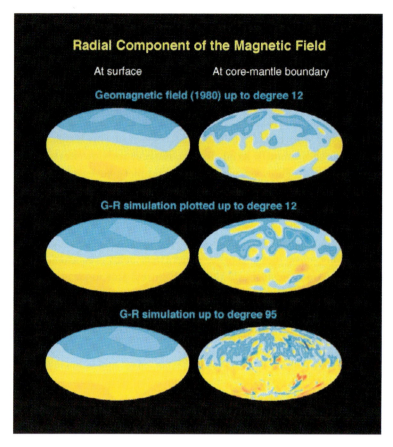

Fig. 3　Comparison of observations and numerical models of the Earth's magnetic field at the surface (left side) and the CMB (right side), from Glatzmaier [2002]. Top panel shows observations for wavelengths >3300 km; middle panel shows numerical results for same wavelength range; bottom panel shows results for wavelengths > 400 km. Note that the short-wavelength signals present at the CMB are not visible at the surface. Reprinted, with permission, from the *Annual Review of Earth and Planetary Sciences*, Volume 30, ©2002 by Annual Reviews, www.annualreviews.org.

(2001); Coe *et al.* (2000)]. In doing so, further constraints on the physical processes governing dynamo behaviour will be obtained.

Figure 5 shows one such discriminant. Figure 5(a) is a histogram of the strength of the dipolar component of the field, showing that it is generally either strongly negative (S) or positive (N), and that neither orientation is preferred. Figures 5(b–d) show the same data for three numerical

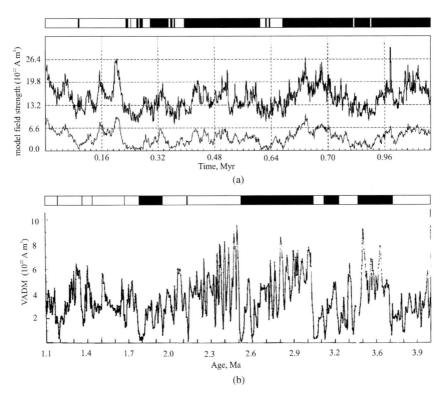

(a)

(b)

Fig. 4   Time-variable behaviour of observed and modelled magnetic field. (a) Model from Kutzner and Christensen [2002]. Bold line is total magnetic field intensity, thin line is field due to dipole component alone. Black and white boxes show episodes of normal and reversed polarity (defined by dipole angles of $<90°$ and $>90°$, respectively). The magnetic diffusion timescale is 160 kyr. Reprinted from *Phys. Earth Planet. Int.* **131**, ©2002, with permission from Elsevier. (b) Observed recent field intensity variations and reversals based on sedimentary cores in the Pacific Ocean, from Valet and Meynadier [1993]. "VADM" stands for virtual axial dipole moment. Reprinted by permission from *Nature* **366**, p. 236, ©1993, Macmillan Publishers Ltd, www.nature.com.

models. Although each has different characteristics, none closely resembles the observations. Other discriminants, such as the frequency content of the magnetic field [Kono and Roberts (2002); Kutzner and Christensen (2002)], and its time variability, may also be applied.

A long-recognised consequence of the dynamo's existence is that the inner core might rotate relative to the outer core due to magnetic torques [e.g. Braginsky (1964)]. Seismological studies, making use of inner core

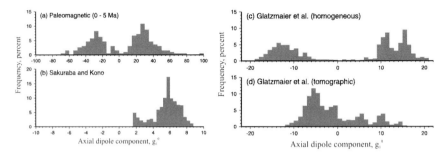

Fig. 5 Comparison of observed frequency distribution of axial dipole component g_1^0 (in μT) with numerical models, from Roberts and Kono [2002]. Reproduced by permission of American Geophysical Union. (a) Observed variation over last 5 Myr. The x-axis gives the strength of the dipole field and whether it is normal (positive) or reversed (negative). (b) Similar plot from numerical model of Kono *et al.* [2000] over 50 kyr. Note that the field never reverses. (c) As for (b) but from Glatzmaier *et al.* [1999] with a homogeneous CMB heat flux and an interval of 0.3 Myr. (d) As for (c) but with a spatially varying core heat flux based on seismic observations.

anisotropy, subsequently appeared to confirm this hypothesis [Song and Richards (1996); Su *et al.* (1996)], though more recent observations have been more equivocal [Song (2003); Souriau and Poupinet (2003); Laske and Masters (1999)]. Similarly, the apparently preferred path taken during polarity reversals can be investigated with numerical models. Models with a core surface heat flux having a minimum in the Pacific showed preferential paths very similar to those inferred [Coe *et al.* (2000); Kutzner and Christensen (2004)], though as noted above, the observations are disputed. The pattern of core heat flux may also influence the frequency of reversals [Glatzmaier *et al.* (1999)]. A particularly important example is the long (> 30 Myr) hiatuses in pole reversals, e.g. during the late Cretaceous and late Carboniferous-middle Permian. This kind of timescale is much longer than characteristic core timescales (O(10^4) yrs) and strongly suggests that the mantle is playing an important role [e.g. Hide (1967); Larson and Olson (1991)], though the details have yet to be worked out. Similarly, the inner core is likely to have an effect on the frequency of magnetic reversals, though there is as yet no agreement on this point [Roberts and Glatzmaier (2001); Gubbins (1999); Sakuraba and Kono (1999)].

As well as comparing them with observations, numerical models can also throw light on aspects of the geodynamo which are not observable at all. For instance, the toroidal component of the magnetic field is not directly observable [though see Jackson (2003)], but may be at least equal in strength to the observable (poloidal) component. Similarly, numerical models may be able to place constraints on how much energy is dissipated by electromagnetic (Ohmic) heating in the core. This heating occurs mainly at short wavelengths, which are not observable at the surface. Understanding the energy requirements of the dynamo is critical to models of how the dynamo has evolved through time, and will be discussed further below.

At this point, it should again be stressed that the model parameters adopted are in some cases a factor of 10^4 different from those applicable to the Earth. The agreement between models and observations is thus somewhat surprising, and suggests that Earth-like dynamos are possible over a large parameter space. What is not yet clear is the extent to which the model results will change as more Earth-like parameters are approached. At this stage, a certain amount of caution needs to be exercised in interpreting numerical model results.

A relatively recent development, which avoids some of the problems inherent in the numerical models, is to simulate aspects of dynamo behaviour using laboratory experiments [Gailitis *et al.* (2002); Muller and Stieglitz (2002)]. While these experiments suffer from their own problems (e.g. the velocity field is generally specified *a priori* by the geometry of the experiment), they sample a different region of parameter space to the numerical models, and one in some ways closer to that of the Earth [Busse (2000)]. A powerful approach is to use laboratory experiments to verify extrapolations made from numerical models [Christensen and Tilgner (2004)]. It is likely that advances in dynamo studies over the next decade will be driven increasingly by laboratory experiments as well as numerical models.

In summary, the last ten years have seen a breakthrough in dynamo studies. Although there remain caveats about the applicability of the parameter space explored, numerical models can now reproduce many aspects of the Earth's dynamo. Discriminating between different models of the basis of observations is likely to further constrain the physics of dynamo generation, for instance, the roles of the mantle and inner core. The numerical models are likely to be increasingly complemented by laboratory studies.

5. The Evolution of the Core and Dynamo

Given the understanding of dynamo generation provided by the models discussed above, it has become possible to investigate the long-term evolution of the dynamo. However, before examining this aspect, we will first discuss the manner in which it is powered at the present day.

5.1. *Present-day heat balance*

Although the distinction is not critical for the dynamo models discussed above, there are two sources of convective motion in the core: Thermal convection, driven by core cooling and latent heat release (as the inner core solidifies); and compositional convection, which arises because the inner core as it freezes expels light elements (Sec. 3). Core solidification thus makes it easier to generate a dynamo, since the solidification provides additional sources of energy.

Whether or not convection occurs depends on the rate at which heat is extracted from the core into the mantle. In the absence of an inner core, convection only occurs if the CMB heat flux somewhere exceeds the adiabatic value, which is the maximum amount which can be transported without convection. It is therefore straightforward to predict whether or not a dynamo operates simply by tracking the CMB heat flux, or equivalently the core cooling rate [e.g. Nimmo and Stevenson (2000)]. However, if an inner core exists, a dynamo might operate even for a sub-adiabatic heat flux, due to the effect of compositional convection. In this situation, it is more convenient to use a criterion based on the rate of change of entropy, rather than energy [Gubbins *et al.* (2003, 2004)]. In this context, the entropy production rate can be thought of as the heat flux divided by a characteristic temperature, giving units of W/K. The entropy production rate also depends on a thermodynamic efficiency factor controlled by the location and temperature of heat sources and sinks. This efficiency factor shows that, for instance, compositional convection is a more efficient way of producing a dynamo than thermal convection. The utility of the entropy approach is that it allows both thermal and compositional effects to be accounted for. An important point is that almost all the entropy production terms are proportional to the rate at which the core is cooling. As a result, more rapid core cooling is more likely to allow a dynamo to operate. The entropy requirement of a minimum core cooling rate, the equivalent of the adiabatic heat flux requirement, arises because the adiabatic entropy term is constant and negative; the positive terms (which arise from core cooling, latent heat release etc.) must outweigh this contribution.

A potentially important driving mechanism for the dynamo is radioactive decay of heat producing elements within the core. The entropy contribution in this case depends not on the core cooling rate, but on the rate of radioactive heat production. The most likely radioactive species to be present in the core is potassium-40, with a half life of 1.3 Gyr. Although early experimental results suggested that the partitioning of ^{40}K into the core was negligible [Chabot and Drake (1999)], more recent results [Gessmann and Wood (2002); Murthy *et al.* (2003); Lee *et al.* (2004)] have found that significant partitioning is in fact likely to occur. The inferred abundance of potassium in the silicate mantle is slightly lower than elements (such as sodium) which have similar condensation temperatures [e.g. Lodders and Fegley (1998)], suggesting that partitioning into the core is acceptable on geochemical grounds. Unfortunately, the uncertainties are large enough to preclude useful geochemical constraints [e.g. Lassiter (2004)]. Nonetheless, it will be argued below that potassium could have played a major role in the history of the geodynamo [Nimmo *et al.* (2004)].

While calculating the rate of entropy production within the core is straightforward, the excess (positive) entropy required to drive the dynamo is unknown. The entropy production required depends on the amount of Ohmic dissipation in the core, which occurs at small (< 100 km) length scales. These length scales are not observable at the surface, because of upwards attenuation. Nor are such length scales readily achieved in simulations, for reasons discussed above. Finally, dissipation depends on both the toroidal and poloidal fields, only the latter of which can be observed. Recent Ohmic dissipation estimates fall in the range 0.2–2 TW [Buffett (2002); Gubbins *et al.* (2003); Roberts *et al.* (2003); Christensen and Tilgner (2004)], equivalent to an excess entropy production rate required to drive the dynamo of 40–400 MW/K. Although there are large uncertainties in these values, the entropy production rate must exceed zero for a dynamo to operate.

In Sec. 3, we argued that the various parameters describing the core's temperature structure and composition are reasonably well known. Given such a set of parameters, it is possible to calculate the various entropy production terms as a function of core cooling rate, or heat flux out of the core. Figure 6 shows how the rate of entropy production varies as a function of the heat flux out of the core, both for a set of core parameters appropriate to the present-day Earth, and for a situation in which the inner core has not yet formed.

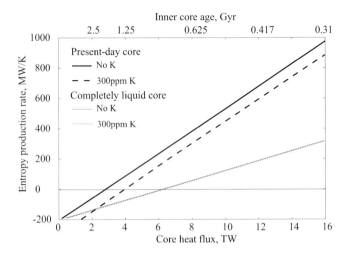

Fig. 6 Variation in entropy production rate with CMB heat flux. Core temperature structure and parameters are from Nimmo *et al.* [2004]. Bold lines are for present-day core with inner core radius 1220 km; thin lines are for completely liquid core. Dashed lines have 3 TW of heating from radioactive decay of potassium-40. The entropy production rate must exceed zero for a dynamo to be possible. The inner core age is inversely proportional to the CMB heat flux.

When an inner core is present, positive contributions to entropy production arise from cooling, latent heat release and gravitational energy; the latter two arise from inner core solidification. All these contributions are proportional to the core cooling rate; radioactive decay within the core is another potential source of entropy production, and is simply proportional to the rate of radioactive heating. The net rate of entropy production, the amount available to drive the dynamo, is obtained by subtracting the (constant) adiabatic contribution. For a present-day, potassium-free core, a CMB heat flux of < 2.5 TW (core cooling rate of ≈ 12 K/Gyr) results in a negative net entropy contribution and, therefore, no dynamo (Fig. 6). A higher core cooling rate generates a higher net entropy production rate; it also means that the inner core must have formed more recently. This is a tradeoff that we return to below. Prior to the formation of an inner core, a significantly larger heat flux (> 6 TW) was required to maintain the dynamo, because neither latent heat nor gravitational energy were then available. Figure 6 also shows that for the same CMB heat flux, less entropy is available to drive a dynamo if radioactive heating is important. This is because such heating has a lower thermodynamic efficiency than that associated with latent heat and gravitational energy release.

Figure 6 shows that the existence of a dynamo at the present day places constraints upon the CMB heat flux, and thus the rate at which the core is cooling. The rate of core cooling is determined by the ability of the mantle to remove heat. Importantly, independent estimates on this cooling rate exist, based on our understanding of mantle behaviour.

One approach to estimating the heat flux across the base of the mantle relies on the conduction of heat across the bottom boundary layer. The temperature at the bottom of this layer (the core) arises from extrapolating the temperature at the ICB outwards along an adiabat, and is about 4100 K (Fig. 1c). The temperature at the top of the layer is obtained from extrapolating the mantle potential temperature inwards along an adiabat, and is about 2700 K (Fig. 1c). The thickness of the bottom boundary layer, based on seismological observations, is 100–200 km. For likely lower mantle thermal conductivities, the resulting heat flux is probably in the range 6–12 TW (Buffett, 2003). Values for the CMB heat flux based on the inferred contribution from rising convective plumes [Davies (1988); Sleep (1990)] are a factor of 2–4 smaller, but are probably underestimates [Labrosse (2002); Bunge (2005)]. Figure 6 shows that a heat flux in the range 6–12 TW results in a net entropy production rate of 200–700 MW/K, likely enough to drive the dynamo.

The inferred value of the CMB heat flux has two implications. Firstly, it is a significant fraction of the heat flux at the Earth's surface, 42 TW [Sclater *et al.* (1980)]. This result may help to explain the long-standing paradox that the mantle is getting rid of heat at about twice the rate at which it is being generated by radioactive decay [Breuer and Spohn (1993)]. Secondly, the inner core age implied by this heat flux (assumed constant) is young, about 0.6 Gyrs for the values used in Fig. 6. In practice, of course, the core heat flux will vary with time; investigating the time evolution of the core and mantle is the subject of the next section.

5.2. *Thermal evolution of the Earth*

As discussed above, there is evidence that a dynamo similar to that at the present day has existed through the bulk of Earth history. It is therefore natural to enquire whether the prolonged life of the dynamo places constraints on the thermal evolution of the Earth. Anticipating the results of the sections below, it has recently become clear that the constraints are quite strong: generating a dynamo requires relatively rapid cooling of the core, while producing an inner core of the correct present-day size requires relatively slow core cooling [Buffett (2002); Gubbins *et al.* (2003)]. The parameter space which allows these two opposing constraints to be satisfied is relatively restricted, and in particular appears to require both a young inner core ($<\approx 1.5\,\text{Gyr}$) and, less certainly, significant ($O(100\,\text{ppm})$) potassium in the core.

As discussed above, powering a dynamo requires the core cooling rate to exceed a given value. The core cooling rate depends on the rate at which the mantle extracts heat from the core. The ability of the mantle to extract heat depends, in turn, on the rate at which the mantle is cooling, and thus the behaviour of the near-surface boundary layer. Plate tectonics on the Earth is an efficient way of cooling the mantle; other planets, in which lateral motion of the surface material does not occur, probably cool much more slowly. This link, between the top 100 km of the Earth's mantle, and the behaviour of the dynamo, is both suprising and of fundamental importance. It also means that the evolution of the Earth as a whole has to be investigated in order to investigate the evolution of the core.

Modelling the thermal evolution of the Earth is a challenging problem. Although 3D numerical mantle convection models can be run, doing so for 4.5 Gyr is not yet possible. Alternatively, parameterised evolution schemes [e.g. Butler and Peltier (2000)] can be adopted, which consider only globally-averaged properties and thus run very much faster, allowing a proper exploration of parameter space. The disadvantage of this approach is that complications, such as compositional layering or vertical viscosity variations, are less easy to include.

Figure 7 shows one such parameterised thermal evolution model, which generates a present-day thermal structure similar to Fig 1(c) while permitting a dynamo throughout Earth history. Figure 7(a) shows the temperature evolution of the core and mantle, and demonstrates the slow cooling regulated by radioactive decay in the mantle. The kink in the central temperature at 3.5 Gyr is due to the inner core starting to solidify. Figure 7(b) shows the evolution of the heat fluxes with time. The model present-day surface heat flux matches the observed value, and the CMB heat flux is 9 TW, in agreement with the arguments presented above. The core heat flux is high early on because of the presence of 400 ppm potassium, the effect of which is discussed below. Figure 7(a) also shows the net entropy production rate as a function of time, which is always positive, indicating a dynamo could have operated over the whole of Earth history. Figure 7(b) shows the inner core

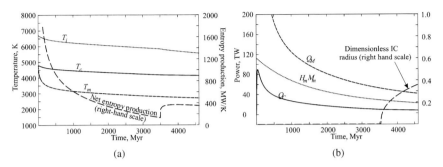

Fig. 7 Parameterized thermal evolution model, modified from Nimmo *et al.* [2004], with 400 ppm potassium in the core. (a) Temperature variation and entropy production rate as a function of time. T_m is the mantle temperature at the CMB, T_c is the core temperature at the CMB, T_i is the temperature at the centre of the planet, or at the ICB if an inner core exists. The kink in T_i at 3500 Myr is due to the onset of inner core solidification. (b) Heat output and inner core size as a function of time. Q_M is the surface heat loss, $H_m M_m$ the mantle contribution from radioactive decay and Q_C the core heat loss. The inner core size is normalised by the core radius.

growth history, demonstrating that it is young (1.1 Gyr) and at the correct present-day size. The entropy production rate increases when the inner core solidifies, due to additional release of latent heat and compositional convection. Prior to inner core formation, the dynamo was maintained by the relatively rapid cooling rate of the core, plus radioactive decay.

The above model produces results compatible with our understanding of present-day Earth structure and geodynamo history. However, it does so mainly because of the presence of 400 ppm potassium in the core. Similar models run without potassium generally result in an inner core which is much too large. This is because the heat released by the potassium reduces the rate at which the core cools and the inner core grows. In the absence of potassium, the core cooling rate has to be significantly reduced to generate an inner core of the correct present-day size. However, a lower core cooling rate and an absence of potassium means a reduction in the rate of entropy production (Fig. 6). There is thus a tradeoff between getting the correct inner core size (which requires slow cooling) and generating enough entropy to drive the dynamo (which requires rapid cooling).

This tradeoff is shown explicitly in Fig. 8, which plots the mean entropy production rate against the present-day inner core size. Except at large inner core sizes, increasing the entropy production rate also results in a larger inner core. Adding potassium to the core shifts the curves to higher

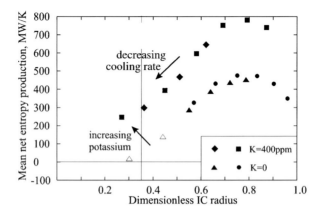

Fig. 8 Tradeoff between time-averaged entropy production and present-day inner core (IC) size, normalised by core radius, from Nimmo *et al.* [2004]. Open symbols have a minimum rate of entropy production < 0, indicating an at least temporary cessation of the dynamo. Different points are for different mantle viscosity structures (and hence core cooling rates). Increased cooling rates lead to higher entropy production and larger IC sizes. Adding potassium (K) allows smaller inner cores for the same entropy production rate. Vertical line denotes present day IC size.

rates of entropy production for a given inner core size, because potassium both delays core solidification and is an additional entropy source. The curves demonstrate that none of the models lacking potassium are able to match both the entropy and the inner core size requirements simultaneously. It also turns out that none of the models with a correct present-day inner core size resulted in a core older than 1.5 Gyr.

The results presented here depend on a large number of parameters, many of which are poorly known. Furthermore, as discussed above, the parameterised calculations are unlikely to capture the full complexity of convection in the Earth. Nonetheless, the two main results — that the inner core is young, and that potassium is likely present in the core — are relatively robust. For instance, appealing to initially hotter core temperatures (rather than potassium) to delay inner core formation fails because the mantle is more efficient at cooling the core at higher temperatures.

Other authors have derived similar results using different techniques. For instance, Buffett [2002] found that obtaining an ancient inner core required a present-day heat flux across the CMB much lower than that inferred (Fig. 1c). To solve this problem he posited a significant amount of radioactive heat production, either within the bottom mantle boundary layer or in the core. Both Roberts *et al.* [2003] and Labrosse *et al.* [2001] examined the evolution of the core for specified CMB heat fluxes, and concluded that an inner core age of 1 ± 0.5 Gyrs was most likely in the absence of any radiogenic heating.

In summary, whether or not a dynamo operates is ultimately controlled by the ability of the mantle to extract heat. There appears to be general agreement that both the present-day thermal structure of the Earth, and the maintenance of a dynamo, are compatible with a present-day CMB heat flux of about 9 ± 3 TW [Buffett (2003); Labrosse and Macouin (2003); Nimmo *et al.* (2004)]. This heat flux implies an inner core age of < 1.5 Gyr. Parameterised thermal evolution models suggest that maintaining a dynamo over Earth history while producing an inner core of the correct size is difficult (Fig. 8), because the dynamo requires rapid core cooling, while the small inner core requires slow core cooling. A possible resolution of this paradox, which is supported by experimental results, is the presence of $O(100)$ ppm potassium in the core, generating 1.5–3 TW of radioactive heating at the present day.

6. Other Silicate Bodies

Although the bulk of this paper has focussed on the Earth, many of the principles discussed can be applied equally well to other silicate bodies. In this section, we discuss some useful generalisations of the principles; describe briefly the data on cores of other planets; and suggest how these data might be interpreted. A useful summary of current understanding is in Stevenson [2003].

Figure 6 showed that sustaining a dynamo depends mainly on the rate of core cooling, and Sec. 5.2 argued that the core cooling rate is ultimately controlled by the rate at which the mantle can extract heat. The Earth is the only silicate body which currently exhibits plate tectonics; the other terrestrial planets do not have mobile plates, and as a result mantle cooling (and thus core cooling) is likely to be less rapid.

In the absence of an inner core, heat must be extracted at a rate exceeding the adiabatic core heat flux. This adiabat depends on gravity, and thus the size of the planet. However, the mantle's ability to extract heat also depends on gravity, though less strongly. This simple analysis suggests that, other things being equal, it is easier to maintain dynamos in small bodies than larger ones at the same temperature. On the other hand, since larger bodies take longer to cool than small ones, and are likely to begin at higher temperatures, a dynamo (if present) is likely to persist for longer in a larger body. Of course, there are numerous additional complications, notably the increase in mantle viscosity (and decreasing mantle heat flux) with pressure, and the possible presence of an inner core. Nonetheless, this analysis suggests that a dynamo in a Moon- or Ganymede-sized body ($g \approx 1 \, \mathrm{m\,s^{-2}}$) should be relatively easy to maintain, while mantle cooling on an Earth-sized body lacking plate tectonics is likely too sluggish to allow a dynamo to operate.

The role of gravity is also important because it controls the relative slopes of the core adiabat and melting curve. For the Earth, the adiabat is shallower than the melting curve (Fig. 1c); but for Mars, the two curves are roughly parallel [Williams and Nimmo (2004)], and for even smaller bodies the adiabat may become steeper. Since the intersection of these curves determines the location of the inner core, it is clear that varying the gravity can have a dramatic effect on inner core behaviour.

The discussion of the Earth's dynamo showed that the role of contaminants in the core is important. In the solar system at large, the most significant contaminant is likely to be sulphur, which is both abundant and has a strong tendency to partition into iron at low pressures. Sulphur has two important effects. Firstly, at low pressures it can dramatically reduce the melting temperature of iron [Fei *et al.* (1997)]. Secondly, a core which is initially rich in sulphur ($> 21 \, \mathrm{wt\%}$) will expel a dense fluid as it solidifies, which is the opposite case to that for the Earth's core. This behaviour is a consequence of the iron-sulphur phase diagram [Fei *et al.* (1997)]. The solidifying material will have a comparable density to that of the initial fluid [Kavner *et al.* (2001); Sanloup *et al.* (2002)].

The effects of gravity and sulphur can be combined into a single diagram to generate four possible scenarios for the core (Fig. 9). The top left panel depicts the situation for Earth, where the adiabat is shallower than the melting curve, and the fluid expelled from the inner core is low density. The top right curve is similar, except here the fluid expelled from the inner core is high density (this situation applies to a sulphur-rich core). In this case, compositional convection will not occur, and the probability of generating a dynamo will be significantly reduced. The bottom left panel has an adiabat steeper than the melting curve, and a light fluid. Here core solidification will start at the outer core boundary. The solid core material is presumed to sink, re-melting as it does so and generating compositional convection; the fluid released during solidification will be stably stratified at the outer core boundary. Finally, the lower right panel shows a similar situation but

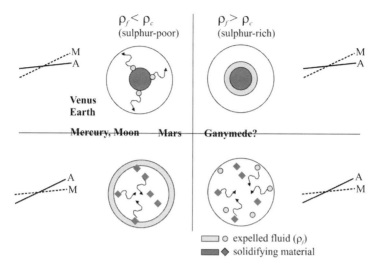

Fig. 9 Different potential geodynamo regimes (see text). Labelled lines denote core melting curve (M) and adiabat (A). Fluid outer core and fluid released during core solidification have densities ρ_c and ρ_f, respectively. The case of $\rho_c < \rho_f$ corresponds to a sulphur-rich scenario. For the sulphur-rich case we are assuming that the solid material is denser than the outer core fluid, which is uncertain.

with a dense fluid. In this case, both fluid and (re-melting) solid will tend to sink, generating vigorous convection.

It is clear that these different scenarios will have very different implications for core and dynamo evolution. Unfortunately, only one scenario has been studied in any detail. Although this scenario is likely appropriate to Earth and Venus, other bodies (especially Ganymede) may lie in quite different regimes.

6.1. *Observations and deductions*

Figure 10 shows a selection of silicate bodies of interest. As explained in Sec. 2, density and moment of inertia data suggest that all possess cores, and tidal observations suggest that many have cores which are at least partially liquid. More interesting are the available magnetic observations. The Earth, Ganymede [Kivelson *et al.* (2002)], Mercury [Connerney and Ness (1988)] and possibly Io [Kivelson *et al.* (2001)] have predominantly dipolar magnetic fields at the present day, which are likely the result of active dynamos. The Moon [Hood *et al.* (2001)] and Mars [Acuna *et al.* (1999)]

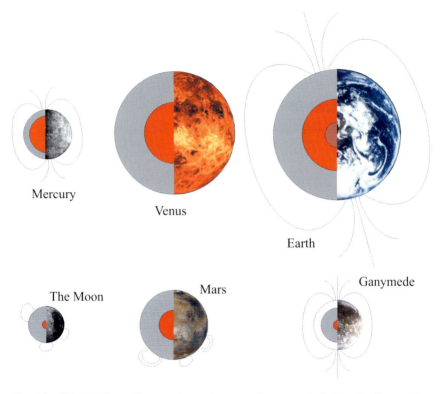

Fig. 10   **Illustration of internal structures and magnetic fields of silicate solar system bodies. Objects are drawn to scale; internal structures are based on information from Table 1. Magnetic fields are schematic, but reflect relative magnitudes and orientations.**

do not have global fields now, but show local magnetic anomalies which are likely the result of an ancient dynamo. In the case of Mars, these crustal anomalies are enormous — an order of magnitude stronger than their terrestrial equivalents. Venus does not possess a global field [Russell (1980)], and the surface temperatures are too high to retain magnetic anomalies.

How might these disparate observations be explained? The case of Venus is relatively straightforward: As suggested by the analysis above, an Earth-size planet which lacks plate tectonics is likely cooling too slowly to allow generation of a dynamo [Nimmo (2002)]. The ancient dynamo on Mars suggests early, rapid cooling; possible explanations for this are an early episode of plate tectonics [Nimmo and Stevenson (2000)], overturn of an initially unstably stratified mantle [Elkins-Tanton *et al.* (2003)], or an initially hot core [Williams and Nimmo (2004)]. With the exception of plate tectonics, similar arguments probably apply to the Moon [e.g. Stegman *et al.* (2003); Collinson (1993)], although an alternative not requiring a dynamo is local magnetisation by impact-generated plasmas [Hood and Huang (1991)].

Mercury is more puzzling. It is not clear that a dynamo is the only way of generating its magnetic field [Schubert *et al.* (1988); Aharonson *et al.* (2004)]. If a dynamo is operating, it is hard to understand how: Mercury has been geologically inactive for 4 Gyr and must be cooling sluggishly at present [Hauck *et al.* (2004)]. Potassium is not an attractive explanation because it is probably too volatile to have been present when Mercury was forming; tidal heating may be an option but depends on very poorly known parameters [Schubert *et al.* (1988)].

The dynamo of Ganymede is equally poorly understood. Firstly, Ganymede may be sulphur-rich, in which case it probably occupies a different regime of parameter space to the other terrestrial planets (Fig. 9). Secondly, the strong background magnetic field of Jupiter may have important effects [Sarson *et al.* (1997)]. Thirdly, Ganymede's thermal evolution was probably drastically influenced by an episode of tidal heating midway through its history [Showman and Malhotra (1997)], with consequences for the dynamo which remain obscure.

It is clear that our understanding of planetary dynamos remains rudimentary. This is in part due to the absence of data, especially time-resolved data, compared to the Earth. But it is also true that comparatively little theoretical effort has been devoted to understanding dynamos which may operate in quite different regimes from the familiar terrestrial one (Fig. 9). The acquisition of new data is likely to be a time-consuming and expensive process; it is to be hoped that swifter progress will be made in the theoretical understanding of planetary dynamos.

7. Conclusions

In the last decade, there have been three main advances in our understanding of the Earth's core and dynamo. Firstly, a combination of improved experimental and numerical techniques have allowed much tighter constraints to be placed on the density and melting behaviour of iron and iron compounds, and thus on the likely properties of the core. Secondly, numerical models generating realistic-looking dynamos have been achieved; these models make it possible to use the time-dependent behaviour of the Earth's magnetic field as a constraint on the behaviour of the core and mantle. Finally, the first two advances now allow the evolution of the core and dynamo to be investigated. Preliminary results suggest that the inner core is a young ($< 1.5\,\mathrm{Gyr}$) feature, and that part of the energy driving the dynamo may be provided by radioactive decay within the core.

The next decade is likely to see a change in emphasis. Despite the success of numerical models up to now, it is likely that future studies of both core properties and dynamo behaviour will be increasingly influenced by experimental results. As the crucial parameters become more tightly constrained, investigating the coupled core-mantle evolution problem will develop as a major area of interest. Observational constraints on the Earth's magnetic field are unlikely to improve significantly, but data on planetary magnetic fields will be dramatically expanded. The MESSENGER spacecraft will comprehensively characterise Mercury; the Dawn spacecraft will investigate the apparently magnetised asteroid Vesta; and sample return from both the Moon and Mars may take place. As the planetary observations improve, significant theoretical effort will need to be devoted to dynamos which may behave in very different ways to the Earth's. Thus, four centuries after the first publication in geomagnetism, this field shows no signs of dissipating.

Acknowledgements

DA acknowledges support from the Royal Society and the Leverhulme Trust; FN the Royal Society and NSF-EAR 0309218. We thank Dave Gubbins and an anonymous reviewer for careful reviews.

References

Acuna, M. H., *et al.* (1999) Global distribution of crustal magnetization discovered by the Mars Global Surveyor MAG/ER experiment. *Science* **284**, 790–793.

Aharonson, O., Zuber, M. T. & Solomon, S. C. (2004) Crustal remanence in an internally magnetized non-uniform shell: A possible source for Mercury's magnetic field? *Earth. Planet. Sci. Lett.* **218**, 261–268.

Alfè, D. (2003) First-principles simulations of direct coexistence of solid and liquid aluminum. *Phys. Rev. B* **68**, 064423.

Alfè, D., Gillan, M. J. & Price, G. D. (1999) The melting curve of iron at the pressures of the Earth's core from *ab initio* calculations. *Nature* **401**, 462–464.

Alfè, D., Gillan, M. J. & Price, G. D. (2000) Constraints on the composition of the Earth's core from *ab initio* calculations. *Nature* **405**, 172–175.

Alfè, D., Price, G. D. & Gillan, M. J. (2001) Thermodynamics of hexagonal-close-packed iron under Earth's core conditions. *Phys. Rev. B* **64**, 045123.

Alfè, D., Gillan, M. J. & Price, G. D. (2002a) Complementary approaches to the *ab initio* calculation of melting properties. *J. Chem. Phys.* **116**, 6170–6177.

Alfè, D., Price, G. D. & Gillan, M. J. (2002b) Iron under Earth's core conditions: Liquid-state thermodynamics and high-pressure melting curve from *ab initio* calculations. *Phys. Rev. B* **65**, 165118.

Alfè, D., Gillan, M. J. & Price, G. D. (2002c) *Ab initio* chemical potentials of solid and liquid solutions and the chemistry of the Earth's core. *J. Chem. Phys.* **116**, 7127–7136.

Alfè, D., Gillan, M. J. & Price, G. D. (2002d) Composition and temperature of the Earth's core constrained by combining *ab initio* calculations and seismic data. *Earth Planet. Sci. Lett.* **195**, 91–98.

Alfè, D., Gillan, M. J. & Price, G. D. (2003) Thermodynamics from first principles: Temperature and composition of the Earth's core. *Min. Mag.* **67**, 113–123.

Anderson, J. D., *et al.* (1996) Gravitational constraints on the internal structure of Ganymede. *Nature* **384**, 541–543.

Bachelet, G. B., Hamann, D. R. & Schluter, M. (1982) Pseudopotentials that work — from H to Pu. *Phys. Rev. B* **26**, 4199–4228.

Barton, C. E. (1989) Geomagnetic secular variation. *The Encyclopedia of Solid Earth Sciences*, D. K. James, (ed.), Van Nostrand Reinhold, New York, pp. 560–577.

Belonoshko, A. B., Ahuja, R. & Johansson, B. (2000) Quasi — *Ab initio* molecular dynamic study of Fe melting. *Phys. Rev. Lett.* **84**, 3638–3641.

Bloxham, J. (2000) Sensitivity of the geomagnetic axial dipole to thermal core-mantle interactions. *Nature* **405**, 63–65.

Boehler, R. (1993) Temperatures in the Earth's core from melting-point measurements of iron at high static pressures. *Nature* **363**, 534–536.

Boehler, R. (2000) High-pressure experiments and the phase diagram of lower mantle and core materials. *Rev. Geophys.* **38**, 221–245.

Braginsky, S. I. (1964) Magnetohydrodynamics of the Earth's core. *Geomag. Aeron.* **4**, 898–916.

Breuer, D. & Spohn, T. (1993) Cooling of the Earth, Urey ratios, and the problem of potassium in the core. *Geophys. Res. Lett.* **20**, 1655–1658.

Brown, J. M. & McQueen, R. G. (1986) Phase-transitions, Gruneisen parameter and elasticity for shocked iron between 77 GPa and 400 GPa. *J. Geophys. Res.* **91**, 7485–7494.

Buffett, B. A. (2002) Estimates of heat flow in the deep mantle based on the power requirements for the geodynamo. *Geophys. Res. Lett.* **29**, 1566.

Buffett, B. A. (2003) The thermal state of Earth's core. *Science* **299**, 1675–1677.

Bunge, H. P. (2005) Low plume excess temperature and high core heat flux inferred from non-adiabatic geotherms in internally-heated mantle circulation models. *Phys. Earth Planet. Int.* **153**, 3–10.

Busse, F. H. (2000) Homogeneous dynamos in planetary cores and in the laboratory. *Ann. Rev. Fluid Mech.* **32**, 383–408.

Butler, S. L. & Peltier, W. R. (2000) On scaling relations in time-dependent mantle convection and the heat transfer constraint on layering. *J. Geophys. Res.* **105**, 3175–3208.

Chabot, N. L. & Drake, M. J. (1999) Potassium solubility in metal: The effects of composition at 15 kbar and 1900 degrees C on partitioning between iron alloys and silicate melts. *Earth Planet. Sci. Lett.* **172**, 323–335.

Christensen, U. R., & Olson, P. (2003) Secular variation in numerical geodynamo models with lateral variations of boundary heat flow, *Phys. Earth Planet. Int.* **138**, 39–54.

Christensen, U. R. & Tilgner, A. (2004) Power requirement of the geodynamo from ohmic losses in numerical and laboratory dynamos. *Nature* **429**, 169–171.

Clement, B. M. (2004) Dependence of the duration of geomagnetic polarity reversals on site latitude. *Nature* **428**, 637–640.

Coe, R. S., Hongre, L. & Glatzmaier, G. A. (2000) An examination of simulated geomagnetic reversals from a palaeomagnetic perspective. *Phil. Trans. R. Soc. London A* **358**, 1141–1170.

Collinson, D. W. (1993) Magnetism of the Moon — a lunar core dynamo or impact magnetization. *Surv. Geophys.* **14**, 89–118.

Connerney, J. E. P. & Ness, N. F. (1988) Mercury's magnetic field and interior. *Mercury*, F. Vilas *et al.* (eds.), pp. 494–513, Univ. Ariz. Press, Tucson.

Davies, G. F. (1988) Ocean bathymetry and mantle convection 1. Large-scale flow and hotspots. *J. Geophys. Res.* **93**, 10467–10480.

Dormy, E., Valet, J. & Courtillot, V. (2000) Numerical models of the geodynamo and observational constraints. *Geochem. Geophys. Geosyst.* **1**, doi:10.1029/2000GC000062.

Dreizler, R. M. & Gross, E. K. U. (1990) *Density Functional Theory*, Springer-Verlag.

Dunlop, D. J. & Yu, Y. (2004) Intensity and polarity of the geomagnetic field during Precambrian time. *Geophys. Monogr.* **145**, J. E. T. Channell *et al.* (eds.), pp. 85–100, Amer. Geophys. Union.

Dziewonksi, A. M. & Anderson, D. L. (1981) Preliminary reference Earth model. *Phys. Earth Planet. Int.* **25**, 297–356.

Elkins-Tanton, L. T., Parmentier, E. M. & Hess, P. C. (2003) Magma ocean fractional crystallization and cumulate overturn in terrestrial planets: Implications for Mars. *Meteorit. Planet. Sci.* **38**, 1753–1771.

Fei, Y. W., Bertka, C. M. & Finger, L. W. (1997) High-pressure iron sulfur compound, Fe3S2, and melting relations in the Fe-FeS system. *Science* **275**, 1621–1623.

Frenkel, D. & Smit, B. (1996) *Understanding Molecular Simulation,* (Academic Press).

Gailitis, A. *et al.* (2002) Laboratory experiments on hydromagnetic dynamos. *Rev. Modern Phys.* **74**, 973–990.

Gessman, C. K. & Wood, B. J. (2002) Potassium in the Earth's core. *Earth Planet. Sci. Lett.* **200**, 63–78.

Glatzmaier, G. A. & Roberts, P. H. (1995) A 3-dimensional self-consistent computer simulation of a geomagnetic field reversal. *Nature* **377**, 203–209.

Glatzmaier, G. A. *et al.* (1999) The role of the Earth's mantle in controlling the frequency of geomagnetic reversals. *Nature* **401**, 885–890.

Glatzmaier, G. A. (2002) Geodynamo simulations — How realistic are they? *Ann. Rev. Earth Planet. Sci.* **30**, 237–257.

Gubbins, D. (1999) The distinction between geomagnetic excursions and reversals. *Geophys. J. Int.* **137**, F1–F3.

Gubbins, D. *et al.* (2003) Can the Earth's dynamo run on heat alone? *Geophys. J. Int.* **155**, 609–622.

Gubbins, D. *et al.* (2004) Gross thermodynamics of two-component core convection. *Geophys. J. Int.* **157**, 1407–1414.

Hauck, S. A. *et al.* (2004) Internal and tectonic evolution of Mercury. *Earth Planet. Sci. Lett.* **222**, 713–728.

Hide, R. (1967) Motions of the Earth's core and mantle, and variations of the main geomagnetic field. *Science* **157**, 55–56.

Hohenberg, P. & Kohn, W. (1964) Inhomogeneous electron gas. *Phys. Rev.* **136**, B864–B871.

Hollerbach, R. (1996) On the theory of the geodynamo. *Phys. Earth Planet. Int.* **98**, 163–185.

Hood, L. L. *et al.* (2001) Initial mapping and interpretation of lunar crustal magnetic anomalies using Lunar Prospector magnetometer data. *J. Geophys. Res.* **106**, 27825–27839.

Hood, L. L. & Huang, Z. (1991) Formation of magnetic anomalies antipodal to lunar impact basins — 2-dimensional model calculations. *J. Geophys. Res.* **96**, 9837–9846.

Jacobs, J. A. (1998) Variations in the intensity of the Earth's magnetic field. *Surv. Geophys.* **19**, 139–187.

Jackson, A. (2003) Intense equatorial flux spots on the surface of the Earth's core. *Nature* **424**, 760–763.

Jeffreys, H. (1929) *The Earth*, Cambridge University Press.

Kavner, A., Duffy, T. S. & Shen, G. Y. (2001) Phase stability and density of FeS at high pressures and temperatures: implications for the interior structure of Mars. *Earth Planet. Sci. Lett.* **185**, 25–33.

Khan, A. *et al.* (2004) Does the Moon possess a molten core? Probing the deep lunar interior using results from LLR and Lunar Prospector. *J. Geophys. Res.* **109**, E09007.

Kivelson, M. G. *et al.* (2001) Magnetized or unmagnetized: Ambiguity persists following Galileo's encounters with Io in 1999 and 2000. *J. Geophys. Res.* **106**, 26121–26135.

Kivelson, M. G., Khurana, K. K. & Volwerk, M. (2002) The permanent and inductive magnetic moments of Ganymede. *Icarus* **157**, 507–522.

Kohn, W. & Sham, L. J. (1965) Self-consistent equations including exchange and correlation effects. *Phys. Rev.* **140**, A1133–A1138.

Kono, M., Sakuraba, A. & Ishida, M. (2000) Dynamo simulation and palaeosecular variation models. *Phis. Trans. R. Soc. London A* **358**, 1123–1139.

Kono, M. & Roberts, P. H. (2002) Recent geodynamo simulations and observations of the geomagnetic field. *Rev. Geophys.* **40**, 1013.

Konopliv, A. S. & Yoder, C. F. (1996) Venusian k(2) tidal Love number from Magellan and PVO tracking data, *Geophys. Res. Lett.* **23**, 1857–1860.

Kuang, W. L. & Bloxham, J. (1997) An Earth-like numerical dynamo model. *Nature* **389**, 371–374.

Kutzner, C. & Christensen, U. R. (2002) From stable dipolar towards reversing numerical dynamos. *Phys. Earth Planet. Int.* **131**, 29–45.

Kutzner, C. & Christensen, U. R. (2004) Simulated geomagnetic reversals and preferred virtual geomagnetic pole paths. *Geophys. J. Int.* **157**, 1105–1118.

Labrosse, S., Poirier, J. P. & Le Mouel, J. L. (2001) The age of the inner core. *Earth Planet. Sci. Lett.* **190**, 111–123.

Labrosse, S. (2002) Hotspots, mantle plumes and core heat loss. *Earth Planet. Sci. Lett.* **199**, 147–156.

Labrosse, S. & Macouin, M. (2003) The inner core and the geodynamo. *Comptes Rendus Geosci.* **335**, 37–50.

Laio, A. *et al.* (2000) Physics of iron at Earth's core conditions. *Science* **287**, 1027–1030.

Laj, C. *et al.* (1991) Geomagnetic reversal paths. *Nature* **351**, 447.

Larson, R. L. & Olson, P. (1991) Mantle plumes control magnetic reversal frequency. *Earth Planet. Sci. Lett.* **107**, 437–447.

Laske, G. & Masters, G. (1999) Limits on differential rotation of the inner core from an analysis of the Earth's free oscillations. *Nature* **402**, 66–69.

Layer, P. W., Kroner, A. & McWilliams, M. (1996) An Archean geomagnetic reversal in the Kaap Valley pluton, South Africa. *Science* **273**, 943–946.

Lassiter, J. C. (2004) Role of recycled oceanic crust in the potassium and argon budget of the Earth: Toward a resolution of the "missing argon" problem. *Geochem. Geophys. Geosyst.* **5**, Q11012.

Lee, K. K. M., Steinle-Neumann, G. & Jeanloz, R. (2004) *Ab-initio* high-pressure alloying of iron and potassium: Implications for the Earth's core. *Geophys. Res. Lett.* **31**, L11603.

Lodders, K. & Fegley, B. (1998) *The Planetary Scientist's Companion*, Oxford University Press.

Love, J. J. (2000) Statistical assessment of preferred transitional VGP longitudes based on palaeomagnetic lava data. *Geophys. J. Int.* **140**, 211–221.

Ma, Y. Z. *et al.* (2004) *In situ* X-ray diffraction studies of iron to Earth-core conditions. *Phys. Earth Planet. Int.* **143**, 455–467.

Masters, G. & Gubbins, D. (2003) On the resolution of density within the Earth. *Phys. Earth Planet. Int.* **140**, 159–167.

Masters, T. G. & Shearer, P. M. (1990) Summary of seismological constraints on the structure of the Earth's core. *J. Geophys. Res.* **95**, 21691–21695.

McElhinny, M. W. & Senanayake, W. E. (1980) Paleomagnetic evidence for the existence of the geomagnetic field 3.5 Ga ago. *J. Geophys. Res.* **85**, 3523–3528.

McKenzie, D. & Bickle, M. J. (1988) The volume and composition of melt generated by extension of the lithosphere. *J. Petrol.* **29**, 625–679.

McMillan, D. G. (2001) A statistical analysis of magnetic fields from some geodynamo simulations. *Geochem. Geophys. Geosyst.* **2**, doi:10.1029/2000 GC000130.

Margot, J. *et al.* (2004) Earth-based measurements of planetary rotational states. *Eos Trans.* AGU **85**(47), abs. G33A-02.

Montelli, R. *et al.* (2004) Finite-frequency tomography reveals a variety of plumes in the mantle. *Science* **303**, 338–343.

Muller, U. & Stieglitz, R. (2002) The Karlsruhe dynamo experiment. *Nonlinear Process. Geophys.* **9**, 165–170.

Murakami, H. *et al.* (2004) Post-perovskite phase transition in MgSi03. *Science* **304**, 855–858.

Murray, C. D. & Dermott, S. F. (1999) *Solar System Dynamics*, Cambridge University Press.

Murthy, V. M., van Westrenen, W. & Fei, Y. W. (2003) Experimental evidence that potassium is a substantial radioactive heat source in planetary cores. *Nature* **323**, 163–165.

Nakagawa, T. & Tackley, P. J. (2004) Effects of thermo-chemical mantle convection on the thermal evolution of the Earth's core Earth planet. *Sci. Lett.* **220**, 107–119.

Nguyen, J. H. & Holmes, N. C. (2004) Melting of iron at the physical conditions of the Earth's core. *Nature* **427**, 339–342.

Nimmo, F. (2002) Why does Venus lack a magnetic field? *Geology* **30**, 987–990.

Nimmo, F. *et al.* (2004) The influence of potassium on core and geodynamo evolution. *Geophys. J. Int.* **156**, 363–376.

Nimmo, F. & Stevenson, D. (2000) Influence of early plate tectonics on the thermal evolution and magnetic field of Mars. *J. Geophys. Res.* **105**, 11969–11979.

Ochi, M. M., Kageyama, A. & Sato, T. (1999) Dipole and octapole field reversals in a rotating spherical shell: Magnetohydrodynamic dynamo simulation. *Phys. Plasmas* **6**, 777–787.

Oganov, A. R. & Ono, S. (2004) Theoretical and experimental evidence for a post-perovskite phase of $MgSiO_3$ in Earth's D' layer. *Nature* **430**, 445–448.

Parr, R. G. and Yang, W. (1989) *Density-Functional Theory of Atoms and Molecules*, Oxford Science Publications.

Perdew, J. P., Burke, K. & Ernzerhof, M. (1996) Generalized gradient approximation made simple. *Phys. Rev. Lett.* **77**, 3865–3868.

Prevot, M. & Camps, P. (1993) Absence of preferred longitude sectors for poles from volcanic records of geomagnetic reversals. *Nature* **366**, 53–57.

Roberts, P. H. & Glatzmaier, G. A. (2001) The geodynamo, past, present and future. *Geophys. Astrophys. Fluid Dyn.* **94**, 47–84.

Roberts, P. H., Jones, C. A. & Calderwood, A. (2003) Energy fluxes and Ohmic dissipation in the Earth's core. *Earth's Core and Lower Mantle*, C. A. Jones *et al.* (eds.), Taylor & Francis.

Russell, C. T. (1980) Planetary magnetism. *Rev. Geophys.* **18**, 77–106.

Sanloup, C. *et al.* (2002) Physical properties of liquid Fe alloys at high pressure and their bearings on the nature of metallic planetary cores. *J. Geophys. Res.* **107**, 2272.

Sarson, G. R. *et al.* (1997) Magnetoconvection dynamos and the magnetic fields of Io and Ganymede. *Science* **276**, 1106–1108.

Sakuraba, A. & Kono, M. (1999) Effect of the inner core on the numerical solution of the magnetohydrodynamic dynamo. *Phys. Earth Planet. Int.* **111**, 105–121.

Schubert, G. *et al.* (1988) Mercury's thermal history and the generation of its magnetic field. *Mercury*, pp. 429–460, F. Vilas *et al.*, (eds.) Univ. Ariz. Press.

Sclater, J. G., Jaupart, C. & Galson, D. (1980) The heat flow through oceanic and continental crust and the heat loss of the Earth. *Rev. Geophys. Space Phys.* **18**, 269.

Shearer, P. & Masters, G. (1990) The density and shear velocity contrast at the inner core boundary. *Geophys. J. Int.* **102**, 491–498.

Shen, G. Y. *et al.* (1998) Melting and crystal structure of iron at high pressures and temperatures. *Geophys. Res. Lett.* **25**, 373–376.

Showman, A. P. & Malhotra, R. (1997) Tidal evolution into the laplace resonance and the resurfacing of Ganymede. *Icarus* **127**, 93–111.

Sleep, N. H. (1990) Hotspots and mantle plumes — some phenomenology. *J. Geophys. Res.* **95**, 6715–6736.

Song, X. D. & Richards, P. G. (1996) Seismological evidence for differential rotation of the Earth's inner core. *Nature* **382**, 221–224.

Song, X. (2003) Three-dimensional structure and differential rotation of the inner core. *Earth's Core, Geodynamics Series, Amer. Geophys. Union* **31**, 45–64.

Souriau, A. & Poupinet, G. (2003) Inner core rotation: a critical appraisal. *Earth's Core, Geodynamics Series, Amer. Geophys. Union* **31**, 65–82.

Stegman, D. R. *et al.* (2003) An early lunar core dynamo driven by thermochemical mantle convection. *Nature* **421**, 143–146.

Stein, S. & Wysession, M. (2000) *Seismology*, Blackwell Publishers.

Stevenson, D. J. (1990) Fluid dynamics of core formation. In Origin of the Earth, pp. 231–250, H. E. Newsom and J. E. Jones (eds.), Oxford University Press.

Stevenson, D. J. (2003) Planetary magnetic fields. *Earth Planet. Sci. Lett.* **208**, 1–11.

Su, W. J., Dziewonski, A. M. & Jeanloz, R. (1996) Planet within a planet: Rotation of the inner core of earth. *Science* **274**, 1883–1887.

Tonks, W. B. & Melosh, H. J. (1993) Magma ocean formation due to giant impacts. *J. Geophys. Res.* **98**, 5319–5333.

Tsuchiya, T. *et al.* (2004) Phase transition in $MgSiO_3$ perovskite in the earth's lower mantle. *Earth Planet. Sci. Lett.* **224**, 241–248.

Valet, J. P. & Meynadier, L. (1993) Geomagnetic field intensity and reversals during the past 4 million years. *Nature* **366**, 234–238.

Valet, J. P. (2003) Time variations in geomagnetic intensity. *Rev. Geophys.* **41**, 1004.

Vočadlo, L. & Alfè, D. (2002) *Ab initio* melting curve of the fcc phase of aluminum. *Phys. Rev. B* **65**, 214105.

Wang, Y. & Perdew, J. P. (1991) Correlation hole of the spin-polarized electron gas, with exact small-wave vector and high-density scaling. *Phys. Rev. B* **44**, 13298–13307.

Williams, J. G. *et al.* (2001) Lunar rotational dissipation in solid body and molten core. *J. Geophys. Res.* **106**, 27933–27968.

Williams, J. P. & Nimmo, F. (2004) Thermal evolution of the Martian core: Implications for an early dynamo. *Geology* **32**, 97–100.

Williams, Q. *et al.* (1987) The melting curve of iron to 250 GPa — a constraint on the temperature at Earth's center. *Science* **236**, 181–182.

Williams, Q. & Garnero, E. J. (1996) Seismic evidence for partial melt at the base of the Earth's mantle. *Science* **273**, 1528–1530.

Yoder, C. F. *et al.* (2003) Fluid core size of Mars from detection of the solar tide. *Science* **300**, 299–303.

Yoo, C. S. *et al.* (1993) Shock temperatures and melting of iron at Earth core conditions. *Phys. Rev. Lett.* **70**, 3931–3934.

SECTION 3

APPLIED EARTH SCIENCE

Giant Catastrophic Landslides

Christopher R. J. Kilburn

Benfield UCL Hazard Research Centre
Department of Earth Sciences
University College London
Gower Street, London WC1E 6BT, UK

When whole mountainsides collapse, they feed giant landslides that travel kilometres within minutes. Their size and speed prevent effective hazard mitigation after collapse. Risk reduction therefore depends on advance warning of collapse, as well as assessment of how far such a landslide might travel. Early studies invoked special mechanisms to explain catastrophic collapse and runout. It is now apparent, however, that their core behaviour can be explained in terms of common physical processes, from accelerating crack growth before failure to pressurised granular flow during transport. Nevertheless, as mountainous districts become more populated, new data are required to enhance current methods of evaluating the threat from giant catastrophic collapse.

1. Catastrophe in the Mountains

On Wednesday, 9th October 1963, Longarone awoke to brilliant sunshine along the Piave Valley in Italy's northeastern Alps [Merlin (1997)]. By the end of the day, the town ceased to exist. At 10:42 in the evening, a wall of water 70 m high crashed through the settlement, sweeping away almost every building and extinguishing more than 2000 lives [Merlin (1997); Müller (1964); Hendron and Patton (1985)].

The cause of the tragedy was a collapsing mountainside, which failed behind the new Vajont (also spelt "Vaiont") dam two kilometres away. Designed to impound some 150 million m^3 of water, the dam was at the vanguard of civil engineering and rose 262 m above the floor of the Vajont river, which ran through a narrow gorge into the Piave (Fig. 1).

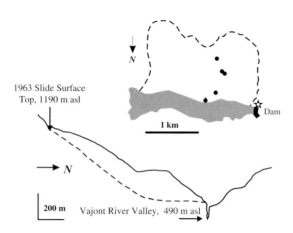

Fig. 1   Cross-section through Mt. Toc before its collapse on 9 October 1963. Broken line shows failure plane. *Inset.* Map of the M-shaped scar produced by the collapse. Locations of monitoring deformation stations (*filled circles*) and seismic station (*star*) are also shown. Modified from Kilburn and Petley [2003].

Filling of the reservoir had begun in February 1960. Within eight months, an M-shaped crack, almost 2 km long and 500–600 m above the valley floor, had opened across Mt. Toc, which formed the southern flank of the reservoir next to the dam. During the following three years, the crack continued to widen at rates of millimetres a day. The rate increased with the depth of water in the reservoir, raising the belief that movement of Mt. Toc could be controlled by regulating the water level. This expectation, together with economic and political pressures, encouraged the drive to test the reservoir at full capacity by the Autumn of 1963.

Once again, the rate of slope movement increased as the water level rose, reaching centimetres a day during early September. In a bid to slow the movement, drainage of the reservoir was started at the end of the month. On this occasion, however, the acceleration continued, exceeding 20 cm a day by October 8th [Hendron and Patton (1985)]. Finally, at 22:39 on October 9th, Mt. Toc collapsed. Within 45 seconds, 270 million m³ of rock had crashed into the reservoir. After sweeping 245 m up the north flank of the reservoir, a wave of water overtopped the dam by more than 100 m. Minutes later, it had overwhelmed Longarone and its neighbours, Pirago, Villanova, Rivalta and Fae, erasing them all from the face of the Earth.

2. The Threat from Sturzstroms

Although initiated by human activity, the collapse of Mt. Toc was not an unusual geological event. Giant volumes of rock fail catastrophically several times a decade, mostly in young mountain ranges and at volcanoes. The resulting landslides have normally acquired minimum velocities of 100–$200\,\mathrm{km\,h^{-1}}$ after collapse; unless trapped by topography, as happened at Mt. Toc, they have the potential to travel large distances and to wipe out entire communities [Voight (1978)]. Such behaviour emerges when the collapse volume exceeds between 1 and 10 million cubic metres [Hsü (1975); Melosh (1987)]. Subaerial landslides have maximum recorded volumes approaching $30\,\mathrm{km^3}$; submarine landslides may be more than 10 times larger. The mechanical energy released is commonly between 10^{14} and $10^{17}\,\mathrm{J}$ on land and up to at least $10^{19}\,\mathrm{J}$ beneath the sea; for comparison, earthquakes of Richter Magnitude 8–9 release some 10^{17}–$10^{18}\,\mathrm{J}$. The implied rates of energy release place giant, catastrophic landslides among the most powerful natural hazards on Earth.

The rapid final acceleration to collapse and large subsequent velocities have been cited as evidence that sturzstroms are produced under unusual conditions. As a result, previous studies of sturzstrom emplacement have focussed on exotic mechanisms for increasing their mobility with respect to the properties of smaller landslides. Recently, however, it has become clear that exotic mechanisms, although they may operate on occasion, are not essential for explaining catastrophic collapse and sturzstrom runout. Rather, the behaviour of such landslides can be explained in terms of well-known physical processes, so offering the prospect of developing reliable models for forecasting collapse and the area vulnerable to destruction.

3. Characteristics of Sturzstrom Deposits

Few direct observations exist of sturzstrom emplacement: Most eyewitnesses have been victims or were fleeing for their lives at the time. Reccurring features are the suddenness of slope failure and the high speed of the landslide. Additional features from emplaced deposits include:

(1) Deposits tend to appear as wide sheets, with peripheries that are often lobate or divided into tongues (Fig. 2); the surface area of a deposit is normally greater than that of the collapse scar, indicating that the landslide has spread during travel [Hsü (1975)].

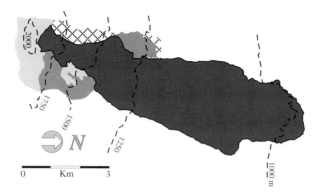

Fig. 2   The subaerial Blackhawk landslide (*black*), 150 km from Los Angeles,
California. About 300 million m^3 collapsed 18 000 years ago to form a lobate
tongue 10–30 m thick, 2 km wide and 7 km long that spread over the desert
in the Lucerne Valley from mountains of gneiss (*dark grey*), sandstone and
limestone (*light grey*) and breccia and conglomerate (*cross-hatched*). Modified
from Shreve [1968].

(2) Deposits usually preserve their pre-failure stratigraphy, such that a pre-
failure sequence of, for example, gneiss over limestone would yield a
deposit of broken gneiss on top of disrupted limestone; evidently, mixing is rare between different levels of a sturzstrom [Erismann (1979);
Erismann and Abele (2001)].
(3) The surfaces of deposits are fragmented, with debris ranging from fine
grains to blocks the size of a house (Fig. 3) [Voight (1978)].

Together, these characteristics suggest that (1) sturzstroms do not
behave as rigid blocks, but can effectively flow during emplacement;
(2) deformation is concentrated within restricted horizons, against which
adjacent layers can move with only a small amount of internal deformation, so inhibiting mixing between layers; and (3) fragmentation may exert
an important control on dynamic behaviour. Although these key features
have long been recognised, uncertainty has continued on the factors that
favour such circumstances [see Melosh (1987); Kilburn (2001)]. Much of this
uncertainty has been encouraged by the view that sturzstroms are exotic
landslides and so require exotic mechanisms to explain their behaviour.
An alternative view, which will be explored here, is that sturzstroms are a
natural result of granular fluids being deformed along restricted horizons
[Campbell (1989); Iverson (1997); Kilburn and Sørensen (1998); Kilburn
(2001)].

Fig. 3   The Köfels deposit exposed (*left*) along the River Ötztaler Ache and (*right*) near its toe. (*Photos*: C.R.J. Kilburn.)

4. The Evolution of Sturzstroms

By focussing on final deposits, the list of sturzstrom characteristics reflects the state of a landslide as it comes to rest. Occasionally, a deposit may also yield clues as to how a sturzstrom evolves. An excellent example is the sturzstrom deposit at Köfels, in the Austrian Tyrol [Erismann and Abele (2001); Sørensen and Bauer (2003)] (Fig. 3). The 2.5 km³ landslide is the largest in the crystalline Alps and was emplaced about 8700 years ago (estimated from radiocarbon dates of deformed wood in landslide sediment [Sørensen and Bauer (2003)]). The collapse removed the peak of the Fundus crest, at about 2400 m above sea level, and carried material across the Ötz valley, some 1300 m below, before climbing 500 m up the facing valley wall into the Horlach hanging valley. Covering 13 km², the sturzstrom dammed the River Ötztaler Ache and reached just over 5 km from the Fundus crest.

The river has since cut a new path through the middle of the landslide, where the deposit was originally some 300–400 m thick. It is now flanked in some sections by cliffs of augengneiss, which from a distance appear to be intact (Fig. 3) but which, when viewed close-up, are seen to be criss-crossed by subvertical and subhorizontal hairline fractures, centimetres to decimetres apart. Remarkably, adjacent pieces have not moved with respect to each other. Except for the fracturing, therefore, the main body of at least the proximal half of the deposit was emplaced with minimal internal deformation. In contrast, the toe of the landslide, which is tens of metres thick, is a fragmental deposit with debris ranging from grains to blocks up to tens of metres across (Fig. 3).

The key point here is that most of the Köfels sturzstrom did not break into loose debris during collapse and runout. Complete disruption occurred only towards the distal end of the landslide. It is possible that, should runout have continued further, the whole landslide might have become disrupted to yield a completely fragmented final deposit. The corollary is that even for deposits which are giant masses of debris, the parent sturzstrom need not have spent most of its emplacement in such a condition. Wholesale disruption, therefore, cannot be an essential feature of sturzstrom behaviour. Hence, if the state of the main body is not crucial, then conditions at its base must exert the primary control on how far a sturzstrom can travel.

Unfortunately, basal layers are rarely exposed in sturzstrom deposits. The few published descriptions show a wide range of features, including:

(1) Mud that was under sufficient pressure to inject itself upwards between gaps in the overlying landslide [McGuire *et al.* (2002a)].
(2) Mud that was entrained by the sturzstrom as it eroded the ground [Dutton (2004)].
(3) Carbonate horizons that have been calcined, with the original calcium carbonate having been broken down into calcium oxide and carbon dioxide, of which the latter has escaped [Hewitt (1988)].
(4) Rare glassy layers that indicate rock melting [Erismann and Abele (2001)].

Although very different, the descriptions suggest that basal layers can be expected to consist of fragmented rock, which is probably fine grained and at elevated temperature (sufficient in the extreme to disassociate carbonate rock or to induce melting), mixed together with a pressurised fluid (e.g. water, steam or CO_2). In other words, the basal layers of sturzstroms are expected to be pressurised granular fluids. Such requirements provide useful constraints on dynamical models for explaining how far a sturzstrom can travel.

5. The Importance of Sturzstrom Volume

Sturzstroms tend to travel greater distances as their volume increases. The nature of the volumetric control reflects the dynamics of sturzstrom emplacement; it is also important for evaluating which districts are at risk,

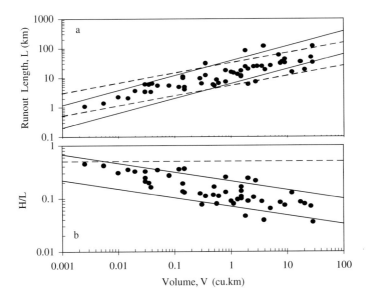

Fig. 4 Variation of subaerial sturzstrom runout length (L) and energy line (H/L) with deposit volume (V). In *a*, the solid lines show an L-$V^{1/2}$ relation and the broken lines an L-$V^{1/3}$ trend; both trends are consistent with observation. In *b*, the solid lines show and H/L-$V^{-1/6}$ relation and the broken line an H/L trend independent of volume (in this case, $H/L = 0.5$). In *a* and *b*, the solid lines correspond to viscous deformation in a basal layer, whereas the broken lines correspond to sliding or to collisional or turbulent flow in the basal layer.

since the volume of an unstable slope may be known before collapse. Arguments have been made that runout length (L) increases with either the cube root [Davies (1982); Dade and Huppert (1998)] or with the square root [Kilburn and Sørensen (1998); Kilburn (2001)] of volume (V). As shown in Fig. 4, field data are sufficiently scattered to accommodate either relation. Fortunately, a second relation exists between sturzstrom volume and runout that can help not only to identify the preferred L-V trend, but also to indicate a landslide's preferred mode of resistance to motion.

The second volume relation concerns a sturzstrom's energy gradient, or energy line, which measures how rapidly a landslide loses energy with distance from its source [Hsü (1975); Malin and Sheridan (1982)]. For simplicity in field measurement, the energy line is defined as H/L, the ratio of vertical to horizontal distance between the top of the landslide scar and the toe of the deposit (Fig. 5). Thus defined, the energy line has become

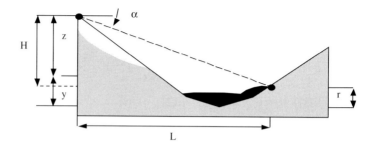

Fig. 5 Definition sketch for dimensions used in describing sturzstrom deposits. Analyses that use H/L as a measure friction should properly use the corresponding distances between the landslide's centre of mass before and after collapse. However, the adjusted values do not significantly change the logarithmic H/L-V trend in Fig. 4.

important for practical and theoretical reasons. First, H/L is the tangent of the angle α linking the endpoints. When drawn on a topographic cross-section, the point where the energy line intersects the ground marks the runout distance of the sturzstrom, independent of the intervening topography [Hsü (1975); Malin and Sheridan (1982)] (Fig. 5); such a simple procedure is obviously helpful for hazard evaluation, especially when under stress during an emergency, provided that H/L is known before collapse.

Second, if it is assumed, as has commonly been the case, that a sturzstrom behaves like a rigid sliding block, it is easy to show that, if the base of the block maintains a constant frictional resistance during collapse and runout, H/L is equivalent to the coefficient of sliding friction, μ, between the base of the landslide and the ground [Hayashi and Self (1992)]. Experiments indicate that μ is about 0.5–0.6 for rock and some clay [Hoek and Bray (1981)]. Sturzstroms, however, are characterised by H/L from about 0.5 to about 0.03, the value tending to become smaller as volume increases (Fig. 4).

The friction coefficient describes the fraction of total surface area that is in contact between adjacent surfaces; it is less than 1 between uneven surfaces, because gaps are maintained where opposing irregularities are unable to lock together. By equating H/L with μ, therefore, early analyses sought to explain the small values of H/L for sturzstroms by invoking mechanisms that reduce a landslide's contact with the ground. Initial explanations focussed on the support of a pressurised fluid, including trapped air, vapour produced by frictional heating of landslide pore water, and gases released by the thermal disassociation of parent rock (these mechanisms are discussed in Melosh (1987) and Kilburn (2001) and references therein). A popular alternative was acoustic fluidisation [Melosh (1987)], according to which the irregular motion of a sturzstrom, treated as a fragmented body, generates pressure (acoustic) waves that allow adjacent surfaces to become temporarily separated. Further models have sought to reduce H/L by appealing to basal melting [Erismann (1979); Erismann and Abele (2001)] or to an additional source of mechanical energy, such as stored strain energy, that drives sturzstroms further than expected by conventional models, even without changing μ [Davies and McSaveney (2004)].

Although each explanation has its particular attraction, none has accounted for the volume-related trends in Fig. 4. A major obstacle has been the often implicit assumption that sturzstroms should slide as rigid blocks. This obstacle disappears if, instead of rigid sliding, sturzstroms are considered to runout over basal layers that behave as granular fluids (Sec. 4); indeed, such an approach can retain the importance of a pressurised basal fluid while accounting for the trends in Fig. 4.

6. Resistance to Motion in Basal Granular Fluids

The trends in Fig. 5 suggest that sturzstroms may be emplaced under similar dynamic conditions. It is possible that these conditions can be satisfied by more than one process. Rather than speculate on a particular process for reducing friction, therefore, it is more rewarding first to identify constraints on the essential dynamic conditions; the constraints can later be used to investigate the effectiveness of a specific process.

Field observations (Sec. 4) are consistent with the basal layers of sturzstroms behaving as pressurised granular fluids. Several sources may contribute to the fluid phase (Secs. 4 and 5). Their relative importance is not discussed here; instead, it is assumed that a fluid phase is normally present. As a sturzstrom advances, it leaves behind material in its wake, some of which comes from the basal layer. For similar dynamic conditions to be maintained, therefore, the basal layer must be replenished with new material during transport, and possible sources are (1) material entrained from the head and sides of a spreading sturzstrom, including the air around it, (2) material reaching the base by falling down from the main body, and (3) material entrained from the ground. Although a different combination of sources may be important to any given sturzstrom (and, indeed, at different stages during its emplacement), the key condition is that the basal layer is maintained, independent of the specific source of replenishment.

Energy in the basal granular fluid may be dissipated by inelastic collisions between fragments, by solid friction between fragments, and by the deformation of any fluid between the fragments [Iverson (1997)]. At low rates of deformation, fragments collide weakly and most energy is lost by grains sliding passed each other. At higher rates, the control on energy loss depends on the condition and amount of any interstitial fluid: When the fluid is dilute or in small quantities, energy losses are dominated by collisions among fragments; otherwise, energy is consumed mainly in deforming the interstital fluid, which may deform in a laminar or turbulent manner.

The three potential controls on energy loss — sliding, colliding or fluid deformation — depend in distinctly different ways on the conditions of emplacement (Appendix). Thus, sliding resistance increases with the pressure (and, hence, thickness) of the overlying landslide, but decreases with increasing fluid pressurisation; collisional energy losses and turbulent fluid resistance increase with the square of the landslide's velocity; and laminar fluid resistance increases with the viscosity of the interstitial fluid and with landslide velocity. Both thickness and velocity can be linked to sturzstrom volume, so that the trends between L, H/L and V (Fig. 5) can be used to infer the dominant resisting mechanism.

7. Quantifying Sturzstrom Runout

Sturzstrom advance is controlled by how resisting stresses consume the potential energy released during collapse (through conversions of potential to kinetic energy and of kinetic to frictional energy). By equating the total amounts of potential and resisting energy involved (Appendix), runout length is expected to increase in proportion with drop height H and volume, but in inverse proportion to the basal resisting stress τ and basal area of the landslide. By geometry, H and A will increase with $V^{1/3}$ and $V^{2/3}$ respectively (Appendix). The final volumetric control on runout length and energy line therefore depends on its influence also on the resisting stress.

The large velocities attained by sturzstroms additionally suggest that friction is not significant during collapse, when the landslide accelerates, but becomes important during runout from the collapse scar. In this case, a simple conversion of potential to kinetic energy during collapse indicates that landslide velocity varies in proportion to $V^{1/6}$ (Appendix). Accordingly, the resisting stress will be proportional to $V^{1/3}$ for sliding (through landslide thickness) and for collisional energy loss and turbulent flow (through the square of the velocity), but proportional to $V^{1/6}$ for viscous resistance during laminar flow (through velocity alone).

Combining the scaling relations then gives runout length proportional to $V^{1/3}$ for sliding, collisional and turbulent flow in the basal layer, but proportional to $V^{1/2}$ for viscous deformation. Both sets of relations could account for the length-volume trends in Fig. 4. However, when applied to H/L, the stress relations yield no volumetric dependence for sliding, collisional or turbulent flow, but a decrease with $V^{1/6}$ for viscous resistance. The latter trend alone is consistent with observation (Fig. 4). Thus, only viscous-dominated resistance can account for both the length-volume and H/L-volume trends.

The factors favouring a predominantly viscous response are (1) a high fluid pressure (for reducing energy losses by sliding), (2) a small thickness for the basal layer (Appendix), and (3) a large effective viscosity for the interstitial fluid (for raising the viscous stress and preventing turbulent flow). Smaller-volume sturzstroms will also favour lower energy losses from sliding and collisions (through landslide thickness and velocity, respectively). The simplest preferred state for viscous resistance is thus a basal layer containing pressurised mud, a combination consistent with three of the four observations of exposed basal layers given in Sec. 4. The fourth condition, for rock melting, may thus be viewed as a limiting case when viscous heating becomes sufficiently large.

8. Minimum Volume for Sturzstroms

As well as influencing runout conditions, sturzstrom volume is also important for initiating sudden collapse and, hence, for providing the accelerations necessary for sturzstroms to achieve high velocities. Thus, on Earth, the minimum volume for subaerial catastrophic collapse is on the order of millions of cubic metres [Hsü (1975); Melosh (1987)]. Failure normally occurs on slopes greater than 20° and produces scars with geometries similar to those associated with a wide range of other types of landslide [McGuire *et al.* (2002b)]. An essential difference is that, for catastrophic collapse, the zones of failure have *minimum* thicknesses of tens of metres.

Although giant catastrophic collapse may be triggered by large earthquakes [Keefer (1984)] (typically with Richter Magnitudes of 6.5 or more), it may also be preceded by months to decades of accelerating creep [Voight (1978)]. The factors controlling instability, therefore, must be able to evolve slowly towards a critical state, after which a sudden change occurs to promote catastrophic failure; they can also be accelerated by earthquakes, but their operation must not depend on such sources of external energy alone.

By virtue of their rapid onset in frequently remote areas, accelerating movements before sudden failure are rarely monitored. A classic exception is the set of deformation data for Mt. Toc before it collapsed into the Vajont reservoir. Gathered for economic concerns, the data show a hyperbolic increase in the rate of movement for at least two months before collapse (Fig. 6). This acceleration raised rates of movement to more than $20\,\mathrm{cm\,d^{-1}}$ during 8 October 1963, the day before catastrophic collapse [Hendron and Patton (1985)]. Extreme guesses at the time might have anticipated maximum velocities during collapse perhaps 10 000 times greater, at about $2\,\mathrm{km\,d^{-1}}$; even so, Mt. Toc wuld have required 6 hours to enter the Vajont reservoir. At such rates no catastrophe was foreseen. In the end, the mountain collapsed with a mean velocity of $40\,\mathrm{km\,h^{-1}}$.

Failure occurred along a clay-rich horizon as much as 200 m below the surface (Fig. 2) after at least a month of heavy rainfall [Müller (1964)]. Increased water pore pressure is thus commonly cited as the key destabilising factor. Indeed, frictional heating of water in the clay, and the consequent decrease in its sliding resistance, has been proposed as the trigger for catastrophic final acceleration [Hendron and Patton (1985); Voight and Faust (1982)]. However, pore pressure alone cannot explain the consistent

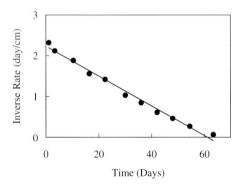

Fig. 6 Inverse rates of horizontal slope movement for the two-month period before the catastrophic collapse of Mt. Toc on 9 October 1963. Linear regression gives $R^2 = 0.99$. A linear inverse rate is equivalent to a hyperbolic increase in deformation rate with time. Modified from Kilburn and Petley [2003].

style of acceleration *for 60 days or more* before collapse, suggesting that an additional factor was important [Kilburn and Petley (2003)].

As it happens, water is also extremely efficient in corroding rock (including clay) under stress [Atkinson (1984); Main *et al.* (1993)]. Corrosion weakens the tips of cracks existing in rock, promoting their growth and eventual coalescence. Initially, the fractured zones remain isolated and the unstable slope deforms slowly because its bulk resistance is controlled by the strength of intact rock. Movement accelerates when the fractured zones begin joining together, so reducing the proportion of intact rock along the future plane of failure; this stage is associated with a hyberbolic increase in the rate of cracking and, hence, also in the rate of slope movement [Kilburn and Petley (2003); Kilburn (2003)].

When the fracture zones finally unite into a single discontinuity, the bulk resistance suddenly decreases to that for sliding along the new failure surface. The drop in resistance may be as much as 20% of the intact rock strength, sufficient to accelerate a landslide to $100 \, \mathrm{km \, h^{-1}}$ within a minute [Kilburn and Petley (2003)], without recourse to frictional heating of pore water (although such heating would reinforce the acceleration). Slow rock cracking has thus the characteristics required for a period of accelerating creep to culminate in catastrophic collapse. Moreover, brittle deformation is expected at depths of tens of metres or more, because such levels commonly consist of brittle bedrock and, even when they contain clay horizons, clay at such pressures can also deform as a brittle material [Petley (1999)].

In contrast, movement at shallower depths typically involves soils and weathered rock that do not deform in a brittle manner and so do not provide a sudden drop in resistance. As a result, shallow landslides rarely have the potential for producing sturzstroms (unless assisted, perhaps, by energy supplied from earthquakes). Observation shows that unstable volumes are commonly 100–1000 times the cube of their mean thickness [McGuire *et al.* (2002b); Kilburn and Petley (2003)]. If the minimum depth for brittle failure is set nominally at 25 m, therefore, the minimum volume expected for a sturzstrom would be in the range 1–10 million $\mathrm{m^3}$, in agreement with field data [Hsü (1975); Melosh (1987)].

Development of a failure surface by slow-cracking has thus the attraction of accounting for long-term hyperbolic accelerations to failure, the rapid onset of catastrophic collapse, and the minimum volume of material required for such behaviour. It is, however, not unique as an explanation. Current investigations are exploring the alternative possibility that catastrophic collapse may also occur after extended intervals of hyperbolic acceleration along an *existing* failure surface [Helmstetter *et al.* (2004)]. In this case, acceleration results from the progressive degradation of the failure surface during motion, leading to a self-feeding decrease in frictional resistance. It may yet emerge, therefore, that hyperbolic accelerations to catastrophe may be reached from more than one starting condition.

9. Implications for Sturzstrom Emplacement

Although rare compared with other classes of landslide [McGuire *et al.* (2002b)] giant, catastrophic collapse and sturtzstrom formation are not the result of unusual physical processes. They are instead produced by common phenomena operating under uncommon conditions. Rather than devise separate interpretations for individual sturzstroms, therefore, it is important to identify the common conditions under which most of them are emplaced. Subsequent modelling can then address any unusual circumstances under which a particular sturzstrom has evolved.

A simple reference model follows from the volumetric controls on sturzstrom formation and runout. The minimum volume for sturzstroms can be explained by the criterion for deep-seated failure in brittle rock. After collapse, deformation is concentrated along the base of the sturzstrom, where the landslide breaks up into a granular mass. Frictional heating raises the pressure of trapped fluids and this allows the basal layer to behave as a viscous fluid. Frictional forces slow down the sturzstrom until basal deformation is dominated by sliding between fragments, which eventually brings the landslide to rest.

The runout distance increases with volume, because as volume increases the potential energy released during collapse increases proportionally more than does the rate of energy loss along the basal layer. Larger sturzstroms will thus tend to have longer emplacement times, as well as larger collapse velocities, so favouring greater runout lengths. Under such boundary-layer flow, the energy line H/L depends on landslide volume and mean density, the viscosity and thickness of the basal layer, and on gravity (Appendix). Improved forecasts of runout distance will therefore require a better understanding of basal layers and more reliable methods for estimating the volume of an instability before collapse. Such advances are crucial. As populations migrate into mountainous regions, so the risk from sturzstroms increases. It is thus imperative that new field and experimental data are acquired to enhance current models of how sturzstroms behave.

10. Appendix

The resisting shear stress τ along the base of a sturzstrom can be expressed as: (1) for sliding, $\tau \propto \rho g h \, (1 - \mathrm{P}^*)$; (2) for collisions and for turbulent fluid deformation, $\tau \propto \rho U^2$ (the constant of proportionality depending on the specific mechanism involved); and (3) for laminar fluid deformation, $\tau \propto \eta U/d$, where ρ, h and U are the sturzstrom's mean density, thickness and forward velocity, g is gravitational acceleration, η is the mean effective viscosity of interstitial fluid (which, for simplicity, is assumed to have a newtonian rheology) and d is the mean thickness of the basal layer; P^* is the ratio of interstitial fluid pressure to the overburden pressure, $\rho g h$.

Equating potential and resisting energy yields $L = (\rho g/\tau) \, (H/A)V$, where A is the mean basal area over which the resisting stress τ acts during emplacement. By geometry, $H \propto \{1 + [(y - r)/z]\}V^{1/3}$ (since $z \propto V^{1/3}$) and $A \propto V^{2/3}$, where z is the vertical extent of the collapse scar, y is the vertical drop to the lowest point below the scar reached by the landslide, and r is the vertical runup from the lowest point (Fig. 6). L and H/L can thus be rewritten as $L \propto (\rho g/\tau)[1 + (y - r)/z]V^{2/3} \propto V^{2/3}/\tau$, and $H/L \propto (\tau/\rho g)V^{-1/3} \propto \tau V^{-1/3}$. A simple conversion of potential to kinetic energy during collapse yields $U \propto (gz)^{1/2} \propto V^{1/6}$, because $z \propto V^{1/3}$. Combining this relation with those above for resisting stress yields the volumetric controls expected for different modes of energy loss, as described in Sec. 7.

Acknowledgements

Bill Murphy suggested numerous improvements to the text, which also benefitted from lively discussions with Ken Hsü, Søren Sørensen, David Petley, John Hutchinson, Tim Davies, Mauri McSaveney and Ken Hewitt. May the discussions continue.

References

Atkinson, B. K. (1984) *J. Geophys. Res.* **89**, 4077.

Campbell, C. S. (1989) *J. Geol.* **97**, 653.

Dade, W. B. & Huppert, H. E. (1998) *Geology* **26**, 803.

Davies, T. R. H. (1982) *Rock Mech.* **15**, 9.

Davies, T. R. & McSaveney, M. J. (2004) *Security of Natural and Artificial Rockslide Dams, Extended Abstracts Volume, NATO Advanced Research Workshop*, (eds.) Abdrakhmatov, K., Evans, S. G., Hermanns, R., Scarascia Mugnozza, G. and Strom, A. L., Bishkek, Kyrgyzstan, p. 28.

Dutton, S. (2004) PhD Thesis, University of Luton, UK.

Erismann, T. H. (1979) *Rock Mech.* **12**, 15.

Erismann, T. H. & Abele, G. (2001) *Dynamics of Rockslides and Rockfalls*, Springer, Berlin.

Hayashi, J. N. & Self, S. (1992) *J. Geophys. Res.* **97**, 9063.

Helmstetter, A., Sornette, D., Grasso, J.-R., Andersen, J. V., Gluzman S. & Pisarenko, V. (2004) *J. Geophys. Res.* **109**, doi: 10.1029/2002JB002160.

Hendron, A. J. & Patton, F. D. (1985) *Tech Rept GL-85-5*, US Army Corps of Engineers, Washington DC.

Hewitt, K. (1988) *Science* **242**, 64.

Hoek, E. & Bray, J. W. (1981) *Rock Slope Engineering*, 3rd Edn, Rev, E&FN Spon, London.

Hsü, K. J. (1975) *Geol. Soc. Am. Bull.* **86**, 129.

Iverson, R. M. (1997) *Rev. Geophys.* **35**, 245.

Keefer, D. K. (1984) *Geol. Soc. Am. Bull.* **95**, 406.

Kilburn, C. R. J. (2001) *Paradoxes in Geology*, Briegel, U. and Xiao, W. (eds.), p. 245, Elsevier, Amsterdam.

Kilburn, C. R. J. (2003) *J. Volcanol. Geotherm. Res.* **125**, 271.

Kilburn C. R. J. & Petley, D. N. (2003) *Geomorphology* **54**, 21.

Kilburn, C. R. J. & Sørensen, S. A. (1998) *J. Geophys. Res.* **103**, 17877.

Main, I. G., Sammonds, P. R. & Meredith, P. G. (1993) *Geophys. J. Int.* **115**, 367.

Malin, M. C. & Sheridan, M. F. (1982) *Science* **217**, 637.

McGuire, W. J., Day, S. J. and Kilburn, C. R. J. (2002a) *Proceedings of International Landslide Symposium*, United Nations and Kyoto University, p. 691.

McGuire, W. J., Mason, I. & Kilburn, C. (2002b) *Natural Hazards and Environmental Change*, Arnold, London.

Melosh, H. J. (1987) *Rev. Eng. Geol.* **VII**, 41.

Merlin, T. (1997) *Sulla Pelle Viva*, Cierre Edizioni, Verona.

Müller, L. (1964) *Felsmech. Ingenieurgeol.* **2**, 148.

Petley, D. N. (1999) *Geol. Soc. London Spec. Pub.* **158**, 61.

Shreve, R. L. (1968) *Spec. Paper Geol. Soc. Am.* **108**, 1.

Sørensen, S. A. and Bauer, B. (2003) *Geomorphology* **54**, 11.

Voight, B. & Faust, C. (1982) *Géotechnique* **32**, 43–54.

Voight, B. (ed.), (1978) *Rockslides and Avalanches. 1, Natural Phenomena*, Elsevier, Amsterdam.

Remote Monitoring of the Earthquake Cycle Using Satellite Radar Interferometry

Tim J. Wright

School of Earth and Environment, University of Leeds
Leeds, LS2 9JT, UK
t.wright@see.leeds.ac.uk

The earthquake cycle is poorly understood. Earthquakes continue to occur on previously unrecognised faults. Earthquake prediction seems impossible. These remain the facts despite nearly a hundred years of intensive study since the earthquake cycle was first conceptualised. Using data acquired from satellites in orbit 800 km above the Earth, a new technique, radar interferometry (InSAR) has the potential to solve these problems. For the first time, detailed maps of the warping of the earth's surface during the earthquake cycle can be obtained with a spatial resolution of a few tens of metres and a precision of a few millimetres. InSAR does not need equipment on the ground or expensive field campaigns, so it can gather crucial data on earthquakes and the seismic cycle from some of the remotest areas of the planet. In this article, I review some of the remarkable observations of the earthquake cycle already made using radar interferometry, and speculate on breakthroughs that are tantalisingly close.

1. The Earthquake Cycle

In the small hours of 17 August, 1999, the largest earthquake to hit Turkey in 60 years devastated the town of Izmit and the surrounding region. More than 18 000 people lost their lives and 250 000 people were made homeless. The earthquake cost the Turkish economy around $6 billion. The Izmit earthquake, like most large continental earthquakes, literally tore the Earth — the pulse emanating from the epicentre at a speed of 11 000 km/h ruptured a 130 km section of a major fault in the earth's crust

in just 40 seconds. The region immediately to the south of the fault moved
to the west when compared with the northern side; walls, roads, railway
lines, rivers that once crossed the fault in a straight line were given a new
step, up to 5 m in magnitude (Fig 1). This was not the first earthquake to
have occurred in this area; historical records show that very similar events
occurred in 1719 and 1894 [Ambraseys and Jackson (2000)]. These earthquakes occurred on the North Anatolian Fault, an extremely active fault
that accommodates the westward escape of Turkey as it is squeezed between
the northward motion of the Arabian plate and the Eurasian plate, which
blocks its path (Fig. 2). It is this repetition of earthquakes on the same
fault that we refer to as the earthquake cycle.

In its simplest form, the idea of the earthquake cycle was developed by
Harry Fielding Reid, to explain observations of the San Francisco Earthquake of 1906, associated with an average of 4–5 m surface slip along 450 km
of the San Andreas Fault [Scholz (1990)]. Reid examined precise survey data
from a triangulation network that spanned the fault. It had been carefully
measured in the 1880s, and again immediately after the earthquake. For the
first time, these observations revealed the surface deformation caused by
an earthquake. Points on the southwest side of the San Andreas Fault had
moved to the northwest during the earthquake, compared to points on the
other side of the fault. When these displacements were extrapolated onto
the fault, they matched the offsets observed at the ground break. Importantly, the magnitude of movement of the survey points decayed rapidly
away from the fault, so that they were small at distances of 20 km or more.

Furthermore, on comparing survey data from the 1860s with that of the
1880s, Reid noticed that the Farallon lighthouse, a point a long way to the
southwest of the fault, had moved in a northwesterly direction in this period
when compared to points on the other side of the fault. These observations
led Reid to propose his *elastic rebound* model of the earthquake cycle [Fig. 3;
Reid (1910)], in which "distant forces"[a] gradually strain the region around
the fault, building up elastic energy, typically over a period of hundreds to
thousands of years, until frictional forces on the fault surface are overcome
and breaking point is reached. The elastic energy is then suddenly and
catastrophically released; the regions on either side of the fault rebound,
and the ground breaks along the fault in an earthquake. The cycle then

---

[a]This was before the idea of plate tectonics had been developed, so Reid's model did not
suggest a driving mechanism.

Fig. 1   Surface rupture of the Izmit earthquake. (a) Newly completed accommodation blocks, unfortunately located directly on top of the North Anatolian Fault, but fortunately yet to be occupied. The surface rupture and sense of motion is shown by the dashed lines, with the large white arrow indicating the 4 m offset in the wall to the complex as it crosses the rupture; (b) A newly created 2–3 m step in the Istanbul to Ankara mainline railway [Photographs courtesy of Aykut Barka (Istanbul Technical University)].

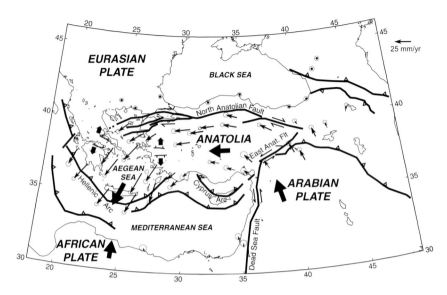

Fig. 2   Tectonic setting of Turkey. Velocity arrows were measured by the
Global Positioning System (GPS) during the interval 1988–1998 and are
shown relative to Eurasia [McClusky *et al.* (2000)]. The Anatolian block is
caught between the northward motion of Arabia and Eurasia, and is forced
westwards. The major faults are shown.

begins again. Over long periods of time, this *stick-slip* behaviour results in
the large cumulative fault offsets observed in the geological record: Around
75 km for the North Anatolian Fault [Armijo *et al.* (1999)], an offset that
would take something like 15 000 Izmit-like earthquakes and over 3 million
years to accrue at present rates of deformation.

    We now know that Reid's elastic rebound model is too simplistic.
Because is ignores changes in the properties of rocks with depth, it cannot
explain why interseismic deformation is focussed on the fault that eventually ruptures, or why rapid deformation often occurs in the immediate aftermath of an earthquake (*postseismic deformation*). There is broad agreement
that most continental earthquakes occur in the so-called seismogenic crust,
typically the upper 10–20 km, and that this behaves elastically, just as in
Reid's model. Below this, where rocks are hotter, the material properties
and behaviour of continental crust are still controversial. Observations of
surface deformation at various stages in the earthquake cycle, many now
coming from InSAR, are beginning to place bounds on competing models.

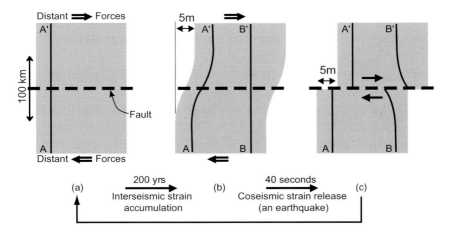

Fig. 3 A schematic representation of Reid's *elastic rebound* model of the earthquake cycle. (a) Map view of area spanning a hypothetical fault, in the instant after the last earthquake. (b) The same area, 200 years later. The profile A–A′, straight at the beginning of the cycle, has become curved. This is known as interseismic strain accumulation. Note that the magnitude of the warping is vastly exaggerated in this diagram. (c) 40 seconds later, after an earthquake. A–A′ is once more a straight line, but this time with a 5 m step at the fault. B–B′, straight immediately before the earthquake, is now curved with an offset of 5 m at the fault, decaying as large distances from the fault. The timings and displacements are representative of a typical earthquake, such as the 1999 Izmit event.

2. Satellite Radar Interferometry

The majority of today's remote sensing satellites operate at optical wavelengths, primarily recording light originating from the Sun that makes it to the satellite's sensors having been reflected or scattered off the Earth's surface. Radar antennae, such as those carried by the European Space Agency's Earth Resources Satellites, ERS-1&2, are different: They actively illuminate the Earth, recording the backscattered waves, and because radar wavelengths (typically 5–25 cm) are around 100 000 times longer than the wavelength of visible light, radar travels through clouds. Radar satellites can therefore operate night and day, and in all weather conditions.

None of the radar satellites currently in orbit was designed to measure the Earth's deformation. Nevertheless, because they are illuminating the Earth with controlled, coherent radar waves, radar interferometry is possible. In essence, interferometry works by ignoring the amplitude of the

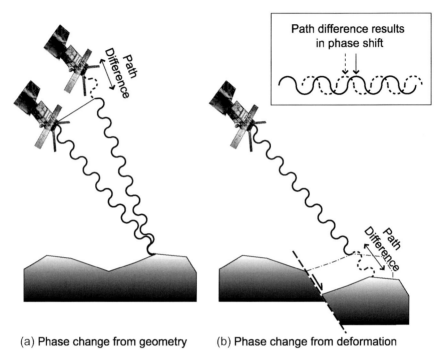

(a) **Phase change from geometry**   (b) **Phase change from deformation**

Fig. 4   Schematic representation of radar interferometry: (a) Two satellites image the same point on the ground at different times but from different positions, creating a phase shift; (b) A point on the ground imaged from the same point in space, before and after an earthquake. The phase change induced is directly related to the component of surface deformation in the look-direction of the satellite (the path of the radar wave). Note that this image is not to scale: ERS-1&2 orbit at 780 km with a wavelength of 5.6 cm.

waves that return to the satellite's antenna. Instead, the phase[b] of the wave is used (Fig. 4). We know the wavelength of the radar waves, the phase of the waves when they left the satellite, and the phase of the waves returning from a particular patch of ground (or pixel in the radar image). The distance from the satellite to the ground is simply a large unknown number of whole wavelengths (around 30 million for ERS-1&2), plus a known fraction of that wavelength, determined from the difference in phase between the waves that leave and those that return.

---

[b]The phase of a wave describes its position within the wave cycle — i.e. if it is at a peak, or trough, or somewhere in between.

This does not initially seem like a very practical method for measuring distances, particularly when we also know that there is a random phase shift added to each measurement when the wave bounces off the ground. But imagine returning to exactly the same position in space at a different time. If nothing has changed then the phase measured for each pixel will be identical to that measured previously. On the other hand, if the distance between the ground and the satellite changes between the times, perhaps due to an earthquake, then the phase measured at the satellite will change. By creating images of these phase changes, it is possible to map deformation with a precision of a small fraction of the radar wavelength: A few millimetres for the 5.6 cm wavelength of ERS-1&2. In practice, the two radar images are unlikely to have been acquired from exactly the same position, introducing additional phase shifts due to the orbital separation and surface topography.[c] Using an elevation model of the target area and precise orbital models, most of these phase signals can be removed, leaving those caused by surface deformation.

Figure 5 shows an interferogram of northwest Turkey, constructed from two ERS-2 radar images acquired 35 days apart, before and after the Izmit earthquake. Each of the coloured interference fringes is equivalent to a 28 mm contour (half the ERS wavelength) of surface deformation in the satellite's line of sight (LOS). The phase measurements are relative, so to calculate the deformation at the fault you simply count the fringes from the edges of the interferogram to the fault at the centre (approximately 25 on each side) and multiply by 28 mm. In this case, the area immediately south of the fault moved more than 70 cm closer to the satellite and the area to the north moved by a similar amount, but away from the satellite. Knowing that the LOS of ERS-2 is nearly along the fault, but 23° from the vertical, and that this earthquake produced largely horizontal motion, the 1.4 m LOS offset implies a horizontal fault offset of ca. 4 m; exactly what was observed in the field.

Because none of the current crop of InSAR satellites was designed to do interferometry, the fact that we can do it at all is remarkable. The technique is a fortuitous byproduct; hence, current satellite design is not optimised for InSAR, and it is not always the first priority of the mission planners.

---

[c]By using one antenna on the Shuttle and another at the end of a 60 m long retractable boom, NASA and NIMA's Shuttle Radar Topography Mission collected topographic interferograms for 80% of the Earth's land surface, taking just 10 days in February 2000. These data have been used to produce a high-resolution global topographic data set.

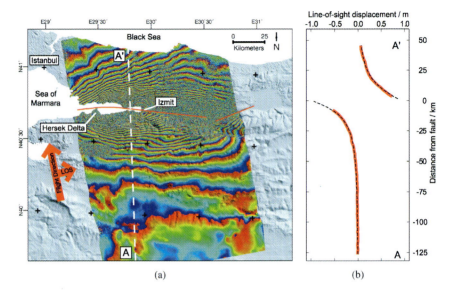

(a)   (b)

Fig. 5   (a) Radar interferogram mapping the deformation field caused by
the 1999 Izmit (Turkey) earthquake whose rupture is shown by the red line.
Each coloured interference fringe is equivalent to a 28 mm contour of surface
displacement in the satellite's line of sight (LOS; red arrow); (b) LOS displacements along profile A–A′, revealing a characteristic *elastic rebound* form
(red line). The dashed black line is a profile through a simple elastic model
of the earthquake.

Images are not acquired as often as we would like and satellite orbits are
generally not sufficiently well steered.

Furthermore, InSAR is only possible if the character of the ground surface does not change between image acquisitions. Otherwise, there is a
change in the random phase contribution due to the interaction between
the radar waves and the ground, resulting in a meaningless phase change
measurement (termed incoherence). Bare rock and man-made structures
often remain coherent for long periods of time, but C-band (ca. 6 cm wavelength) interferograms of forested areas, for example, can be incoherent
even if the images were only acquired a day apart.[d]

Also troublesome is the Earth's changing atmosphere: Water vapour
concentrations, in particular, distort the phase ruler, causing phase shifts

[d]This is due to multiple scattering off leaves and small branches whose positions are
unstable.

that can be confused with deformation. A thundercloud, for example, can cause phase changes equivalent to ground motions of up to 10 cm. In addition, a single interferogram can only measure surface deformation in the LOS of the satellite. The true, 3D character of the motion is lost, and we only have information from a single dimension. This can cause ambiguities in our interpretation of interferograms that lead to uncertainties in physical models. Despite these limitations, the radars carried by ERS-1 and ERS-2 in particular, have produced many important observations of the earthquake cycle.

Although this paper covers the application of satellite radar interferometry to the coseismic and interseismic phases of the earthquake cycle, InSAR has been used to measure postseismic deformation. InSAR has also imaged volcanic eruptions, the inflation of volcanoes as new magma fills buried magma chambers, and even the thermal contraction of erupted lava flows. Land subsidence due to mining, water extraction and oil wells has been measured remotely using InSAR, along with landslides and even Antarctic ice flows. For more comprehensive information on how InSAR works, and details of these other applications, refer to excellent review papers by Massonnet and Feigl [1998] and Bürgmann *et al.* [2000].

3. Coseismic Deformation: Images of Earthquakes

For most of the 20th century, seismology provided the only way of studying the majority of continental earthquakes. By listening to the seismic waves emanating from an earthquake, seismologists can determine the location, magnitude and type of earthquake that occurred. However, there are often large uncertainties in some of the earthquake parameters for shallow crustal earthquakes: The depth of faulting is often poorly resolved with seismology, and, except for very large earthquakes, the distribution of slip on the fault cannot be reliably determined. The 1992 Landers (California) earthquake furnished the first image of an earthquake's deformation field, and it graced the front cover of *Nature* [Massonnet *et al.* (1993)]. Since then, InSAR has been used to map the deformation resulting from more than 40 earthquakes. This may not sound like a large number, but conventional surveying techniques had captured the deformation of less than 15 earthquakes before 1992.

The location, magnitude and type of an earthquake can be determined from its deformation field by posing the question "what type of earthquake could have resulted in this deformation?". This is a classic geophysical inverse problem. Given knowledge of the earthquake fault orientation and distribution of slip, it is straightforward to calculate the deformation that we would expect to observe. Doing the reverse is difficult and computer intensive, and methods for doing this are still evolving. Nevertheless, InSAR is gradually breaking seismology's earthquake monopoly, in many cases providing vital information that was not available from seismology.

The Izmit earthquake showed how important InSAR has become. A third of the surface rupture of this magnitude 7.5 earthquake was offshore in the Gulf of Izmit, and field geologists were unable to quantify the magnitude of slip that occurred there (Fig. 5). Not only this, they concluded that the earthquake had terminated east of the prominent Hersek Delta, which crosses most of the Gulf, 35 km west of Izmit [e.g. Barka (1999)]. InSAR data showed that over 1 m of slip must have occurred beyond the Hersek Delta for at least 10 km [Wright et al. (2001b)].

The location of this termination is a crucial variable for determining the future seismic hazard for the Istanbul area. The Izmit earthquake was the seventh in a westward sequence of earthquakes that have unzipped over 1000 km of the North Anatolian Fault since 1939, in a relentless march towards Istanbul. The last major earthquakes on the North Anatolian Fault south of Istanbul were in 1509 and 1766 [Ambraseys and Jackson (2000)], leaving a ca. 160 km long seismic gap, and a potential disaster on an even bigger scale than that of 1999.

Tom Parsons and colleagues at the US Geological Survey calculated the probability of strong shaking in the Istanbul area after the Izmit earthquake. They required a historical record of earthquakes in the region, and an accurate slip model of the Izmit earthquake. The latter was important because although the Izmit earthquake provided an additional push to the areas at either end of the fault, increasing the seismic hazard there. Using InSAR data alone, we quickly produced a preliminary slip model. When combined with the long gap since the last earthquake, Parsons et al. [2000] estimated a 62% chance of strong shaking for greater Istanbul in the next 30 years: One of the highest probabilities for any fault zone in the world.

4. Interseismic Deformation

Earthquakes cannot occur without the build up of elastic strain. It is in mapping the accumulation of this interseismic strain that InSAR offers the most potential as a medium-range forecast tool. Note that I carefully avoided the term *prediction*, as this implies something like a 2-day earthquake warning. An earthquake forecast would give the likelihood of an earthquake occurring over a certain time period (e.g. "There is a 62% chance of strong shaking in greater Istanbul in the next 30 years"). These medium-range forecasts are vital because they enable civil defence agencies to prepare communities through education, rebuilding and retrofitting programs.

Measuring the build up of elastic strain between earthquakes (interseismic strain) using InSAR is not straightforward. The strain rates are extremely small — the North Anatolian Fault, for example, moves at around 24 mm/yr horizontally and it therefore takes around three years to create a single interference fringe. Interseismic fault creep, where slip continues to the surface, produces a discontinuity that is relatively straightforward to observe in interferograms [e.g. Rosen *et al.* (1998)]. In contrast, interseismic deformation associated with faults that are locked at the surface is typically distributed on a length scale of 30–150 km and therefore much harder to distinguish from atmospheric and orbital errors.

In an ideal world, we would look at interferograms spanning a very long time interval, but, in areas such as Turkey, interferograms with time intervals of larger than 2 years are generally incoherent. We are therefore restricted to shorter-period interferograms, but these contain such a small deformation signal that they tend to be swamped by noise. To overcome this problem requires the use of multiple interferograms to amplify the tectonic signal and reduce the noise. By summing several interferograms, I was able, with colleagues, to extract the pattern of strain accumulation across the North Anatolian Fault [Fig. 6; Wright (2000); Wright *et al.* (2001a)]. The stacked interferogram is effectively an image of the gradual build up of elastic energy in a ca. 70 km wide zone across the North Anatolian Fault. This energy will eventually be released in an earthquake. The image is also a direct observation of plate tectonics in action, revealing the relative motion of Anatolia with respect to the Eurasian Plate.

Elsewhere, Peltzer *et al.* [2001] have used similar methods to measure the deformation of the San Andreas Fault Zone in southern California. My colleagues and I have also used InSAR to show that the present-day rate of strain accumulation on the major faults in western Tibet is lower than expected from geological observations [Wright *et al.* (2004)].

With current satellites, it is only possible to use InSAR to measure strain accumulation if ground conditions are optimal. In particular, vegetation cover must be relatively low. With future satellite technology, it will be possible to build-up a time-varying map of strain across most of the

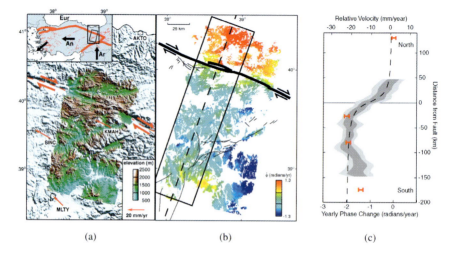

(a)   (b)   (c)

Fig. 6   (a) Topographic and tectonic map of the eastern end of the North
Anatolian Fault. The coloured area is an elevation model calculated from a
1-day interferogram: the North Anatolian Fault (dashed red line) can clearly
be seen cutting through the landscape. Arrows are GPS-determined velocities relative to the Eurasian plate (McClusky *et al.*, 2000); (b) Stacked interseismic interferogram, converted to a yearly phase change ($\dot{\phi}$). Positive phase
changes (warm colours) indicate a relative increase in distance to the satellite; (c) Phase profile perpendicular to the North Anatolian Fault (along
the dashed line in (b)). The grey bands delimit the 1- and 2-sigma error
bounds, with red bars the GPS velocities. Phase changes predicted by an
elastic model are plotted as a dashed line.

planet, which could used to construct accurate medium-range earthquake
forecasts.

5. A Look Into the Future

The immediate future for InSAR looks good with the work of the ERS satellites being continued by ESA's Envisat, launched in 2002, and launches of
the Japanese ALOS satellite and Canadian Radarsat-2 planned this year.
These satellites will extend the lifespan of this new technology, but none of
them will greatly improve upon current capabilities. That requires something new: A dedicated InSAR mission, targeted at earthquake and volcanic hazards. Such a mission would have onboard GPS to control and
measure the satellite's orbit to a very high precision; it would operate at
L-band (20 cm) wavelength to ensure better coherence in vegetated areas; it
would collect data on every satellite pass, enabling detailed time series to be

created, atmospheric noise to be reduced, and faults with low slip rates to be monitored/identified; it would acquire data from several look directions allowing 3D displacements to be recovered; it would provide data cheaply and quickly to the scientific community. Proposed missions with these specifications are being considered by NASA and ESA, and there are reasonable prospects that one will be launched within the next 5–10 years.

Many of the last decade's significant earthquakes occurred on faults that had not previously been recognised as major faults of their regions [e.g. Northridge, CA (1994); Kobe, Japan (1995); Athens, Greece (1999); Bam, Iran (2003)]. It is here that InSAR could have a big impact, by mapping strain accumulation globally and producing reliable medium-range earthquake forecasts. The human argument for such a mission is compelling: A tenth of the world's population lives in areas classified as having medium to high seismic hazard by the Global Seismic Hazard Assessment Program. Earthquake fatalities are highest in developing countries, which cannot afford ground-based monitoring equipment.

The economic argument is also simple. The 1994 Northridge earthquake in Los Angeles caused total property damage estimated at $20 billion. This would have been greater were it not for an intense program of hazard mitigation activities over the previous two decades. Rebuilding or retrofitting structures to protect them from earthquakes is relatively cheap compared to the cost of rebuilding after an earthquake. For example, the US Federal Emergency Management Agency estimate the cost of retrofitting bridges to be just 22% of the cost if they are destroyed by earthquakes, and this does not take into account the cost to the local economy of the temporary loss of infrastructure. Although the cost of a dedicated InSAR mission is high (ca. 150–250 million), a city saved from extensive earthquake damage after an InSAR forecast led to a major retrofitting program might consider the price tag cheap. A satellite-based system is also much cheaper than attempting to make similar measurements using ground based techniques such as continuous GPS.[e]

A dedicated InSAR mission in the next ten years, and perhaps a constellation of Earth monitoring satellites in the next 25 years, will lead to a vastly improved understanding of the physics of the earthquake cycle, a complete time-varying map of the Earth's strain and reliable earthquake forecasts. Ultimately, this will save lives.

---

[e]Covering just the populated areas of the planet at risk from earthquakes with continuous GPS instruments spaced on a 15 km grid would cost ca. 1 billion.

Acknowledgements

TJW is supported by a Royal Society University Research Fellowship. Thanks to Aykut Barka for providing figures, and Barry Parsons, Jenefer Brett, and an anonymous reviewer for comments that helped improve the manuscript. Apologies to those whose work could not be included in this brief review of a rapidly expanding field.

References

Ambraseys, N. & Jackson, J. (2000) Seismicity of the Sea of Marmara (Turkey) since 1500. *Geophys. J. Int.* **141**(3), 1–6.

Armijo, R., Meyer, B., Hubert, A. & Barka, A. (1999) Westward propagation of the North Anatolian fault into the northern Aegean: Timing and kinematics. *Geology* **27**(3), 267–270.

Barka, A. (1999) The 17 August 1999 Izmit earthquake. *Science* **285**(5435), 1858–1859.

Bürgmann, R., Rosen, P. & Fielding, E. (2000) Synthetic Aperture Radar interferometry to measure Earth's surface topography and its deformation. *Ann. Rev. Earth. Planet. Sci.* **28**, 169–209.

Massonnet, D. & Feigl, K. L. (1998) Radar Interferometry and its application to changes in the earth's surface. *Rev. Geophys.* **36**(4), 441–500.

Massonnet, D., Rossi, M., Carmona, C., Adragna, F., Peltzer, G., Feigl, K. & Rabaute, T. (1993) The displacement field of the Landers earthquake mapped by radar interferometry. *Nature* **364**, 138–142.

McClusky, S., Balassanian, S., Barka, A., Demir, C., Ergintav, S., Georgiev, I., Gurkan, O., Hamburger, M., Hurst, K., Kahle, H., Kastens, K., Kekelidze, G., King, R., Kotzev, V., Lenk, O., Mahmoud, S., Mishin, A., Nadariya, N., Ouzounis, A., Paradissis, D., Peter, Y., Prilepin, M., Reilinger, R., Sanli, I., Seeger, H., Tealeb, A., Toksöv, M. & Veis, G. (2000) Global Positioning System constraints on plate kinematics and dynamics in the eastern Mediterranean and Caucasus. *J. Geophys. Res.* **105**(B3), 5695–5719.

Parsons, T., Toda, S., Stein, R., Barka, A. & Dieterich, J. (2000) Heightened odds of large earthquakes near Istanbul: An interaction-based probability calculation. *Science* **228**, 661–665.

Peltzer, G., Crampé, F., Hensley, S. & Rosen, P. (2001) Transient strain accumulation and fault interaction in the Eastern California shear zone. *Geology* **29**(11), 975–978.

Reid, H. F. (1910) The mechanics of the earthquake. *The California Earthquake of 18 April, 1906: Report of the State Earthquake Investigation Commission, 2.* Carnegie Institution, Washington.

Rosen, P., Werner, C., Fielding, E., Hensley, S., Buckley, S. & Vincent, P. (1998) Aseismic creep along the San Andreas Fault northwest of Parkfield, CA measured by radar interferometry. *J. Geophys. Res.* **25**(6), 825–828.

Scholz, C. H. (1990) *The Mechanics of Earthquakes and Faulting.* Cambridge, Cambridge University Press.

Wright, T. J. (2000) *Crustal Deformation in Turkey from Synthetic Aperture Radar Interferometry.* D.Phil. Thesis, University of Oxford, Oxford, UK.

Wright, T. J., Parsons, B. E. & Fielding, E. J. (2001a) Measurement of interseismic strain accumulation across the North Anatolian Fault by satellite radar interferometry. *Geophys. Res. Lett.* **28**(10), 2117–2120.

Wright, T. J., Fielding, E. J. & Parsons, B. E. (2001b) Triggered slip: Observations of the 17 August 1999 Izmit (Turkey) earthquake using radar interferometry. *Geophys. Res. Lett.* **28**(6), 1079–1082.

Wright, T. J., Parsons, B., England, P. C. & Fielding, E. (2004) InSAR observations of low slip rates on the major faults of Western Tibet. *Science* **305**, 236–239.

Human Influence on the Global Geochemical Cycle of Lead

Dominik J. Weiss* and Malin E. Kylander

Department of Earth Science and Engineering
Imperial College London, London SW7 2AZ, UK

Department of Mineralogy, The Natural History Museum
London SW7 5BD, UK
*d.weiss@imperial.ac.uk

Matthew K. Reuer

Department of Environmental Sciences
Colorado College, Colorado Springs, CO 80903, USA

The global biogeochemical cycles of several trace metals are presently dominated by human activities, a result of the nature and magnitude of historical resource consumption. Lead has been mined since ancient times, often as a by-product of silver extraction, and has one of the longest associations with man of all heavy metals [Nriagu (1983)]. As of 1983, human activities accounted for an estimated 97% of the global mass balance of lead [Nriagu and Pacyna (1988)]. At that time as well as today, most of the lead was derived from leaded gasoline [Nriagu (1990)], where it was used as an anti-knock agent.

New estimates of anthropogenic sources of lead suggest that the overall burden of anthropogenic lead emissions has decreased but new pollution sources (e.g. China, Mexico) have become important [Pacyna and Pacyna (2001); Pacyna et al. (1995)] leaving anthropogenic Pb emission to remain a global problem and leaded gasoline as the main source. In addition, as metals are not biodegradable, the Pb in the environment has accumulated over the decades and its fate and pathways within the ecosystem need to be investigated.

In a manner similar to chlorofluorocarbons and radionuclides derived from atomic testing, the release of lead into the environment represents an inadvertent geochemical tracer experiment, providing new insights into its fate and transport within marine and terrestrial systems. There have been several review papers and books discussing lead, its historical place in society and its impact on human and environmental health [Boutron (1995); Needleman (1997); Nriagu (1983); Nriagu (1989b); Nriagu (1990); Reuer and Weiss (2002); Shotyk and Le Roux (2005); Weiss *et al.* (1999)] and the reader is encouraged to refer to these works as well.

1. Getting the Lead Out: Sampling and Analysis

How is lead accurately measured in natural samples? The enormous amount of lead in the environment would suggest that analyses are fairly routine; this is not the case, however. As emitted lead is dispersed across the large integrated area of the earth's land surfaces and oceans it is diluted to low concentrations that require rigorous control of sampling and analytical protocols. It was not until a few decades ago that much attention was paid to contamination occurring during sampling and measurements. One of the great contributions of Patterson and co-workers at Caltech was the introduction of stringent clean laboratory methods needed to assess lead's biogeochemical cycles and human impacts [Patterson and Settle (1976)]. To reduce contamination, advances in the field of uranium-lead chronology were adopted and improvements on data quality were immediate. For example, in High Sierra freshwater lakes Hirao and Patterson first measured lead concentrations of $0.300 \, \mathrm{ng\,g^{-1}}$ but improved sampling protocols lowered this value to $0.015 \, \mathrm{ng\,g^{-1}}$. This 20-fold difference underscores the large possible artefacts generated during sampling [Hirao and Patterson (1974)]. New standards for laboratory work have been established with much trace metal work being performed in 'clean laboratories'. In these types of laboratories it is common that the air is filtered using high efficiency particulate air (HEPA) filters, workers wear protective clothing and shoes, ultra-pure chemicals are used and all laboratory ware is stringently cleaned using acid baths.

An in-depth description of all the different analytical techniques used to measure lead concentrations, isotope ratios and organo-Pb species in environmental samples is surely beyond the scope of this review but we want to give a small overview focusing on mass spectrometers, arguably the most widely used technique for trace lead analysis (concentration and isotope ratios). Mass spectrometer arrangements are composed of three fundamental parts: an ion source; a mass analyser to separate ions generated based on their mass/charge ratio; and an ion collector to count the ions. Thermal ionisation mass spectrometers (TIMS) using filaments for ionisation deserve first mention as they have been pivotal in producing the first high precision and accurate lead isotope and concentration data in environmental and geological samples [Albarède and Beard (2004) Thirlwall (1997); Vallelonga *et al.* (2002)]. The development of various plasma based mass-spectrometry techniques (inductive coupled plasma — or ICP — mass spectrometers) in the 1980s combined with different mass separation and detection methods (single and multiple detection with variable mass resolution) allowed not only precise and accurate measurements of elemental concentrations and isotope ratios often similar to TIMS but also a larger sample throughput. Such mass spectrometer developments include quadropole (Q)-ICP-MS [Weiss *et al.* (2000)], sector field (SF)-ICP-MS [Krachler *et al.* (2004); Schwikowski *et al.* (2004)] or multicollector (MC)-ICP-MS [Rehkaemper *et al.* (2001)]. Figure 1 shows representative

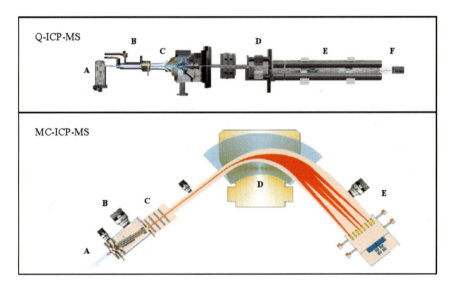

Fig. 1   Cross sections of two different mass spectrometry arrangements: A quadrupole inductively coupled plasma-mass spectrometer (Q-ICP-MS) and a single focussing multiple collector (MC-ICP-MS). In the Q-ICP-MS, the sample is introduced by a nebuliser (A) and enters the plasma and torch (B). The analyte passes through sampling cone and skimmer (C), ion lenses (D), the quadrupole mass analyser (E) to reach the electron multiplier (F) [Agilent Technologies (2005)]. In the single focussing MC-ICP-MS, the sample leaves the ion source (A), enters a collision cell (B), passes the lens stack (C), is bent by the magnetic sector mass analyser (D) and counted by the Faraday cup collectors (E) [modified from GV Instruments (2005)].

mass spectrometer arrangements for a Q-ICP-MS and a single focussing MC-ICP-MS.

The determination of organo-Pb species requires quite different instrumentation due to the chemical form of the lead [Heisterkamp and Adams (1999)]. A detailed recent review of the various analytical methods for lead and other trace metals in environmental samples is given elsewhere [Hill *et al.* (2002)].

2. Lead Isotopes as Tools for Source Identification

Lead isotopes are used in cosmological, geological and environmental investigations that involve age dating and source tracing. Lead has four stable isotopes: ^{204}Pb, ^{206}Pb, ^{207}Pb and ^{208}Pb. The latter three are the end members of ^{238}U, ^{235}U and ^{232}Th decay series (thus called radiogenic), respectively while ^{204}Pb is not generated by radioactive decay. The age and geological history of the reservoir and the initial U-Th-Pb concentrations of the parent rock determine the isotopic composition of lead. The amount of daughter isotope is equal to the difference between the original amount of parent isotope and the amount of parent isotope left after a certain passing of time at a given decay rate. Table 1 lists the different isotopes, their relative abundance, parent isotopes, half life and decay constants. When there is a separation between parent and daughter isotopes, as, for example, during ore genesis and the formation of galena (PbS), the production of a given lead isotope ends and a distinct 'isotope signature' is created [Dickin (1995)]. With the number of factors affecting the amount of daughter isotope created, the fact that there are three parent nuclides and one non-radiogenic isotope, considerable variations in lead isotope ratios are generated.

Table 1   Geochemical features of lead's four stable isotopes including parent nuclides and their radioactive decay rate (taken from [Dickin (1995)].

Isotope	Atomic mass (amu)	Relative abundance (%)	Parent isotope	Parent half life (years)	Decay constant, λ (year^{-1})
^{204}Pb	203.9730	1.4	—	—	—
^{206}Pb	205.9744	24.1	^{238}U	4.5×10^9	1.55125×10^{-10}
^{207}Pb	206.9759	22.1	^{235}U	77.1×10^8	9.8485×10^{-10}
^{208}Pb	207.9766	52.4	^{232}Th	1.4×10^{10}	4.9475×10^{-11}

The application of lead isotopes to trace anthropogenic sources was demonstrated already four decades ago. Measuring stable lead isotope variability within leaded petrol, North American coals and aerosols, Chow *et al.* [1975] made the critical observation that a large isotopic range is found in natural and synthetic materials, e.g. a 14% range in leaded petrol ^{206}Pb/^{207}Pb between Bangkok, Thailand (1.072) and Santiago de Chile, Chile (1.238) [Chow *et al.* (1975)]. Since isotopic differences are preserved from petrol exhausts and other industrial sources to aerosols, tracing of source inputs to the atmosphere is possible.

The lead isotope tracer technique has proven repeatedly as powerful tool in environmental research. For example, the assessment of high lead concentrations in petrol contaminated groundwater in South Carolina (USA) showed that the dissolved lead derived from sediment particulates of the native aquifer material and not from leaded petrol spills [Landmeyer *et al.* (2003)]. Investigations into the lead isotope variations in environmental and biological samples in Broken Hill, Australia, one of the largest Pb-Zn-Ag mines in the world, showed that in individual cases, lead from petrol and paint was the major source and not the ore body itself [Gulson *et al.* (1994)]. First attempts have also been made to use lead isotope geochemistry to model and predict lead input and output in soils [Semlali *et al.* (2004)], which is crucial to estimate the future behaviour of historical lead deposited in the environment.

3. 'Natural' Lead in the Environment

To understand the impact of anthropogenic lead in the environment, the 'natural' geochemical cycles and biogeochemistry of lead must be fully understood.

3.1. *Natural lead in the terrestrial environment*

Lead is generally associated with massive sulphide deposits; it is a 'chalcophile' element similar to copper, cadmium, zinc and silver. It occurs in the crystal structures of rock forming silicates (e.g. K-feldspar) and oxides of common crustal rocks [Klein and Hurlbut (1999); Krauskopf and Bird (1995)]. With respect to the mineralogical forms, the primary mineral of lead in nature is galena (PbS, Fig. 2) and its oxidation products like plattnerite (PbO_2), cerussite ($PbCO_3$), and anglesite ($PbSO_4$) are important.

Most lead is found in the lithosphere (soils and sediments: $\sim 5 \times 10^{19}$ g), followed by the hydrosphere ($\sim 10^{16}$ g) and biosphere ($\sim 10^{12}$ g). The main

Fig. 2   **An example of coarse, cubic galena (PbS), the principal ore mineral of lead, intergrown with pyrrhotite (iron sulphide). Sample taken from Black Mountain Pb-Zn-(Cu-Ag) mine in the northern Cape Province, South Africa.**

natural source of lead in sediments and soils is dust from rock weathering. Weathering of igneous and magmatic rocks results in accumulation of lead particularly in the clay fraction and in ferruginous components. A review of the basic geochemical processes and mechanisms affecting the chemistry of lead in soils (in particular with respect to bioavailability and solubility) is given elsewhere [Hettiarachchi and Pierzynski (2004)].

Important insights with respect to the isotope systematics of lead in soils during weathering were gained from field studies [Hansmann and Koeppel (2000); Teutsch *et al.* (2001)] and dissolution experiments of fresh granite and soil samples [Harlavan and Erel (2002)]. The latter suggested (see Fig. 3) that during the early stages, lead is preferentially released from accessory phases (i.e. allanite, sphene, apatite) which results in higher ^{206}Pb/^{207}Pb values and different REE patterns in solution compared to rock values [Harlavan and Erel (2002)]. A very recent intriguing study used the lead isotopic composition of river sediments from the earth's major river basins, from old cratonic to young orogenic areas and from subarctic to tropical climates, to estimate the lead isotopic composition of the average upper crust, the source of 'natural' soil dust. The calculations based on flux weighted averages of particulate lead gave values of 19.07, 15.74 and 39.35 for ^{206}Pb/^{204}Pb, ^{207}Pb/^{204}Pb and ^{208}Pb/^{204}Pb [Millot *et al.* (2004)].

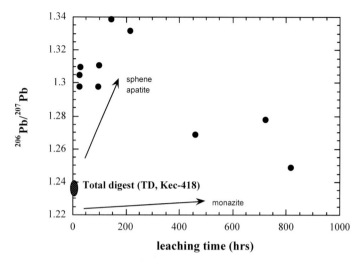

Fig. 3 $^{206}Pb/^{207}Pb$ versus leaching time (in hrs) of acid leach fractions of a granite (sample Kec/TF-418) along with the average ratios of the total digest (TD) and isotopic trends for monazite, sphene, and apatite (taken from [Harlavan and Erel (2002)].

3.2. *Natural lead in the marine environment*

In the ocean's surface waters, lead's geochemical cycle reflects its atmospheric source and particle reactivity. Dissolved lead accounts for approximately 90% of the total open ocean concentration, and a significant fraction (50–70%) of dissolved lead is complexed to organic ligands. Inorganic complexes dominate the remaining dissolved lead. It is estimated that $PbCO_3$ accounts for 55% of dissolved inorganic fraction, followed by $PbCl_2$ (11%), $Pb(CO_3, Cl_3)$ (10%) and $PbCl^+$ (7%) [Whitfield and Turner (1980)]. The remaining 17% includes chloride complexes along with additional sulphate and hydroxide complexes. These complexes are rapidly scavenged by biological particles in the open ocean, with surface ocean residence time of approximately two years [Bacon *et al.* (1976)].

Atmospheric deposition and particle scavenging results in upper-ocean lead seasonality, including summer aerosol lead accumulation within a warm, stratified mixed layer, and deep winter mixing diluting the accumulated lead [Wu and Boyle (1997)]. Trace metals, such as lead, with atmospheric sources and surface ocean residence times of less than a few years show substantial spatial and temporal variability in the surface mixed layer and upper thermocline due to seasonal cycles, storm or eddy events and the changing strength of source emissions [Boyle *et al.* (1986); Boyle *et al.* (1994)]. This variability can significantly obscure long-term trends as surface water and measurements near Bermuda showed [Boyle *et al.* (1986); Boyle *et al.* (1994); Véron *et al.* (1993)].

Most of our knowledge on the geochemical controls of lead in the deep sea is derived from analyses of lead isotopes of deep sea sediments around the world including hydrogenic Fe-Mn crusts [Ling *et al.* (2005); von Blanckenburg *et al.* (1996)]. Lead isotope records of the Ceara Rise (western Atlantic) and Sierra Leone Rise (eastern Atlantic) cores show a clear glacial-interglacial cyclicity, reflected by alternating unradiogenic lead and radiogenic lead. The glacial-interglacial lead isotopic contrast is observed in Pb-Pb space and can be explained in terms of binary mixing — variations along the mixing lines reflecting changes in the relative proportions of the glacial (unradiogenic) and interglacial (radiogenic) lead source(s). The correlation found between lead isotope cycles and other paleoclimate proxies suggests that lead isotopes may be responding to variations in Earth's orbital parameters [Abouchami and Zabel (2003)].

3.3. *Natural changes in atmospheric deposition*

Only a small amount of lead is found in the atmosphere ($\sim 10^{10}$ g) but this reservoir is the major transport pathway of lead within the environment and consequently atmospheric circulation patterns are of prime importance to understand atmospheric deposition. Natural particles of lead are generally associated with larger particle size, rapid deposition rates and short atmospheric residence time. Primary natural lead sources include wind-blown dust, volcanic emission and biogenic particulates. Estimated total emission volumes range between 0.9 and 23.5×10^9 g yr^{-1} [Nriagu (1989a)] (see Table 2).

The present day atmosphere is dominated by anthropogenic lead, and information regarding the pattern, controls and characteristics of natural atmospheric deposition can only be derived from geochemical archives including ice cores, peats and sediments [Boutron (1995); Last and Smol (2001); Mackay *et al.* (2004); Shotyk (1996)]. The longest records of atmospheric deposition have been derived form ice cores giving fascinating insights into the glacial-interglacial variations in lead concentrations and lead isotopes, mineral dust fluxes and volcanogenic lead to ice cores.

Table 2 Estimated natural global annual lead emission (in 10^6 kg y^{-1}) (taken from [Nriagu (1989a)]). The biogenic estimate includes terrestrial and marine sources.

Source	Emission (10^6 kg y^{-1})
Windblown soil	0.3–7.5
Volcanic emission	0.5–6.0
Wild forest fires	0.1–3.8
Biogenic particulates	0.0–3.3
Sea Salt spray	0.0–2.8
Total	**0.9–23.5**

3.4. *Glacial-interglacial variations in lead concentrations and lead isotopes*

The Vostok core from Antarctica assessed atmospheric lead deposition from 65 000 to 240 000 yrs BP reaching back to the beginning of the penultimate ice age and the preceding interglacial (isotope stage 7.5) [Hong *et al.* (2003)]. Lead concentrations were highly variable with low values during warm climatic stages and much higher values during cold stages, especially during isotope stage 4.2 and 6.2 to 6.6. Additional work reported from the EPICA Dome C ice core dating back to 220 kyr BP showed that also lead isotopic compositions in Antarctic ice varied with changing climate [Vallelonga *et al.* (2005)]. Figure 4 shows a $^{206}Pb/^{207}Pb$ ratios decrease during glacial periods, with the lowest values occurring during colder climatic periods (stages 2, 4 and 6) and the Holocene. Low lead concentrations were found during the Holocene and the last interglacial (climate stage 5.5) while higher lead concentrations were found during cold climatic periods.

3.5. *Mineral dust fluxes and volcanogenic lead to ice cores*

The Vostok core suggests that virtually 100% of natural lead deposition during cold climate can be accounted for using soil dust and rock, while contributions from volcanoes might have been significant during warm stages [Hong *et al.* (2003)]. The EPICA Dome C core shows also a dominance of soil and rock derived lead for the pre-industrial period [Vallelonga *et al.* (2005)]. This source apportionment, however, contrasts with findings from other Antarctic locations. Matsumoto and Hinkley [2001] suggested that the deposition rate in the pre-industrial ice from coastal west Antarctica

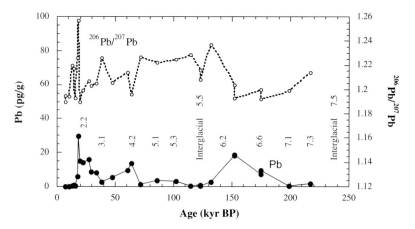

Fig. 4 **A 220-kyr record of lead concentration and ^{206}Pb/^{207}Pb from the EPICA Dome C ice core. Marine isotope stage numbers are also shown (taken from [Vallelonga *et al.* (2005)]).**

was approximately matched by the output rate to the atmosphere by quiescent (non-explosive) degassing of volcanoes worldwide. This conclusion is supported by the isotopic compositions of lead, which are similar to those of a suite of ocean island volcanoes, mostly in the Southern Hemisphere [Matsumoto and Hinkley (2001)].

For Greenland, an ice record from the GRIP core covers the time period between 8250 to 149 100 and it is suggested that lead in the Northern Hemisphere was derived mainly from soil dust during both glacial and interglacial periods [Hong *et al.* (1996)].

Peat bogs are the other terrestrial archive that testifies only atmospheric deposition and work on long term records in Switzerland [Shotyk *et al.* (1998)], Sweden [Klaminder *et al.* (2003)] and Spain [Kylander *et al.* (2005)] showed a similar control of climate on the lead cycling, for example, the Younger Dryas and Saharan Drying increased atmospheric lead fluxes during the Holocene.

4. Anthropogenic Lead

To what extent has lead's natural geochemical cycle been altered by human activities? Total anthropogenic lead emissions exceeded natural sources by an order of magnitude in 1983, with emissions from non-ferrous metal smelting, coal combustion, waste incineration, other industrial sources and leaded

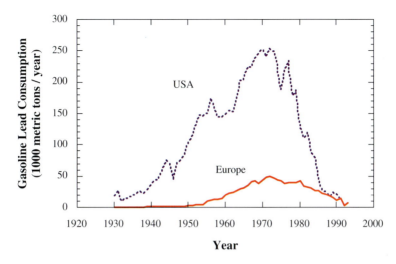

Fig. 5 **Historical leaded petrol consumption in the US and Western Europe between 1930–1990. The nations for Europe include inventories for France, Italy, the United Kingdom and Germany (taken from [Wu and Boyle (1997)]).**

petrol use; the majority of anthropogenic lead released since 1950 comes from the latter [Nriagu and Pacyna (1988)]. The consumption of tetraethyl lead between 1930 and 1990 in the US and Europe, the main global pollution source during that time, is shown in Fig. 5. These economic data demonstrate the large US consumption compared that of countries in Europe and the time-dependent nature of the consumption, which peaked in 1972.

There has been a heated debate about lead usage in gasoline and its effects on environmental and human health. Fascinating historical details are given in two more recent publications [Bertsch-McGrayne (2001); Kitman (2000)]. In particular the impact and achievements of Thomas Midgley (1889–1944), mechanical engineer at General Motors and inventor of organo-Pb additives, and of Clair C. Patterson (1922–1995), who demonstrated the detrimental impact of lead pollution on the environmental health by measuring lead content in 1600 year old Peruvian Indians [Patterson (1965)], are described.

The phasing out of leaded petrol in North America and Europe during the 1970s and 1980s significantly reduced human contributions of lead to the environment [Hagner (2000); von Storch *et al.* (2003)]. Demonstrating the success of the lead policies in Germany, von Storch *et al.* [2003] showed that concentrations in leaves and human blood have steadily declined since the early 1980s and the economic repercussion that had been feared from the phase out did not emerge. Instead, the affected mineral, oil and car manufacturing industries were able to adapt without incurring significant extra costs [von Storch *et al.* (2003)].

New accurate and complete emissions inventories for atmospheric lead on a global scale have been produced [Pacyna and Pacyna (2001); Pacyna *et al.* (1995)] updating the first global estimate made by Niriagu and Pacyna [Nriagu and Pacyna (1988)]. Combustion of leaded, low-leaded and unleaded gasoline continues to the major source of lead in the environment, contributing about 74% to the total anthropogenic emissions of this metal in 1995. As seen in Table 3, the largest contributions, 43–44% comes from combustion of gasoline in Asia. The largest contributions from individual countries come from European Russia, China and Mexico, emitting over 8500 tonnes of lead in 1995 [Pacyna and Pacyna (2001)].

4.1. *Aerosol compositions*

An important difference between anthropogenic and natural emissions is that anthropogenic emissions are often the result of high temperature processes, which result in the release of lead into the fine fraction of aerosols. These fine particles can be transported long distances before being deposited, mixing on hemispheric and global scales [Bollhöfer and Rosman (2000); Bollhöfer and Rosman (2001); Bollhöfer and Rosman (2002); Boutron (1995); Simonetti *et al.* (2003)]. Atmospheric lead fluxes follow the primary anthropogenic sources and prevailing winds; with greatest lead deposition occurring in densely populated urban areas but reaching and affecting remote places as well. For example, Duce *et al.* [1991]

Table 3   Worldwide emissions of lead from mobile sources in 1995 (in tonnes, taken from [Pacyna and Pacyna (2001)]).

Continent	Minimum	Maximum
Europe	19 507	19 507
Africa	6 852	11 992
Asia	32 996	44 293
North America	10 414	15 780
South America	4 866	7 270
Australia	2 000	2 000
World Total	**76 635**	**100 842**

found atmosphere-ocean fluxes over the North Atlantic Ocean being five-fold higher (0.08–$1.03\,\mathrm{mg\,m^{-2}\,y^{-1}}$ versus 0.02–$0.2\,\mathrm{mg\,m^{-2}\,y^{-1}}$) than those found over the Pacific Ocean highlighting the significance of North American emissions.

Most studies assessing sources of atmospheric aerosols have been conducted on limited regional scale in urban and remote areas. A classic study around an urban area was conducted in Canada. The $^{206}\mathrm{Pb}/^{207}\mathrm{Pb}$ ratio of integrated Canadian and US aerosols demonstrated that 24–43% of the anthropogenic lead aerosols close to Toronto were derived from the US [Sturges and Barrie (1987)]. A more recent study conducted in Sicily showed substantial 'natural' lead pollution (e.g. from volcanoes) can occur, accounting up to 30% at Mt. Etna and 80% at Vulcano Island [Monna *et al.* (1999)]. An early study in a remote area has been conducted at the tropical South Pacific island of American Samoa. At this site, Patterson and Settle constructed an atmospheric mass balance demonstrating that natural volcanic emissions and soil dusts accounted for about 1% of the total lead fluxes [Patterson and Settle (1987)].

Three recent studies by Bollhöfer and Rosman [2000–2002] defined the extent to which lead isotopic ratios in aerosols vary on a global scale. They showed first that there are significantly different lead isotope ratios on a regional scale, for example, in the Northern Hemisphere, during the 1990s, the least radiogenic compositions were found in France and Spain ($^{206}\mathrm{Pb}/^{207}\mathrm{Pb}$ between 1.097 and 1.142) and the most radiogenic in the United States ($^{206}\mathrm{Pb}/^{207}\mathrm{Pb}$ between 1.173 and 1.231). Second, countries where leaded petrol is still marketed (e.g. Asia, Africa and Eastern Europe), automobile emissions dominate the lead isotope signature. In the US, however, the impacts of the phasing out of leaded petrol are already detectable. Whereas in the early 1990s the $^{206}\mathrm{Pb}/^{207}\mathrm{Pb}$ ratio was fairly uniform across the country, recent samples show that east (1.173–1.231) and west (1.159–1.188) coast aerosols differed. The reason for this is unclear but one suggestion is that there is long distance transport of aerosols from Asia that are characterised by high lead concentrations and low $^{206}\mathrm{Pb}/^{207}\mathrm{Pb}$ ratios [Bollhöfer and Rosman (2000); Bollhöfer and Rosman (2001)]. Time series from 38 globally distributed sites revealed significant seasonal variations at sampling sites close to Eastern Europe that probably reflect an enhanced westward transport of pollution in winter. They also showed that the temporal variability in Canada and North America is now larger than before due to decreased airborne lead levels coupled with an increase in industrial sources. Temporal variations on mainland Australia are comparatively

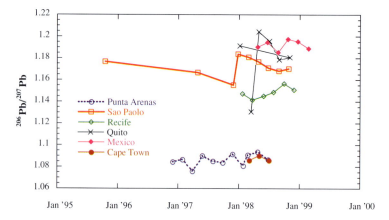

Fig. 6 ^{206}Pb/^{207}Pb ratio time series in South America (Punta Arenas, Sao Paolo, Recife, Quito), South Africa (Cape Town) and Mexico between 1995 and 1999 (taken from [Bollhöfer and Rosman (2002)]).

small with a typical range of 0.2% in ^{206}Pb/^{207}Pb ratio and isotopic ratios that indicate that leaded petrol is still a major source. Figure 6 shows the long-term data measured at selected sites in South America, South Africa and Mexico. The ratios at Punta Arenas and South Africa show similar ^{206}Pb/^{207}Pb ratios and indicate a common supplier of alkyllead. There is also a notable increase of ^{206}Pb/^{207}Pb ratio at both sites compared to data from 1994 [Bollhöfer and Rosman (2000)], which could be due to a relative increase from industrial sources because of the slow phasing out of leaded gasoline in both regions. In Recife in north eastern Brazil, seasonal variations are noticeable with low ratios during the southern autumn and higher ratios in spring. lead isotopes in aerosols collected in Mexico City are very stable and show variation of below 0.3%.

4.2. Lead in the marine system

Historical atmospheric lead fluxes have directly affected seawater lead concentrations. The first papers describing the dominance of anthropogenic lead in ocean surface waters were published in the early 1980s using lead concentration and isotope ratios measurements along vertical depth profiles and in surface water in the Pacific and North Atlantic ocean [Flegal and Patterson (1983); Flegal et al. (1984); Schaule and Patterson (1981); Schaule and Patterson (1983)]. Consequent monitoring of surface waters near Bermuda and the western North Atlantic showed a threefold decline in lead concentrations from 1971 to 1987, with continuing but notably slower decline in the 1990s [Reuer *et al.* (2003); Shen and Boyle (1988b); Wu and Boyle (1997)]. This reduction was concurrent with a 20-fold decrease in leaded petrol consumption in the US from 1979 to 1993 and this apparent difference reflects the amount of emitted lead reaching the North Atlantic and its subsequent admixture within the subtropical gyre (Fig. 7). A similar decrease in lead concentrations has been detected in the Mediterranean [Alleman *et al.* (2000); Nicolas *et al.* (1994)].

Historically, the US has consumed lead with high ^{206}Pb/^{207}Pb signatures (notably Missouri lead) whereas European nations used low ^{206}Pb/^{207}Pb lead. This dissimilarity has been apparent in North Atlantic surface water measurements: westerly atmospheric transport results in high seawater ^{206}Pb/^{207}Pb near North America and reduced ^{206}Pb/^{207}Pb ratios to the south most likely reflect north-easterly European fluxes [Hamelin *et al.* (1997); Véron *et al.* (1994); Weiss *et al.* (2003)]. Isotopic boundaries observed between the tropics and sub-tropics agree with lead concentration gradients, with the southerly advection of low lead concentration waters from the South Atlantic [Véron *et al.* (1994); Weiss *et al.* (2003)]. With respect to North Atlantic surface waters, the presence of American lead has

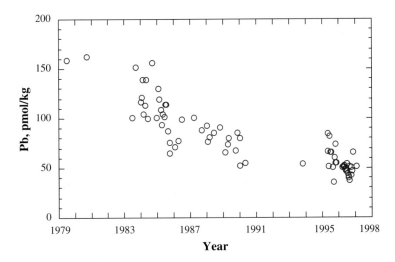

Fig. 7 **Lead concentrations in surface water near Bermuda, 1979–1997 showing samples analysed at the MIT and Cal Tech laboratories** [Schaule and Patterson (1983); Véron *et al.* (1993); Wu and Boyle (1997)].

been detected over the entire north and central North Atlantic [Véron *et al.* (1994)] and the subtropical north-eastern Atlantic [Hamelin *et al.* (1997)]. In two more recent contributions, lead isotopes were further applied to elucidate the role of oceanic circulation on contaminant distribution in the South Atlantic [Alleman *et al.* (2001a); Alleman *et al.* (2001b)].

With the passage of time from the 1972 maximum, the lead tracer has been transported into intermediate and deep water given their greater ventilation ages, allowing stable lead isotopic compositions to be employed as tracers of abyssal mixing. For example, vertical concentration profiles in the sub arctic North Atlantic showed spatial gradients in the isotopic signature which are consistent with the thermohaline circulation pattern of the different water masses in that region and their discrete isotopic signatures [Véron *et al.* (1999)]. Figure 8 shows three distinct $^{206}Pb/^{207}Pb$ ratios at the southern location in the Faroe Bank Channel at the base of the Norwegian sea. The ratio (1.179) in the surface water is comparable to that of the aerosols (1.176 ± 0.004) collected at the relatively proximate site in Mace Head Ireland. The ratio (1.184–1.185) markedly increases in the subsurface (130–430 m) immediately below with the salinity maximum characteristics of the North Atlantic Drift (NAD) flowing into the region. The ratios (1.174–1.175) decrease with the freshening of deeper (750–810 m) waters associated with the formation of Iceland-Scotland Overflow Waters (ISOW). Likewise, it was shown that advective transport of lead into the deep abyssal waters is facilitated through the formation of North Atlantic Deep Water [Alleman *et al.* (1999)].

5. Reconstructing the Anthropogenic Impact

If past variability is known one can place modern observations in a historical context, determine the spatial impact of multiple anthropogenic sources, assess the relative importance of different emission mechanisms, and assess the impact of policy measures. Terrestrial records provide atmospheric flux estimates to a single locality, whereas marine records integrate multiple sources within a larger, advective reservoir. Several terrestrial environmental archives are used in making historical reconstructions of atmospheric deposition, namely polar ice, peat and lacustrine sediment cores.

5.1. *Evidences from ice core records*

The results from the first of these studies, conducted at Camp Century, Greenland [Murozumi *et al.* (1969)], showed that anthropogenic lead was

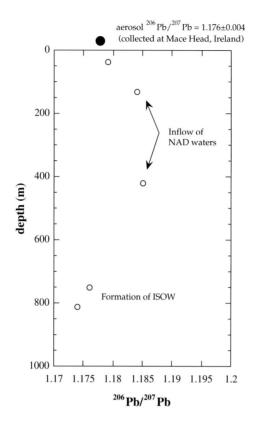

Fig. 8 Vertical profiles of ^{206}Pb/^{207}Pb isotope ratios at the depths around the Iceland-Scotland Ridge (61.4 N, 08.4 E). This profile shows the contamination of the surface waters by relatively local emissions of industrial lead aerosols with a characteristic lead isotopic composition, inflowing North Atlantic Drift (NAD) waters with a contrasting isotopic composition, and the underlying Iceland-Scotland Overflow Waters (ISOW) with a ratio comparable to that of the surface waters which contributes to its formation (see text for details, taken from [Véron *et al.* (1999)]).

detectable in the Arctic; lead concentrations were shown to have increased from less than 0.01 μg g^{-1} in 800 B.C. to greater than 0.200 μg g^{-1} by 1965. The greatest increase during this time occurred in the 1940s, in agreement with Eurasian and North American leaded petrol consumption. Additionally, lead concentrations from 800 B.C. to 1750 A.D. were shown to have increased ten-fold, suggesting significant lead pollution prior to the Industrial Revolution. The work on ice and snow has been extensive and an in depth review up to the mid 1990s is given elsewhere [Boutron (1995)].

Recent work in Greenland produced a continuous high-resolution record of lead and other trace elements (1750 and 1998) showing a large, sustained increase in lead deposition after the introduction of leaded gasoline. Non-gasoline lead contamination likely accounted for $> 50\%$ of the cumulative increase in lead deposition and crustal enrichment since industrialisation [McConnell *et al.* (2002)]. After 1970, the mandated emission reductions mentioned above (mainly phasing out of leaded gasoline in most of Europe) led to a $> 75\%$ decline in annual lead flux with most of the decline occurring from 1970 to 1985.

Recent work from Antarctica reported lead and barium concentrations and lead isotopic compositions from firn core and snow pit samples from Victoria Land dating from 1872 AD to 1994 AD [Van de Velde *et al.* (2005)]. Two periods of major lead enrichment were identified: from 1891 to 1908 AD and from 1948 to 1994 AD. The earlier pollution event is attributed to lead emissions from non-ferrous metal production and coal combustion in the Southern Hemisphere and is in excellent agreement with coincident pollution inputs reported in firn/ice cores from two other regions of Antarctica, at Coats Land and Law Dome. It was calculated that similar to 50% of lead deposited in Victoria Land in 1897 originated from anthropogenic emission sources. The more recent period of lead enhancements, from 1948 to 1994 AD, corresponds to the introduction and widespread use of gasoline alkyl lead additives in automobiles in the Southern Hemisphere, with anthropogenic lead inputs averaging 60% of total lead. Intra- and inter-annual variations in lead concentrations and isotopic compositions were evaluated in snow pit samples corresponding to the period 1991–1994. Substantial variations in Pb/Ba and ^{206}Pb/^{207}Pb ratios were detected but the absence of a regular seasonal pattern for these parameters suggests that the transport and deposition of aerosols to the Antarctic ice sheet are complex and vary from year to year.

Similar comprehensive records for lead isotopes, lead concentrations and organo Pb are reported for several high altitude locations including the Alps [Heisterkamp *et al.* (1999); Rosman *et al.* (2000); Schwikowski *et al.* (2004)] and the Andes [Hong *et al.* (2004)]. The latter record was established using a dated snow/ice core drilled at an altitude of 6542 m on the Sajama ice cap in Bolivia. The analysed sections were dated from the Last Glacial Stage (similar to 22000 years ago), the Mid-Holocene, and the last centuries. The observed variations of crustal enrichment factors (EF_{crust}) for various metals showed contrasting situations. For V, Co, Rb, Sr and U, EF_{crust} values close to unity were observed for all sections, showing that these elements are mainly derived from rock and soil dust. For the other metals including lead, clear temporal trends were observed, with a pronounced increase of EF_{crust} values during the 19th and 20th centuries. This increase shows evidence of metal pollution associated with human activity in South America. For lead an important contribution was from gasoline additives. For metals such as Cu, Zn, Ag and Cd, important contribution was from metal production activities, with a continuous increase of production during the 20th century in countries such as Peru, Chile and Bolivia.

5.2. *Evidences from peat and sediment records*

A historical reconstruction using a peat core from the Jura Mountains in Switzerland [Shotyk *et al.* (1998)] clearly established the time from which local human impacts on the natural lead cycle began. Here significant increases in the Pb/Sc ratio occurred at 3000 radiocarbon years before present and lead isotope data confirmed this change was anthropogenically induced, a product of Phoenician and Greek mining. The lead flux in a sample dated to 1979 $(15.7 \, mg \, m^{-2} \, yr^{-1})$ was found to be 1570 times the natural background value $(0.01 \, mg \, m^{-2} \, yr^{-1})$; this is significantly higher than the Greenland anthropogenic to natural ratio.

Sediment records are numerous — in North America [Gallon *et al.* (2005); Graney *et al.* 1995] and in Europe alike [Kober *et al.* (1999); Renberg *et al.* (1994)]. Sediment records clearly support elevated mid-latitude lead fluxes during the industrialisation (e.g. increase in metal smelting and steel manufacture) and modernisation (coal combustion, electrification) of Europe and North America, followed by the introduction and phase out of leaded petrol [Bindler *et al.* (1999)].

5.3. *Evidences from marine records*

In terms of marine records, surface coral records and marine sediments have been used to place additional constraints on anthropogenic fluxes to the world oceans.

Measuring the Pb/Ca ratio of annually banded corals from Bermuda and the Florida Straits, Shen and Boyle [1987, 1988a] estimated an 11-fold increase in anthropogenic lead fluxes to the western North Atlantic from 1884 $(5.2 \, nmol \, mol^{-1})$ to 1971 $(56.7 \, nmol \, mol^{-1})$ [Shen and Boyle (1987); Shen and Boyle, (1988a)]. This trend was subsequently interrupted by the introduction and phase out of leaded petrol. Reuer *et al.* [2003] reported recently an extended coral proxy record and seawater times series

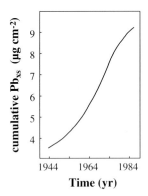

Fig. 9   Average inventory of the input of anthropogenic lead over the North Atlantic (taken from [Véron *et al.* (1987)]).

for the western North Atlantic, showing lead isotope ratios analysed by MC-ICP-MS and using surface coral cores collected from North Rock, Bermuda in 1983 and surface seawater collected at station S [Reuer *et al.* (2003)]. The results show decreasing $^{206}Pb/^{207}Pb$ from 1886 to 1922, diminished $^{206}Pb/^{207}Pb$ variability from 1922 to 1968 and $^{206}Pb/^{207}Pb$ and $^{207}Pb/^{206}Pb$ maxima from 1968 to 1990. The multi decadal reduction of $^{206}Pb/^{207}Pb$ is consistent with increased combustion of North American coal and other lead emissions throughout the Industrial Revolution and agrees with historical anthropogenic lead variability [Reuer *et al.* (2003)].

Significant lead pollution has been documented in surface sediments collected in the north-eastern [van Geen *et al.* (1997); Véron *et al.* (1987)] and north-western Atlantic shelf [Hamelin *et al.* (1997); Hamelin *et al.* (1990)]. The record of Véron *et al.* [1987] suggests that about up to half of the atmospheric pollutant lead which was introduced since the beginning of the Industrial Revolution had accumulated in North Atlantic sediments by the 1980s. Figure 9 shows an average inventory of the input of anthropogenic lead into the North Atlantic, derived from sediment concentrations [Véron *et al.* (1987)]. Lead isotope measurements identified Europe as the dominant source for the northeast and North America for the northwest.

6. Future Steps

Many of the examples given above give credit to the success of phasing out of leaded petrol, so where do we go from here? Is the work done? The answer is that the work is far from done.

At present tetraethyl lead is still used in petrol in the developing world and puts the lives of many people at risk of lead poisoning. This problem is only likely to escalate given that the car population is increasing rapidly. In addition, metal smelting still pose persistent health risks on local to regional levels [Spiro *et al.* (2004)] and ingestion of leaded paint by children remains a significant health problem [Ahamed *et al.* (2005)].

Systematic observations of the anthropogenic lead transient have constrained the global nature of atmospheric pollution. Anthropogenic lead will provide many future insights regarding its elemental flux and provenance, and our current knowledge is incomplete at best. The observations shown here are spatially and temporally biased, often focusing on the Northern Hemisphere, given the location of the principal anthropogenic sources during the 20th century. Limited data presently exists from the Southern Hemisphere and regions emerging from economic development despite known hemispheric-scale effluxes. The transient nature of anthropogenic lead will also provide key insights as the passage of time the lead 'dye' will be transported to older marine and terrestrial sinks. Future geochemical studies of anthropogenic lead and other elements will address the impact of human activities on the global environment and how the Earth dynamically partitions anthropogenic signals into the atmosphere, continents and oceans.

Acknowledgements

The authors would like to thank the many people who helped to shape their research into the environmental geochemistry of lead including Jan Kramers, Bill Shotyk, Antonio Martinez, Bernd Kober, Ed Boyle, Rick Keyser, Victor Köppel, Baruch Spiro, Matthew Thirlwall, Mark Rehkämper, Jerome Nriagu, Yigal Erel, Francois Monna.

A special thanks goes to Mike Warner at Imperial College London, Andy Fleet and Terry Williams at the Natural History Museum for their continuing support of the research facilities. We also thank Peter Sammonds for great editorial handling and one anonymus reviewer for the helpful comments on an earlier version of the paper.

DJW thanks in particular Barry Coles, Thomas Mason, Alla Dolgopolova, John Chapman, Kate Peel and Jo Muller for excellent work at the ICL laboratories.

DJW dedicates this contribution to Stöff and Dicle.

References

Abouchami, W. & Zabel, M. (2003) Climate forcing of the Pb isotope record of terrigenous input into the Equatorial Atlantic. *Earth Planet. Sci. Lett.* **213** (3–4), 221–234.

Ahamed, M., Verma, S., Kumar, A. & Siddiqui, M. K. J. (2005) Environmental exposure to lead and its correlation with biochemical indices in children. *Sci. Total. Environ.* **346**, 48–55.

Albarède, F. & Beard, B. L. (2004) Analytical methods for non-traditional isotopes. *Geochemistry of Non-Traditional Stable Isotopes*, edited by Johnson, C. M., Beard, B. L. and Albarède, F., pp. 113–152, Mineralogical Society of America, Washington.

Alleman, L. Y., Church, T. M., Ganguli, P., Véron, A. J., Hamelin, B. & Flegal, A. R. (2001a) Role of oceanic circulation on contaminant lead distribution in the South Atlantic. *Deep-Sea Res. II* **48**, 2855–2876.

Alleman, L. Y., Church, T. M., Véron, A. J., Kim, G., Hamelin, B. & Flegal, A. R. (2006b) Isotopic evidence of contaminant lead in the South Atlantic troposphere and surface waters. *Deep-Sea Res. II* **48**, 2811–2827.

Alleman, L. Y., Hamelin, B., Véron, A. J., Miquel, J.-C. & Heussner, S. (2000) Lead sources and transfer in the coastal Mediterranean: Evidence from stable lead isotopes in marine particles. *Deep-Sea Res. II* **47**, 2257–2279.

Alleman, L. Y., Véron, A. J., Church, T. M., Flegal, A. R. & Hamelin, B. (1999) Invasion of the abyssal North Atlantic by modern anthropogenic lead. *Geophys. Res. Lett.* **26**(10), 1477–1480.

Bacon, M. P., Spencer, D. W. & Brewer, P. G. (1976) $^{210}Pb/^{226}Ra$ and $^{210}Po/^{210}Pb$ disequilibria in seawater and suspended particulate matter. *Earth Planet. Sci. Lett.* **32**, 277–296.

Bertsch-McGrayne, S. (2001) *Prometheans in the Lab — Chemistry and the Making of the Modern World*, 243 pp., McGrawth Hill, New York.

Bindler, R., Brannvall, M. L., Renberg, I., Emteryd, O. & Grip, H. (1999) Natural lead concentrations in pristine boreal forest soils and past pollution trends: A reference for critical load models. *Environ. Sci. Technol.* **33**(19), 3362–3367.

Bollhöfer, A. & Rosman, K. J. R. (2000) Isotopic source signature for atmospheric lead: The Southern Hemisphere. *Geochim. Cosmochim. Acta* **64**, 3251–3262.

Bollhöfer, A. & Rosman, K. J. R. (2001) Isotopic source signatures for atmospheric lead: The Northern Hemisphere. *Geochim. Cosmochim. Acta* **65**(11), 1727–1740.

Bollhöfer, A., & Rosman, K. J. R. (2002) The temporal stability in lead isotope signatures at selected sites in the Southern and Northern Hemisphere. *Geochim. Cosmochim. Acta* **66**(8), 1375–1386.

Boutron, C. F. (1995) Historical reconstruction of the Earth's past atmospheric environment from Greenland and Antarctic snow and ice cores. *Environ. Rev.* **3**, 1–28.

Boyle, E. A., Chapnick, S. D., Shen, G. T. & Bacon, M. P. (1986) Temporal variability of lead in the western North Atlantic. *J. Geophys. Res.* **91**, 8573–8593.

Boyle, E. A., Sherrell, R. M. & Bacon, M. P. (1994) Lead variability in the western North Atlantic Ocean and central Greenland: Implications for the search for decadal trends in anthropogenic emissions. *Geochim. Cosmochim. Acta* **58**(15), 3227–3238.

Chow, T. J., Snyder, C. B. & Earl, J. L. (1975) Isotope ratios of lead as pollutant source indicators, *UN, FAO and IAEA Symp. IAEA-SM-191/4*, pp. 95–105, International Atomic Energy Agency, Vienna.

Dickin, A. P. (1995) *Radiogenic Isotope Geology*, 475 pp., Cambridge University Press, Cambridge.

Duce, R. A., Liss, P. S., Merill, J. T., Atlas, E. L., Buat-Ménard, P., Hicks, B. B., Mille, J. M., Prospero, J. M., Arimoto, R., Church, T. M., Ellis, W., Galloway, J. N., Hansen, L., Jickels, T. D., Knap, A. H., Reinhardt, K. H., Schneider, B., Soudine, A., Tokos, J. J., Tsunogai, S., Wollast, R. & Zhou, M. (1991) The atmospheric input of trace species to the world ocean. *Global. Biogeochem. Cycles* **5**(3), 193–259.

Flegal, A. R. & Patterson, C. C. (1983) Vertical concentration profiles of Pb in the Central Pacific at 15° N and 20° N. *Earth Planet. Sci. Lett.* **64**, 19–32.

Flegal, A. R., Schaule, B. K. & Patterson, C. C. (1984) Stable isotopic ratios of lead in surface waters of the Central Pacific. *Mar. Chem.* **14**, 281–287.

Gallon, C., Tessier, A., Gobeil, C. & Beaudin, L. (2005) Sources and chronology of atmospheric lead deposition to a Canadian Shield lake: Inferences from Pb isotopes and PAH profiles. *Geochim. Cosmochim. Acta* **69**(13), 3199–3210.

Graney, J. R., Halliday, A. N., Keeler, G. J., Nriagu, J. O., Robbins, J. A. & Norton, S. A. (1995) Isotopic record of lead pollution in lake sediments from the northeastern United States. *Geochim. Cosmochim. Acta* **59**, 1715–1728.

Gulson, B., Mizon, K., Law, A., Korsch, M. & Howarth, D. (1994) Sources and pathways of lead in humans from the Broken Hill mining community — an alternative use of exploration methods. *Econ. Geol.* **89**, 889–908.

Hagner, C. (2000) European regulations to reduce lead emissions from automobiles — did they have an economic impact on the German gasoline and automobiles markets. *Regional Environ. Change* **1**, 135–151.

Hamelin, B., Ferrand, J. L., Alleman, L. & Nicolas, E. (1997) Isotopic evidence of pollutant lead transport from North America to the subtropical North Atlantic gyre. *Geochim. Cosmochim. Acta* **61**(20), 4423–4428.

Hamelin, B., Grousset, F. E. & Sholkovitz, E. R. (1990) Pb isotopes in surficial pelagic sediments from North Atlantic. *Geochim. Cosmochim. Acta* **54**, 37–47.

Hansmann, W. & Koeppel, V. (2000) Lead-isotopes as tracers of pollutants in soils. *Chem. Geol.* **171**, 123–144.

Harlavan, Y. & Erel, Y. (2002) The release of Pb and REE from granitoids by the dissolution of accessory phases. *Geochim. Cosmochim. Acta* **66**(5), 837–848.

Heisterkamp, M. & Adams, F. (1999) *In situ* propylation using sodium tetrapropylborate as a fast and simplified sample preparation for the speciation analysis of organolead compounds using GC-MIP-AES. *J. Anal. Atom. Spectrom.* **14**, 1307–1311.

Heisterkamp, M., Van de Velde, K. Ferrari, C. Boutron, C. F. & Adams, F. (1999) Present century record of organolead pollution in high altitude alpine snow. *Environ. Sci. Technol.* **33**, 4416–4421.

Hettiarachchi, G. M. & Pierzynski, G. M. (2004) Soil lead bioavaliability and *in situ* remediation of lead contaminated soils: A review. *Environ. Progress* **23**(1), 78–93.

Hill, S. J., Arowolo, T. A., Butler, O. T., Chenery, S. R. N., Cook, J. M., Cresser, J. M. & Miles, D. L. (2002) Atomic spectrometry update. Environmental analysis. *J. Anal. Atom. Spectrom.* **17**(3), 284–317.

Hirao, Y. & Patterson, C. C. (1974) Lead aerosol pollution in the High Sierra Overrides natural mechanism which exclude lead from a food chain. *Science* **184**, 989–992.

Hong, S., Candelone, J. P., Turetta, C. & Boutron, C. F. (1996) Changes in natural lead, copper, zinc, and cadmium concentrations in Central Greenland ice from 8,250 to 149,100 years ago: Their association with climatic changes and resultant variations of dominant source contributions. *Earth Planet. Sci. Lett.* **143**, 233–244.

Hong, S., Kim, Y., Boutron, C. F., Ferrari, C. P., Petit, J. R., Barbante, C., Rosman, K. J. R. & Lipenkov, V. Y. (2003) Climate-related variations in lead concentrations and sources in Vostok Antarctic ice from 65,000 to 240,000 years BP. *Geophys. Res. Lett.* **30**(22), 2138.

Hong, S. M., Barbante, C., Boutron, C., Gabrielli, P., Gaspari, V., Cescon, P., Thompson, L., Ferrari, C. F., Francou, B. & Maurice-Bourgoin, L. (2004) Atmospheric heavy metals in tropical South America during the past 22,000 years recorded in a high altitude ice core from Sajama, Bolivia. *J. Environ. Monit.* **6**(4), 322–326.

Kitman, J. L. (2000) The secret history of lead. *The Nation* **270**(11), 11–43.

Klaminder, J., Renberg, I. Bindler, R. & Emteryd, O. (2003) Isotopic trends and background fluxes of atmospheric lead in northern Europe: Analyses of three ombrotrophic boigs from South Sweden. *Global Biogeochem. Cycles* **17**.

Klein, C. & Hurlbut, C. S. (1999) *Manual of Mineralogy*, John Wiley & Sons, New York.

Kober, B., Wessels, M., Bollhofer, A. & Mangini, A. (1999) Pb isotopes in sediments of Lake Constance, Central Europe, constrain the heavy metal pathways and the pollution history of the catchment, the lake and the regional atmosphere. *Geochim. Cosmochim. Acta* **63**(9), 1293–1303.

Krachler, M., Zheng, J., Fisher, D. & Shotyk, W. (2004) Direct determination of lead isotopes in Arctic ice samples at picogram per gram levels using inductively coupled plasma-sector field ms coupled with a high-efficiency sample introduction system. *Anal. Chem.* **76**, 5510–5517.

Krauskopf, K. B. & Bird, J. D. (1995) *Introduction to Geochemistry*, McGraw-Hill, New York.

Kylander, M. E., Weiss, D., Garcia-Sanchez, R., Martinez-Cortizas, A. & Coles, B. J. (2005) Refining the pre-industrial atmospheric lead isotope evolution curve in Europe using an 8,000 years old peat core from NW Spain. *Earth Planet. Sci. Lett.* **240**, 467–485.

Landmeyer, J. E., Bradley, P. M. & Bullen, T. D. (2003) Stable lead isotopes reveal a natural source of high lead concentrations to gasoline-contaminated groundwater *Environ. Geology* **45**(1), 12–22.

Last, W. M. & Smol, J. P. (2001) Tracking environmental change using lake sediments. *Developments in Paleoenvironmental Research*, edited by Last, W. M. and Smol, J. P., pp. 504, Kluwer Academic Publishers, Dordrecht.

Ling, H. F., Jiang, S. Y., Frank, M., Zhou, H. Y., Zhou, F., Lu, Z. L., Chen, X. M., Jiang, Y. H. & Ge, C. D. (2005) Differing controls over the Cenozoic Pb and Nd isotope evolution of deepwater in the central North Pacific Ocean. *Earth Planet. Sci. Lett.* **232**(3–4), 345–361.

Mackay, A., Battarbee, P., Birks, P. & Oldfield, M. (2004) The Holocene, pp. 540, Arnold/Hodder Publisher, London.

Matsumoto, A. & Hinkley, T. K. (2001) Trace metal suites in Antarctic pre-industrial ice are consistent with emissions from quiescent degassing of volcanoes worldwide. *Earth Planet. Sci. Lett.* **186**(1), 33–43.

McConnell, J., Lamorey, G. W. & Hutterli, M. A. (2002) A 250-year high-resolution record of Pb flux and crustal enrichment in central Greenland. *Geophys. Res. Lett.* **29**(23), 2130.

Millot, R., Allègre, C.-J., Gaillardet, J. & Roy, S. (2004) Lead isotope systematics of major river sediments: A new estimate of the Pb isotopic composition of the Upper Continental Crust *Chem. Geol.* **203**, 75–90.

Monna, F., Aiuppa, A., Varrica, D. & Dongarra, G. (1999) Pb isotope composition in lichens and aerosols from eastern Sicily: insights into the regional impact of volcanoes on the environment. *Environ. Sci. Technol.* **33**, 2517–2523.

Murozumi, M., Chow, T. J. & Patterson, C. (1969) Chemical concentrations of pollutant lead aerosols, terrestrial dusts and sea salts in Greenland and Antarctic snow strata. *Geochim. Cosmochim. Acta* **33**(10), 1247–1294.

Needleman, H. L. (1997) Clamped in a straitjacket: The insertion of lead into gasoline. *Environ. Res.* **74**(2), 95–103.

Nicolas, E., Ruiz Pino, D., Buat-Ménard, P. & Béthoux, J. P. (1994) Abrupt decrease of lead concentrations in the Mediterranean: A response to antipollution policy. *Geophys. Res. Lett.* **21**, 2119–2122.

Nriagu, J. O. (1983) *Lead and Lead Poisoning in Antiquity*, John Wiley & Sons, New York.

Nriagu, J. O. (1989a) A global assessment of natural sources of atmospheric trace metals. *Nature* **338**, 47–49.

Nriagu, J. O. (1989b) The history of leaded gasoline. *Heavy metals in the environment*, edited by J.-P. Vernet, pp. 361–366, Page Bros.

Nriagu, J. O. (1990) The rise and fall of leaded gasoline. *Sci. Total Environ.* **92**, 13–28.

Nriagu, J. O. & Pacyna, J. M. (1988) Quantitative assessment of worldwide contamination of air, water, and soils by trace metals. *Nature* **333**, 134–139.

Pacyna, J. M. & Pacyna, E. M. (2001) An assessment of global and regional emissions of trace metals to the atmosphere from anthropogenic sources worldwide. *Environ. Review* **9**, 269–298.

Pacyna, J. M., Scholtz, T. M. & Li, Y.-F. (1995) Global budget of trace metal sources *Environ. Rev.* **3**, 145–159.

Patterson, C. C. (1965) Contaminated and natural Pb environments of man. *Archives Environ. Health* **11**, 344–360.

Patterson, C. C. & Settle, D. M. (1976) The reduction of orders of magnitude errors in lead analysis of biological materials and natural waters by controlling the extend and sources of industrial lead contamination introduced during sample collecting, handling and analysis. *Accuracy in Trace Analysis: Sampling, Sample Handling, Analysis*, edited by P. Lafleur, pp. 321–351, Natl. Bur. Standards Spec. Publ. 422.

Patterson, C. C. & Settle, D. M. (1987) Review of data on eolian fluxes of industrial and natural lead to the lands and seas in remote regions on a global scale. *Mar. Chem.* **22**, 137–162.

Rehkaemper, M., Schoenbaechler, M. & Stirling, C. H. (2001) Multiple collector ICP-MS: Introduction to instrumentation, measurement techniques and analytical capabilities. *Geostand. News.* **25**(1), 23–40.

Renberg, I., Persson, M. W. & Emteryd, O. (1994) Pre-industrial atmospheric lead contamination detected in Swedish lake sediments. *Nature* **368**, 323–326.

Reuer, M. K., Boyle, E. A. & Grant, B. C. (2003) Lead isotope analysis of marine carbonates and seawater by multiple collector ICP-MS. *Chem. Geol.* **200**, 137–153.

Reuer, M. K. & Weiss, D. J. (2002) Anthropogenic lead dynamics in the terrestrial and marine environment. *Phil. Trans. R. Soc. London A* **360**, 2889–2904.

Rosman, K. J. R., Ly, C., Van der Velde, K. & Boutron, C. F. (2000) A two century record of lead isotopes in high altitude Alpine snow and ice. *Earth Planet. Sci. Lett.* **176**(3–4), 413–424.

Schaule, B. K. & Patterson, C. C. (1981) Lead concentrations in the north-east Pacific: Evidence for global anthropogenic perturbations. *Earth. Planet. Sci. Lett.* **54**, 97–116.

Schaule, B. K. & Patterson, C. C. (1983) Perturbations of the natural lead profile in the Sargasso Sea by industrial lead. *Trace Metals in Sea Water*, edited by Wong, C. S., Boyle, E. A., Bruland, K., Burton, D. & Goldberg, E. D., pp. 487–504, Plenum, New York.

Schwikowski, M., Barbante, C. Döring, T., Gaeggler, H. W., Boutron, C., Schotter, U., Tobler, L., Ferrari, C., Cozzi, G. Rosman, K. J. R. & Cescon, P. (2004) Post-17th-century changes of European lead emissions recorded in high-altitude alpine snow and ice. *Environ. Sci. Technol.* **38**(4), 957–964.

Semlali, R. M., Dessogne, J.-B., Monna, F., Bolte, J., Azimi, S., Navarro, N., Denaix, L., Loubet, M., Chateau, C. & Van Oort, F. (2004) Modelling lead input and output in soils using lead isotopic geochemistry. *Environ. Sci. Technol.* **38**, 1513–1521.

Shen, G. T. & Boyle, E. A. (1987) Lead in corals: Reconstruction of historical industrial fluxes to the surface ocean. *Earth Planet. Sci. Lett.* **82**, 289–304.

Shen, G. T. & Boyle, E. A. (1988a) Determination of lead, cadmium and other trace elements in annually-banded corals. *Chem. Geol.* **67**, 47–62.

Shen, G. T. & Boyle, E. A. (1988b) Thermocline ventilation of anthropogenic Pb in the western North Atlantic. *J. Geophys. Res.* **93**, 15715–15732.

Shotyk, W. (1996) Peat bogs archives of atmospheric metal deposition: Geochemical assessment of peat profiles, natural variations in metal concentrations, and metal enrichment factors. *Environ. Rev.* **4**(2), 149–183.

Shotyk, W. & Le Roux, G. (2005) Biogeochemistry and cycling of lead. *Biogeochemcial Cycles of the Elements*, edited by Sigel, A., Sigel, H. and Sigel, R. K. O., Marcel Dekker.

Shotyk, W., Weiss, D., Appleby, P. G., Cheburkin, A. K., Frei, R., Gloor, M., Kramers, J. D., Reese, S. & Van der Knaap, W. O. (1998) History of atmospheric lead deposition since 12,370 C-14 yr BP from a peat bog, Jura Mountains, Switzerland. *Science* **281**(5383), 1635–1640.

Simonetti, A., Gariépy, C. & Carignan, J. (2003) Tracing sources of atmospheric pollution in Western Canada using Pb isotopic composition and heavy metal abundances in epiphytic lichens. *Atmos. Environ.* **37**, 2853–2865.

Spiro, B., Weiss, D. J., Purvis, O. W., Mikhailova, I., Williamson, B., Udachin, V. & Coles, B. J. (2004) Pb isotopes in lichen transplants — transient records of diverse sources around the Karabash smelter, Urals, Russia. *Environ. Sci. Technol.* **38**, 6522–6528.

Sturges, W. T. & Barrie, L. A. (1987) Lead 206/207 isotope ratios in the atmosphere of North America as tracers of US and Canadian emissions. *Nature* **329**, 144–146.

Teutsch, N., Erel, N., Halicz, L. & Banin, A. (2001) Distribution of natural and anthropogenic lead in Mediterranean soils. *Geochim. Cosmochim. Acta* **65**(17), 2853–2864.

Thirlwall, M. F. (1997) Thermal ionisation mass spectrometry (TIMS), *Modern Analytical Geochemistry*, edited by Gill, R., pp. 135–153, Addison Wesley Longman, Singapore.

Vallelonga, P., Gabrielli, P., Rosman, K. J. R., Barbante, C. and Boutron, C. F. (2005) A 220 kyr record of Pb isotopes at Dome C Antarctica from analyses of the EPICA ice core. *Geophys. Res. Lett.* **32**, doi:10.1029/2004GL021449.

Vallelonga, P., Van de Velde, K., Candelone, J. P., Rosman, K. J. R., Boutron, C., Morgan, V. I. & Mackey, D. J. (2002) Recent advances in measurement of Pb isotopes in polar ice and snow at sub-picogram per gram concentrations using thermal ionisation mass spectrometry. *Anal. Chim. Acta* **453**(1), 1–12.

Van de Velde, K., Vallelonga, P., Candelone, J. P., Rosman, K. J. R., Gaspari, V., Cozzi, G., Barbante, C., Udisti, R., Cescon, P. & Boutron, C. (2005) Pb isotope record over one century in snow from Victoria Land, Antarctica. *Earth Planet. Sci. Lett.* **232**, 95–108.

van Geen, A., Adkins, J. F., Boyle, E. A., Nelson, C. H. & Palanques, A. (1997) A 129 yr record of widespread contamination from mining of the Iberian pyrite belt. *Geology* **25**(4), 291–294.

Véron, A. J., Church, T. M., Flegal, A. R., Patterson, C. C. & Erel, Y. (1993) Response of lead cycling in the surface Sargasso Sea to changes in tropospheric input. *J. Geophys. Res.* **98**(C10), 18269–18276.

Véron, A. J., Church, T. M., Patterson, C. C. & Flegal, A. R. (1994) Use of stable isotopes to characterise the sources of anthropogenic lead in North Atlantic surface waters. *Geochim. Cosmochim. Acta* **58**(15), 3199–3206.

Véron, A. J., Church, T. M., Rivera-Duarte, I. & Flegal, A. R. (1999) Stable lead isotopic ratios trace thermohaline circulation in the subarctic North Atlantic. *Deep-Sea Res. II* **46**, 919–935.

Véron, A. J., Lambert, C. E., Isley, A., Linet, P. & Grousset, F. E. (1987) Evidence of recent lead pollution in deep north-east Atlantic sediments. *Nature* **326**, 278–281.

von Blanckenburg, F., O'Nions, R. K. & Hein, J. R. (1996) Distribution and sources of pre-anthropogenic lead isotopes in deep ocean water from Fe-Mn crusts. *Geochim. Cosmochim. Acta* **60**(24), 4957–2963.

von Storch, H. M., Costa-Cabral, M., Hagner, C., Feser, F., Pacyna, J. M., Pacyna, E. M. & Kolb, S. (2003) Four decades of gasoline lead emissions and control policies in Europe: A retrospective assessment. *Sci. Total. Environ.* **311**, 151–176.

Weiss, D., Chavagnac, V., Boyle, E. A., Wu, J. F. & Herwegh, M. (2000) Determination of lead isotope ratios in seawater by quadrupole inductively coupled plasma mass spectrometry after $Mg(OH)_2$ co-precipitation. *Spectrochim. Acta B* **55**(4), 363–374.

Weiss, D. J., Boyle, E. A., Chavagnac, V., Wu, J., Michel, A. and Reuer, M. (2003) Lead isotope evolution of the North Atlantic: Pattern of deposition and source assessment. *J. Geophys. Res.* **108**(C10), 3306 doi:10.1029/2000JC000762.

Weiss, D. J., Shotyk, W. & Kempf, O. (1999) Archives of atmospheric lead pollution. *Naturwissenschaften* **86**, 262–275.

Whitfield, M. & Turner, D. R. (1980) The theoretical studies of the chemical speciation of Pb is seawater, *Lead in the Marine Environment*, edited by Branica, M. and Konrad, Z., Pergamon, New York.

Wu, J. & Boyle, E. A. (1997) Lead in the western North Atlantic Ocean: Completed response to leaded gasoline phaseout. *Geochim. Cosmochim. Acta* **61**(15), 3279–3283.

Natural and Artificial Platinum and Palladium Occurrences World-Wide

Hazel M. Prichard

School of Earth, Ocean & Planetary Sciences
Cardiff University, Main College
Park Place, Cardiff, CF10 3YE, UK
Prichard@cardiff.ac.uk

The introduction of legislation in the western world over the last 20 years or so requires that poisonous exhaust emissions from cars are reduced. Catalytic converters fitted to exhaust systems make these poisonous gases safer. Now nearly half the world's annual production of platinum (Pt) and palladium (Pd) is being used in the manufacture of catalytic converters. As cars travel around our cities these converters lose Pt and Pd which are ejected onto roads. This creates artificial concentrations of Pt and Pd in the urban environment that is a new addition to the global distribution. We know very little about such concentrations but we know much more about Pt and Pd concentrations in geological settings. Pt and Pd occur naturally at the Earth's surface only in a very few rare locations in rocks formed by an unusual combination of geological processes. The major Pt and Pd deposits were formed by crystallisation of magma which concentrated the metals into specific minor rock units within large igneous intrusions. As these precious metals are very rare on the Earth's surface they are economic at concentrations of only a few parts per million. Recent studies, including those described here, show that values of Pt and Pd accumulating in our cities are approaching values found in natural deposits. Certainly Pt and Pd can be located in road dust at road junctions in the cities in the wealthy western world at levels well above natural background values.

1. Platinum and Palladium Distribution in the Earth

During the formation of the Earth the very dense, high melting point Pt and Pd metals would be expected to concentrate in the metal phase that is in the core of the Earth. So the presence of even small amounts of Pt and Pd in the Earth's mantle, which surrounds the core, indicates either an inefficient collection of these elements in the core during its formation so that some remain in the mantle, or transfer of the Pt and Pd from the core into the mantle carried by rising plumes of molten rock [Brandon et al. (1998)] or addition of Pt and Pd by meteorite impact especially during the early history of the Earth [Chou (1978); O'Neill et al. (1995); Palme (1997)].

Today natural Pt and Pd occurrences occur on the Earth's surface as a result of extraction of these elements from the Earth's mantle during unusually extensive mantle melting. The melt produced is injected as molten rock or magma into overlying rocks to crystallise and form igneous rocks. There are three main primary geological settings in which Pt and Pd are concentrated in magmas. These are old continental areas into which magma has been intruded or extruded, younger continental rocks that have been rifted at the initial stage of ocean formation and injected by plumes of magma and areas above plate tectonic collision zones where a down-going oceanic plate sinks beneath another oceanic plate or a continental plate. The igneous rocks, in these three settings, that host concentrations of Pt and Pd may be either formed close to the Earth's surface or exposed by uplift and erosion. If these Pt- and Pd-bearing rocks are eroded by surface processes then sedimentary placer deposits will form where the very dense precious metals accumulate in hollows down stream of the primary source. The tectonic settings of Pt and Pd concentrations world-wide are shown in cartoon form in Fig. 1.

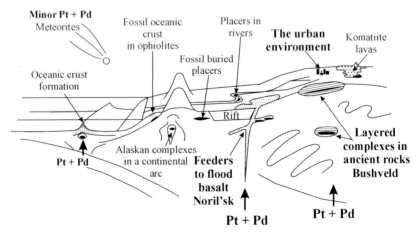

Fig. 1   Sketch (not to scale) showing the tectonic locations of Pt and Pd concentrations.

2. Economic Pt & Pd Concentrations in Natural Deposits

The two major natural economic deposits in the world that produced in 2003 more than 85% of the world's supply of Pt plus Pd are the Bushveld Complex in the Republic of South Africa and the Noril'sk deposits in northern Siberia [Kendall (2003)]. The Bushveld Complex is a large ancient layered igneous complex formed by magma intruding and crystallising as a large body within ancient continental crust. In contrast the Noril'sk deposits occur in feeder sills to flood basalts that form the Siberian traps. These were erupted as the European plate split away from the Asian plate during the initiation of a new ocean of Permo-Triassic age. The magma that formed the flood basalts at Noril'sk originated deep in the Earth's mantle and rose in hot jets of molten material, forming a mantle plume. These two giant Pt and Pd deposits were produced in different geological environments and have very different ages. However in both cases the Pt and Pd are the result of concentration processes that took place in large magmatic systems that resulted from great amounts of mantle melting. These melts formed magmas that crystallised to produce mafic and ultramafic igneous rocks rich in ferromagnesium minerals. The Bushveld Complex is 240 km wide and 350 km long [Cawthorn *et al.* (2002)] and is the largest layered igneous complex exposed on Earth. The Noril'sk sills were the feeder channels for the many cubic km of lava that comprise the largest continental flood basalt system in the world [Wilson (1989)].

In the Bushveld complex Pt and Pd have been concentrated and are extracted from the Merensky Reef which is a coarsely crystalline mafic rock (Fig. 2A) and the UG2 which is a chromitite [Lee (1996)]. Both of these are only 1–2 m thick in a sequence of igneous rocks that is approximately 7–9 km thick and both contain minor Ni-Cu-Fe sulphide minerals. The average values of Pt and Pd in these two reefs are in the order of 5000–10 000 parts per billion (ppb or mg/tonne) [Cawthorn *et al.* (2002)]. They represent very efficient concentration of the precious metals into specific narrow units within the rock sequence during crystallisation of the magma. In Noril'sk Pt and Pd occur in the massive Ni-Cu-Fe-sulphides that efficiently collected the Pt and Pd from the magma as it flowed through feeder sills to the surface where it erupted to form lava flows. The grades of Pt and Pd in the massive sulphides vary considerably from 2000 to 110 000 ppb (Kozyrev *et al.* (2002)]. They are so enriched in some samples that platinum-group minerals can be seen in hand specimen (Fig. 2B).

Fig. 2   Photographs of Pt- and Pd-rich rocks from (A) the Merensky Reef in the Bushveld Complex; Chromite (Ch) forms a sub-horizontal layer cutting across the coarsely crystalline silicate minerals, (S) are Ni-Cu-Fe sulphides, (B) Noril'sk; massive Ni-Cu-Fe sulphide ore containing a white platinum-group mineral (PGM) and (C) Cliff in the Shetland ophiolite; black chromite (Ch) surrounded by Pt- and Pd-bearing green Ni carbonate next to dunite (D). Scale bars represent 1 cm.

Smaller deposits can be economic and one of the best examples of this is the Lac de Isles deposit in Canada which is only 1 km^2. Here ore is extracted from an open pit with grades of 8000 ppb Pd [Watkinson *et al.* (2002)]. Some Pt and Pd are extracted from other layered intrusions including the Stillwater complex in Montana, USA [McCullum (1996)], the Great Dyke in Zimbabwe [Wilson (1996)] and from massive Ni-Cu-Fe-sulphides in the Sudbury igneous complex in Canada that formed after a meteorite impact.

Pt and Pd also can be concentrated in ultramafic magma that erupted to form lava flows and assimilated sulphides from the adjacent sediments. Examples include the Pt and Pd enrichments at Kambalda in Western Australia and Raglan in Cape Smith in Canada [Lesher and Keays (2002)].

Economic concentrations of Pt and Pd sometimes occur in laterites overlying Pt- and Pd-bearing igneous complexes or in placer deposits. Examples of placers include those associated with giant deposits such as the Merensky Reef and smaller but enriched deposits such as in eastern Siberia, the Urals and Columbia [Weiser (2002)].

3. Non-Economic Pt and Pd Concentrations in Natural Occurrences

There are many minor occurrences of Pt and Pd that are not at present economic to extract including concentrations in several layered igneous complexes. Examples include Skaergaard in Greenland and Rum in Scotland, both of which are layered complexes associated with plumes of magma that rose during the opening of the Atlantic Ocean [Andersen *et al.* (2002)]. Oceanic rocks can be created above down-going plates of older oceanic rocks sinking as part of the plate tectonic process. These oceanic rocks are formed by unusually great degrees of mantle melting caused by water driven off the sinking oceanic rocks. This water lowers the melting point of the mantle and the greater degree of melting extracts Pt and Pd which then concentrates in the oceanic rocks during crystallisation of the extracted magma [Prichard *et al.* (1996)]. Slices of fossilised ocean crust that have been emplaced on land by mountain building are known as ophiolite complexes and Pt and Pd concentrations often occur in the ophiolite complexes formed above down going oceanic plates. Magmatic concentrations in ophiolite complexes can reach values of 4000–5000 ppb (e.g. in the Shetland ophiolite complex) but tonnages are insufficient to be economic [Prichard *et al.* (1996)] (Fig. 2C). Similarly magmas produced above oceanic plates going down below continents are Pt enriched [Johan (2002)]. These magmas crystallise to form Pt-rich intrusions that are exposed at the Earth's surface in a number of mountain ranges that formed as ocean plates sank beneath continents. These intrusions are especially well preserved in Alaska and so they are known as Alaskan type complexes.

Away from these geologically rare concentrations the abundances of Pt and Pd are very low. Pt and Pd values in the Earth's mantle are low. Barnes *et al.* [1988] averaged 114 source mantle analyses to produce average modern mantle values of 9.2 parts per billion (ppb or μg/kg or mg/tonne) Pt and 4.4 ppb Pd and similar values are given by Morgan [1986]. The Pt and Pd content of igneous rocks decreases sharply as the silicate content increases. Reliable data is somewhat sparse but generally Pt ranges from 10–20 ppb for ultramafic rocks and 5–10 ppb for mafic rocks. Felsic and intermediate rocks contain ranges of Pd of 0.1–6 ppb [Crocket (1981)]. Sedimentary rocks may contain placers but these are very rare and the vast majority of sedimentary rocks have very low concentrations of these elements. A recent study of 281 sediments resulted in average values of 2.7 ppb Pt and 1.9 ppb Pd and the sediments included lake sediments, marine sediments derived from nearby land and marine shale formed far from land [Terashima *et al.* (2002)].

4. Magmatic Processes that Collect Pt and Pd

The six platinum-group elements (PGE), Os (osmium), Ir (iridium), Ru (ruthenium), Rh (rhodium), Pt and Pd tend to occur together as they share affinities with iron and sulphur. Sulphur saturation and the crystallisation of sulphides from magma are caused by a number of factors. These include cooling and change in composition of the magma during crystallisation and commonly by crystallisation of Fe-rich oxides such as chromite ($FeCr_2O_4$) and magnetite (Fe_3O_4) or by contamination or mixing of magmas [Naldrett (1989)].

The magmatic processes that concentrate PGE tend to be similar in a variety of geological settings from layered intrusions in stable continental areas to ophiolite complexes formed in tectonically active regions. Thus in whatever tectonic setting the magma is generated, Os, Ir and Ru are particularly concentrated with chromitites whereas Pt and Pd are associated with Ni-Cu-Fe-sulphides. If sulphur-saturation was not coincident with chromite crystallisation then there is usually a separation of the Os, Ir and Ru from Pt and Pd [Prendergast (1991); Andersen *et al.* (1998); Auge (1998); Moreno *et al.* (1999)]. If sulphide-saturation of the magma and chromite crystallisation took place at the same time, then all six PGE are concentrated together [Prichard and Lord (1993); Lee (1996)]. Even in this case, it sometimes can be demonstrated mineralogically that Os-, Ir- and Ru-bearing minerals are enclosed in chromite and so started to crystallise early whereas Pt-, Pd- and Rh-bearing minerals are associated with sulphides and are situated in late crystallising silicates that formed between the earlier chromite [Prichard and Tarkian (1988)].

In Ni-Cu-Fe-sulphide-rich magmas Pt and Pd are concentrated in molten sulphide-rich droplets that separate from the silicate magmas in a similar way to oil droplets in water [Naldrett *et al.* (1996); Crocket *et al.* (1997)]. The sulphide droplets may be small (1 cm in diameter) or they may coalesce to form large units that can crystallise to form massive metal-rich ore deposits metres to 10 s of metres across such as at Noril'sk. The result is that Pt and Pd are often associated with Ni-Cu-Fe-sulphides in mafic and ultramafic igneous complexes even if there is only 1–2% Ni-Cu-Fe-sulphide in the rock. The first sulphide liquid to separate from the silicate magma and crystallise tends to concentrate the Pt and Pd and subsequent sulphides are Pt- and Pd-poor.

Once separated from the silicate melt, Pt- and Pd-bearing sulphide liquids crystallise high temperature minerals first. The composition of the remaining liquid sulphide changes, or fractionates because of the separation of a solid of differing composition. Volatile elements such as bismuth (Bi), tellurium (Te), antimony (Sb) and arsenic (As) tend to concentrate in the last sulphide liquid as early crystals contain no volatile elements. PGE often also concentrate in these last sulphide liquids with Bi, Te, Sb and As [Prichard *et al.* (2004a)]. Sometimes, as in the Merensky Reef, Pt and Pd bismuth tellurides occur interstitially with minerals such as quartz that crystallised late from fractionated silicate magma [Prichard *et al.* (2004b)]. Similarly in the Pt- and Pd-enriched horizons in the Freetown layered complex in Sierra Leone Pt-sulphides occur as droplet shaped minerals with magnetite enclosed by amphibole (a mineral that contains hydroxyl ions) that lies between earlier crystallised silicate minerals [Bowles *et al.* (2002)]. Both these examples indicate that the Pt and Pd crystallised late in the silicate crystallisation sequence, with volatile-rich sulphide liquid.

5. Platinum-Group Minerals

Platinum-group elements form a great variety of platinum-group minerals (PGM). Different PGM are often produced by different processes. Therefore the type of PGM found may indicate its origin as, for example, formation during early or late magmatic crystallisation and often the primary PGM have been replaced during subsequent low temperature alteration and/or surface weathering.

Despite its great age the Bushveld Complex is relatively unaltered and undeformed because of its size. Even here the PGE that were magmatically concentrated in Ni-Cu-Fe-sulphide in the Merensky reef are now contained

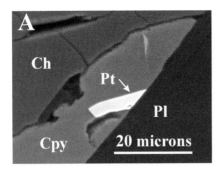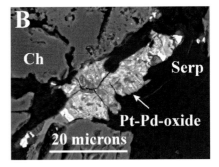

Fig. 3   Photographs taken using a scanning electron microscope of (A) a homogeneous Pt-sulphide with a smooth out line exsolved from a Cu-Fe-sulphide (Cpy) surrounded by chromite (Ch) and plagioclase (Pl) and (B) a mottled inhomogeneous altered Pt-Pd-oxide with a ragged outline enclosed in the low temperature alteration mineral serpentine (Serp) adjacent to chromite (Ch).

in secondary minerals. These include discrete PGM sulphides enclosed by PGE-poor Ni-Cu-Fe-sulphides that are themselves associated with chromite layers. These PGM were produced by re-equilibration during cooling with PGE being expelled (or exsolved) from the crystal structure of their Ni-Cu-Fe-sulphide hosts during cooling after all the magma has crystallised (Fig. 3A) [Prichard et al. (2004b)]. Bi-, Te-, Sb- and As-bearing PGM may have formed either directly by crystallisation from a fractionated sulphide liquid or are the result of later exsolution. In the Shetland ophiolite it is clear that Bi, Te, Sb and As were introduced during low temperature regional metamorphism. The PGE came originally from a magmatic source but have been mobilised to form Pt-, Pd- and Rh-bearing bismuthides, tellurides, antimonides and arsenides and they are surrounded now by low temperature alteration minerals such as serpentine and chlorite [Prichard et al. (1994)].

The association of PGM with low temperature carbonate has been described from a number of localities including the Raglan PGE-bearing massive sulphide deposit in ultramafic lavas in Cape Smith northern Canada. Here Pt and Pd tellurides, antimonides and arsenides occur in carbonates [Seabrook et al. (2004)]. In the Shetland ophiolite, at the very Pt- and Pd-rich locality at Cliff, PGM occur surrounded by Ni carbonate associated with chromite (Fig. 1C). In Jinchuan, a major nickel and PGE deposit in China, Se has been introduced during low temperature alteration to replace palladium-bismuthides by palladium selenides [Prichard *et al.* (2004c)].

The alteration of PGM sulphides, bismuthides, tellurides, arsenides and antimonides to PGE-alloys appears to be the next stage in the alteration sequence. This can be observed where PGM are in contact with secondary minerals such as amphibole and chlorite. Other alloys may also form at this stage including combinations of Au, Ni, Cu, Fe with or without PGE. For example, in the Freetown complex Pt-sulphides alter directly to Pt-Fe alloys [Bowles *et al.* (2002)].

A final stage of alteration, probably caused by surface weathering and oxidation, is the production of PGE oxides (Fig. 3B). Often these are located around other PGM. Now PGE oxides are being described much more commonly [Augé and Legendre (1994); Moreno *et al.* (1999); Ortega *et al.* (2004)] but the processes of their formation are not understood. PGE oxides that have been described so far tend to be poorly crystalline amorphous minerals that may replace earlier minerals.

6. Pt and Pd Mobility at the Earth's Surface

The platinum-group elements were traditionally thought of as inert. However when subjected to alteration and surface processes they are clearly far from inert and show different solubility and mobility in different pH and Eh conditions. The Bacuri complex in the Amazon in Brazil is deformed, altered and has been exposed to tropical weathering. PGE were initially concentrated into Ni-Cu-Fe-sulphides with the chromite during magmatic crystallisation. They were then exsolved to form PGM including, for example, Pd bismuthides. In these rocks PGM occur in veins in the chromite showing that the PGM are being locally remobilised. In the tropically weathered lateritic soils that over lie the complex the Pd has been removed but the Pt remains producing very high Pt/Pd ratios in the laterite (Fig. 4A) [Prichard *et al.* (2001)]. Similarly in the Shetland ophiolite soil pits over the mineralised area revealed that Pt/Pd ratios increase upwards indicating that Pd is being removed from the soil by weathering [Prichard and Lord (1994)]. These two examples from Brazil and Shetland with high rainfall in tropical Brazil and temperate Shetland agree with conventional ideas that suggest that Pd is more mobile during surface alteration with Pt/Pd ratios increasing from fresh mineralised rock into overlying soils [Fuchs and Rose (1974)]. However theory suggests that Pt is more mobile than Pd [Wood (2002)] and analysis of samples from the desert in

Fig. 4 Photographs showing (A) deep lateritic weathering in the Amazon caused by tropical weathering, with a faun coloured weathered unit overlying a more mottled unit extending down to the water, all of which are completely altered with no fresh rock (tropical trees give the scale) compared with (B) rusty rocks and rock talus covered by a very thin soil in the desert of Nevada (man in blue, centre left gives scale).

Nevada, USA, suggests that in very dry conditions Pt is more mobile than Pd with Pt/Pd ratios decreasing in the weathered horizons above the PGE mineralisation in the rock (Fig. 4B). It is possible that the type of weathering is critical in determining the relative solubility of the PGE.

Erosion of natural Pt and Pd concentrations may produce placers and these may be forming today as in Yubdo in Ethiopia [Johan (2002)] or they may occur as fossil placers derived from sources long since completely eroded away. In Brazil fossil Pt and Pd placers are present sporadically in the Espinhaço quartzites which extend for 100s of Km from Bahia in the north to Minas Gerais in the south. These PGE are likely to have been derived from older greenstone belts that contained PGE concentrations in mafic and ultramafic igneous complexes that have been eroded. Some of these placer grains are well rounded (Fig. 5A) and located in conglomerate suggesting mechanical transport of the grains and collection in traps as a consequence of their very high density. In contrast delicate dendritic and botryoidal shaped grains suggest that other PGM may, at least in part, have grown *in situ* by precipitation of the PGE from solution (Fig. 5B) [Bowles (1986)]. PGE can dissolve and they have been identified in natural waters in mineralised areas [e.g. Cook and Fletcher (1993)] and their up take by vegetation has also been recorded [e.g. Hall *et al.* (1990)].

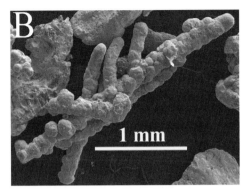

Fig. 5 Photographs taken using a scanning electron microscope of two Pt-Pd alloys derived from placers in the Espinhaço quartzites, Brazil, (A) an alloy that was mechanically transported and rounded during transport and (B) an irregular shaped alloy that probably grew *in situ* by precipitation of Pt and Pd.

The balance of weathering versus erosion of the land's surface varies with climate and relief. Thus in flat tropical areas with high rainfall, deep chemical weathering dominates over erosion. In mountains or hills in temperate zones and deserts rocks are exposed because erosion removes weathered and mechanically dislodged material faster than it can be weathered to form soil. The dispersion and re-concentration of Pt and Pd on the Earth's surface are controlled by their solubility and the mechanical mobility. These depend on the original mineralogy hosting the Pt and Pd and on surface conditions that exist at a particular Pt and Pd occurrence.

7. Pt and Pd in the Urban Environment

Pt and Pd are extremely rare metals. Almost half the world's annual production is being used in catalytic converters fitted to cars to reduce the emissions of poisonous gasses from car exhaust systems. Particles of Pt, Pd and also Rh, are detached from these catalytic converters and land on the roads [Prichard in BA reports (1998); Jarvis *et al.* (2001); Higney *et al.* (2002).] A pilot study in 1996 on road dusts swept from six road junctions in Cardiff (Table 1, Fig. 6) showed that at all the localities Pt and Pd are above the normal background values of less than 1–2 ppb. Values at the six sites varied from 16–126 ppb Pt and 7–99 ppb Pd. The higher values correspond to the busiest roundabouts where traffic flow is high.

Table 1A   Cardiff road dust as swept straight from the road.

Values in ppb	Rh	Pt	Pd
1 A48, Roath roundabout	**22**	**126**	**99**
2 A48, Penylan roundabout	16	98	13
3 M4, Granada services roundabout	**11**	**73**	**21**
4 Cyncoed roundabout	12	72	15
5 Culverhouse cross BMW entrance	**5**	**34**	**10**
6 Traffic lights A48	2	16	7

Table 1B   Roath roundabout road dust.

Sieved into coarse and fine fractions and separated by heavy liquids

Values in ppb	Rh	Pt	Pd	Au	% of sample
Coarse 0.09–1 mm					
Light fraction	4	25	30	3	93
Heavy fraction	220	1679	284	122	7
Fine fraction below 0.09 mm					
Light fraction	5	37	6	6	89
Heavy fraction	111	966	2945	272	11

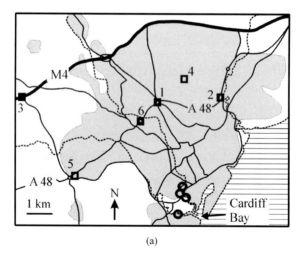

(a)

Fig. 6   (a) A map of Cardiff showing sample locations for road dust collection at major road junctions (squares) and mud samples in the bay (circles); the M4 motorway is shown as a thick black line, and the other roads are shown as thinner black lines. Rivers are shown in dashed lines, Cardiff Bay and the Bristol Channel are filled with horizontal lines and the built up area is shaded grey.

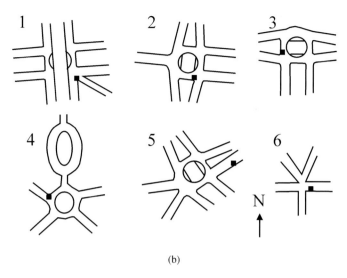

(b)

Fig. 6 (b) A map showing the locations of sample sites (■) at the major road junctions marked on (A). The number of each sample site refers to sample numbers in Table 1.

The Cyncoed roundabout was found to have significantly high Pt and Pd values and, although it does not have the high traffic flow associated with the roundabouts on major route ways and motorways, it is in the centre of a wealthy part of Cardiff. Here new cars have catalytic converters that tend to disintegrate due to temperature fluctuation at the beginning of journeys. The sample with the highest Pt and Pd values at Roath, after density separation, yielded total precious metal values of 4294 ppb in the fine grained, heavy fraction.

From the road sides the PGE are washed down gullies into the complicated network of artificial and natural drainage (Fig. 7) [Laschka and Nachtwey (1997)]. PGE are collecting at points in the urban waste system at concentrations well above normal background levels. The fact that these accumulations of Pt and Pd are being moved through the urban environment is demonstrated by their presence in mud from Cardiff Bay. Here values of 20 ppb Pt plus Pd must have come from artificial sources as there are no nearby natural sources of Pt and Pd. Another example of the mobility of PGE in the urban environment comes from Os isotope evidence in sediments in Massachusets and Cape Cod bays [e.g. Ravizza and Bothner (1996)]. Traces of Pt and Pd contamination have been recorded even in the Greenland ice sheet [Barbante *et al.* (2001)].

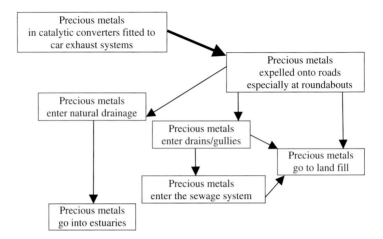

Fig. 7   **Diagram showing possible pathways for the Pt and Pd as they pass through the urban environment.**

8. Knowledge from Natural Pt and Pd Occurrences Applied to the Man-Made Situation

Geological exploration for Pt and Pd in the natural environment includes mineralogical and geochemical analysis of weathered and altered Pt and Pd occurrences exposed on the Earth's surface. Knowledge of the original mineralogy of natural Pt and Pd concentrations and the alteration processes affecting them should be applied to understand the mobility of Pt and Pd in the urban environment. The mineralogy of Pt and Pd in natural occurrences influences the way the different physical and chemical surface processes mobilise Pt and Pd. This also will be the case for Pt and Pd in the urban environment where they will be moved by mechanical processes causing collection by gravity as in placer deposits and by dissolution in acidic conditions and subsequent precipitation. If Pt- and Pd-rich particles from catalytic converters oxidise and disintegrate during surface weathering then they may become less stable and prone to dissolution. An understanding of the Pt and Pd mineralogy, and Eh and pH conditions to which they are subjected, should make it possible to predict where Pt and Pd will concentrate in the urban environment. Little is known about the distribution of Pt and Pd in the urban environment, or the mineralogy of these elements or how they are transported, through our cities in the complicated network of artificial and natural urban drainage systems and further work is necessary. What is clear is that the use of catalytic converters in cars is redistributing

Pt and Pd from the relatively rare natural geological occurrences near the Earth's surface onto road-sides and then into urban drainage systems in many cities world wide.

Acknowledgements

I would like to acknowledge Dr. C. R. Neary and an anonymous referee who have improved the text of this paper with many constructive comments. I would like to thank the many colleagues with whom I have had discussions about the processes described in this paper. Peter Fisher is thanked for providing scanning electron microscope images of PGM. The Royal Society made much of this research possible by funding my Royal Society University Fellowship and subsequent Industrial Fellowship. The research will continue with Royal Society Funding from the Brian Mercer Senior Award for 2004.

References

Andersen, J. C. Ø., Rasmussen, H., Nielsen, T. F. D. & Ronsbo, J. G. (1998) The triple group and Platinova gold and palladium reefs in the Skaergaard intrusion: Stratigraphic and petrographic relations. *Econ. Geol.* **93**, 488–509.

Andersen, J. C. Ø., Power, M. R. & Momme, P. (2002) Platinum-group elements in the Palaeogene North Atlantic Igneous Province. *The Geology, Geochemistry, Mineralogy and Mineral Beneficiation of Platinum-Group Elements* (ed. L. J. Cabri), sp. vol. **54**, pp. 637–668. Can. Inst. Min. Metal. & Pet.

Auge, T. & Legendre, O. (1994) PGE oxides from the Pirogues ophiolitic mineralization. New Caledonia: Origin and significance. *Econ. Geol.* **89**, 1454–1468.

Auge, T., Legendre, O. & Maurizot, P. (1998) The distribution of Pt and Ru-Os-Ir minerals in the New Caledonia ophiolite. *International Platinum* (eds. N. P. Laverov & V. V. Distler), pp. 141–154. Theophrastus publications, Athens.

Barnes, S-J., Boyd, R., Korneliussen, A., Nilsson, L-P., Often, M., Pedersen, R. B. & Robins, B. (1988) The use of mantle normalisation and metal ratios in discriminating between the effects of partial melting, crystal fractionation and sulphide segregation on platinum-group elements, gold, nickel and copper: examples from Norway. *Geo-platinum Symposium Volume*, Prichard, H. M., Potts, J., Bowles, J. F. W. & Cribb, S. J. (eds.), pp. 113–143. Elsevier Applied Science, London and New York.

Brandon, A. D., Walker, R. J., Morgan, J. W., Norman, M. D., & Prichard, H. M. (1998) Coupled 1860s and 1870s evidence for core-mantle interaction. *Science* **280**, 1570–1573.

Barbante, C., Veysseyre, A., Ferrari, C., Van de Velde, K., Morel, C., Capodaglio, G., Cescon, P., Scarponi, G. & Boutron, C. (2001) Greenland snow evidence of large scale atmospheric contamination for platinum, palladium and rhodium. *Env. Sci. & Tech.* **35**, 835–839.

Bowles, J. F. W. (1986) The development of platinum-group minerals in laterites. *Econ. Geol.* **81**, 1278–1285.

Bowles, J. F. W., Prichard, H. M. & Fisher, P. C. (2002) Platinum-group minerals (PGM) in the Freetown Complex, Sierra Leone. Abs. vol. (ed. A. Boudreau), pp. 61–63. *Ext. abs.* 9th *Int. Pt Symp*; Billings, Montana.

Cawthorne, R. G., Merkle, R. K. W. & Viljoen, M. J. (2002) Platinum-group element deposits in the Bushveld complex, South Africa. *The Geology, Geochemistry, Mineralogy and Mineral Beneficiation of Platinum-Group Elements* (ed. L. J. Cabri), sp. vol. **54**, pp. 389–430. Can. Inst. Min. Metal. & Pet.

Chou, C. L. (1978) Fractionation of siderophile elements in the Earth's upper mantle. *Proc. 9th Lunar Planet. Sci Conf.*, pp. 219–230.

Cook, S. J. & Fletcher, W. K. (1993) Distribution and behaviour of Pt in soils, sediments and waters of the Tulameen ultramafic complex, southern British Columbia, Canada. *J. Geochem. Explor.* **46**, 279–308.

Crocket, J. H. (1981) Geochemistry of the platinum-group elements. *In Platinum-Group Elements: Mineralogy, Geology, Recovery* (ed. L. J. Cabri). sp. vol. **23**, pp. 47–64. Can. Inst. Min. Metal. & Pet.: Harpell's press Cooperation, Quebec.

Crocket, J. H., Fleet, M. E. & Stone, W. E. (1997) Implications of composition for experimental partitioning of platinum-group elements and gold between sulphide liquid and basaltic melt. *Geochim. Cos. Acta* **61**, 4139–4149.

Fuchs, W. A. & Rose, A. W. (1974) The geochemical behaviour of platinum and palladium in the weathering cycle in the stillwater complex, Montana. *Econ. Geol.* **69**, 332–346.

Hall, G. E. M., Pelchat, J.-C. & Dunn, C. E. (1990) The determination f Au, Pd and Pt in ashed vegetation by ICP-MS and graphite furnace atomic absorption spectrometry. *J. Geochem. Explor.* **37**, 1–14.

Higney, E., Olive, V., MacKenzie, A. B. & Pulford, I. D. (2002) Isotope dilution analysis of Pt in road dusts from west central Scotland. *Ap. Geochem.* **17**, 1123–1129.

Jarvis, K. E., Parry, S. J. & Piper, J. M. (2001) Temporal and spatial studies of autocatalyst-derived Pt, Rh and Pd and selected vehicle derived trace elements in the environment. *Env. Sci. & Tech.* **35**, 1031–1036.

Johan, Z. (2002) Alaskan-type complexes and their platinum-group element mineralization. *The Geology, Geochemistry, Mineralogy and Mineral Beneficiation of Platinum-Group Elements* (ed. L. J. Cabri), sp. vol. **54**, pp. 669–720. Can. Inst. Min. Metal. & Pet.

Kendall, T. (2003) *Platinum 2003 Interim Review.* 28 pp. Published by Johnson Matthey.

Kozyrev, S. M., Komarova, M. Z., Emelina, L. N., Oleshkevich, O. I., Yakovleva, O. A., Lyalilnov, D. V. & Maximov, V. I. (2002) The mineralogy and behaviour of PGM during processing of the Noril'sk-Talnakh PGE-Cu-Ni ores. *The Geology, Geochemistry, Mineralogy and Mineral Beneficiation of Platinum-Group Elements* (ed. L. J. Cabri), sp. vol. **54**, pp. 757–791. Can. Inst. Min. Metal. & Pet.

Laschka, D. & Nachtwey, M. (1997) Pt in municipal sewage treatment plants. *Chemosphere* **34**, 1803–1812.

Lee, C. A. (1996) A review of mineralization in the Bushveld complex and some other layered intrusions. *Layered Intrusions* (ed. R. G. Cawthorn), pp. 103–147. Elsevier.

Lesher, C. M. & Keays, R. R. (2002) Komatiite-associated Ni-Cu-(PGE) deposits: Geology, mineralogy, geochemistry and genesis. 757–791. *The Geology, Geochemistry, Mineralogy and Mineral Beneficiation of Platinum-Group Elements* (ed. L. J. Cabri), Sp. Vol. **54**, pp. 579–618. Can. Inst. Min. Metal. & Pet.

McCullum, I. S. (1996) The Stillwater complex. *Layered Intrusions* (ed. R. G. Cawthorn), pp. 441–484. Elsevier.

Moreno, T., Prichard, H. M., Lunar, R., Monterrubio, S. & Fisher, P. C. (1999) Formation of a secondary PGM assemblage in chromitites from the Herbeira ultramafic massif in Cabo Ortegal, NW, Spain. *Eur. J. Mineral.* **11**, 363–378.

Morgan J. W. (1986) Ultramafic xenoliths: Clues to the earth's late accretionary history. *J. Geophys. Res.* **91**, 12375–12387.

Naldrett, A. J. (1989) *Magmatic Sulphide Deposits.* 186 pp. Oxford Monographs on Geology and Geophysics **14**.

Naldrett, A. J., Fedorenko, V. A., Asif, M., Shushen, L., Kunlov, V. E., Stekhin, A. I., Lightfoot, P. C. & Gorbachev, N. S. (1996) Controls on the composition of Ni-Cu sulphide deposits as illustrated by those at Noril'sk, Siberia. *Econ. Geol.* **91**, 751–773.

O'Niell, H. St. C., Dingwell, D. B., Borisov, A., Spettel, B. & Palme, H. (1995) Experimental petrochemistry of some highly siderophile elements at high temperatures, and some implications for core formation and the mantle's early history. *Chem. Geol.* **120**, 255–273.

Ortega, L., Lunar, R., Garcia-Palomero, F., Moreno, T., Estevez, J. R. M., Prichard, H. M. & Fisher, P. C. (2004) The Aguablanca Ni-Cu-PGE deposit south western Iberia: Magmatic ore-forming processes and retrograde evolution. *Can. Min. 9th Int. Pt Symp.* **42**(2), 325–350.

Palme, H. (1997) Highly siderophile elements in chondritic meteorites and the nature of the late veneer. *EAG Workshop: The Origin and Fractionation of Highly Siderophile Elements in the Earth's Mantle* (eds. G. Brugmann, J. P. Lorand, & H. Palme), pp. 63–65. Max Planck Institute Chemie, Mainz, Germany.

Prendergast, M. D. (1991) The Wedza-Mimosa platinum deposit, Great Dyke, Zimbabwe: Layering and stratiform PGE mineralization in a narrow mafic magma chamber. *Geol Mag.* **128**, 235–249.

Prichard, H. M. & Tarkian, M. (1988) Pt and Pd minerals from two PGE-rich localities in the Shetland ophiolite complexes. *Can. Min.* **26**, 979–990.

Prichard, H. M. & Lord, R. A. (1993) An overview of the PGE concentrations in the Shetland ophiolite complex. *Magmatic Processes and Plate Tectonics* (eds.) H. M. Prichard, T. Alabaster, N. B. W. Harris and C. R. Neary. pp. 273–294. Sp. Pub. Geol. Soc. London, **76**.

Prichard, H. M. & Lord, R. A. (1994) Evidence for the mobility of PGE in the secondary environment in the Shetland ophiolite complex. *Trans. Inst. Min. & Metal.* B, **103**, 79–86.

Prichard, H. M., Ixer, R. A., Lord, R. A., Maynard, J. & Williams, N. (1994) Assemblages of platinum-group minerals and sulphides in silicate lithologies and chromite-rich rocks within the Shetland ophiolite. *Can. Min.* **32**(2), 271–294.

Prichard, H. M., Lord, R. A. & Neary, C. R. (1996) A model to explain the occurrence of Pt- and Pd-rich ophiolite complexes. *Jnl. Geol. Soc. Lond.* **153**, 323–328.

Prichard, H. M., Sa, H. & Fisher, P. C. (2001) Platinum-group mineral assemblages and chromite composition in the altered and deformed Bacuri complex, Amapa, northeastern Brazil. *Can. Min.* **39**, 377–396.

Prichard, H. M., Hutchinson, D. & Fisher, P. C. (2004a) Petrology and crystallisation history of multi-phase sulphide droplets in a mafic dyke from Uruguay: Implications for the origin of Cu-Ni-PGE-sulphide deposits. *Econ. Geol.* **99**, 365–376.

Prichard, H. M., Barnes, S.-J., Maier, W. D. & Fisher, P. C. (2004b) Variations in platinum-group minerals in a cross-section through the Merensky reef at Impala Platinum: Implications for the mode of formation of the reef. *Can. Min. 9th Int. Pt Symp.* **42**(2), 423–437.

Prichard, H. M., Fisher, P. C., McDonald, I., Zhou, M-F & Wang, C. Y. (2004c) Platinum-group minerals in the Jinchuan intrusion, China. *Recent Advances in Magmatic Ore Systems of Magmatic-Ultramafic Rocks. Abs. vol.* (eds. J. G. Shellnutt, M. F. Zhou & K. N. Pang), pp. 48–49. University of Hong Kong.

Publications resulting from BA talk in Cardiff by H M Prichard on (12th September 1998): Precious metal fall-out may justify mining city streets. *Guardian* Platinum could be extracted from road dust. *Financial Times* Cars paving the streets with Pt. *Independent* Tests show city streets are paved with platinum. *Daily Telegraph* December 1998 paved with platinum. *Chemistry in Britain,* **34**, 17.

Ravizza, G. & Bothner, M. H. (1996) Os isotopes and silver as tracers of anthropogenic metal in sediments from Massachusetts and Cape Cod bays, *Geochim. Cos. Acta* **60**, 2753–2763.

Seabrook, C. L., Prichard, H. M. & Fisher, P. C. (2004) Platinum-group minerals in the Raglan Ni-Cu-(PGE) deposit, Cape Smith, Canada. *Can. Min. 9th Int. Pt Symp.* **42**(2), 485–497.

Terashima, S., Mita, N., Nakao, S. & Ishihara, S. (2002) Platinum and palladium abundances in marine sediments and their geochemical behaviour in marine environments. *Bull. Geol Sur. Japan* **53**(11/12), 725–747.

Watkinson, D. H., Lavigne, M. J. & Fox, P. E. (2002) Magmatic-hydrothermal Cu-and pd-rich deposits in gabbroic rocks from North America. *The Geology, Geochemistry, Mineralogy and Mineral Beneficiation of Platinum-Group Elements* (ed. L. J. Cabri), sp. vol. **54**, pp. 299–319. Can. Inst. Min. Metal. & Pet.

Weiser, T. W. (2002) Platinum-group minerals in placer deposits. *The Geology, Geochemistry, Mineralogy and Mineral Beneficiation of Platinum-Group Elements* (ed. L. J. Cabri), sp. vol. **54**, pp. 721–756. Can. Inst. Min. Metal. & Pet.

Wilson, A. H. (1996) A review of mineralisation in the Bushveld complex and some other layered intrusions. *Layered Intrusions* (ed. R. G. Cawthorn), pp. 365–402. Elsevier.

Wilson, M. (1989) *Igneous Petrogenesis.* 466 pp. Chapman and Hall.

Wood, S. A. (2002) The aqueous geochemistry of PGE: applications to ore deposits. *The Geology, Geochemistry, Mineralogy and Mineral Beneficiation of Platinum-Group Elements* (ed. L. J. Cabri), sp. vol. **54**, pp. 211–249. Can. Inst. Min. Metal. & Pet.

Data Assimilation and Objectively Optimised Earth Observation

David J. Lary* and Anuradha Koratkar

Global Modelling and Assimilation Office
NASA Goddard Space Flight Center, MD, USA
GEST at the University of Maryland Baltimore County, MD, USA
David.Lary@umbc.edu

This chapter describes a vision for a future objectively optimised earth observation system with integrated scientific analysis. The system envisioned will dynamically adapt the what, where, and when of the observations made in an online fashion to maximise information content, minimise uncertainty in characterising the systems state vector, and minimise both the required storage and data processing time for a given observation capability. Higher level goals could also be specified such as the remote identification of sites of likely malaria outbreaks. By facilitating the early identification of potential breeding sites of major vector species before a disease outbreak occurs and identifying the locations for larvicide and insecticide applications. This would reduce costs, lessen the chance of developing pesticide resistance, and minimise the damage to the environment. Here we describe a prototype system applied to atmospheric chemistry with two relatively mature symbiotic components that seeks to achieve this goal. One component is the science goal monitor (SGM), the other is an Automatic code generation system for chemical modeling and assimilation (AutoChem) described online at www.AutoChem.info/. The Science Goal Monitor (SGM) is a prototype software tool to determine the best strategies for implementing science goal driven automation in missions. The tools being developed in SGM improve the ability to monitor and react to the changing status of scientific events. The SGM system enables scientists to specify what to look for and how to react in descriptive rather than technical terms. The system monitors streams of science data to identify occurrences of key events previously specified by the scientist. When an event occurs, the system autonomously coordinates the execution of the scientist's desired goals. The data assimilation system can feed multivariate objective measures to the SGM such as information content and system uncertainty so that SGM can schedule suitable observations given the observing system constraints. The observing system may of course be a sensor web suite of assets including orbital and suborbital platforms. Once the observations are made an integrated scientific analysis is performed which automatically produces a cross-linked web site for easy dissemination and to facilitate investigation of the scientific issues. A prototype is available at www.CDACentral.info/.

1. Introduction

This year, 2004, is the fortieth anniversary of the NASA Nimbus program. The Nimbus satellites, first launched in 1964, carried a number of instruments: Microwave radiometers, atmospheric sounders, ozone mappers, the Coastal Zone Color Scanner (CZCS), infrared radiometers, etc. Nimbus-7, the last in the series, provided significant global data on sea-ice coverage, atmospheric temperature, atmospheric chemistry (i.e. ozone distribution), the Earth's radiation budget, and sea-surface temperature.

What will the earth observing systems of the future look like? Will they be autonomous? This chapter describes one vision for future earth observing systems. New in this vision is the desire for symbiotic communication to dynamically guide an earth observation system. An earth observation system which is not just a single satellite acting on its own but a constellation of satellites, and sub-orbital platforms such as unmanned aerial vehicles [Trego (1994); Zelenka *et al.* (1997); Sujit and Ghose (2004)] (http://uav.wff.nasa.gov/), and ground observations interacting with computer systems used for modeling, data analysis and dynamic observation guidance. Automatic code generation (www.AutoChem.info/) and automatic parallelisation will greatly facilitate the implementation and automatic adaption of the system for different problems and its possible use on a variety of hardware. Automatic documentation of both software and data products facilitate both code maintenance, and the production and quality monitoring of self-consistent analyses. These analyses can be used by scientists to understand and answer major scientific questions, and by policy makers to establish sound policy decisions, thus increasing the accessibility and utility of Earth Science data. Automatic compression minimises both the required cost of storage and dissemination, and the required time for electronic product transfer/download.

Most of the key questions in earth science involve the tracking of dynamically evolving geophysical fields. So it is desirable to make the best use of a given earth observation capability by using an objective dynamic data

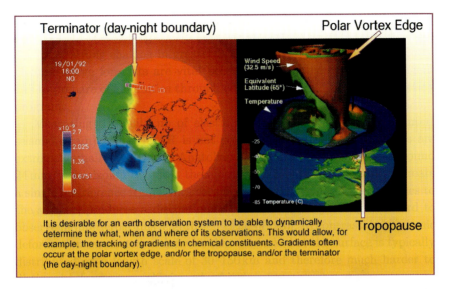

Fig. 1 It would be very useful for an earth observing system to dynamically track evolving features. For example, on the left the sharp gradient in NO (nitric-oxide) at the terminator can be seen. On the right a visualization of both the tropopause and the polar vortex can be seen, both are important mixing barriers.

retrieval control system that dynamically adapts the observations made in an online fashion. This facilitates the dynamic tracking of time-evolving sharp gradients, one example would be those in chemical tracer fields often located at the polar vortex edge, the tropopause and the day-night division. An example of this is shown in Fig. 1. On the left the sharp gradient in NO (nitric-oxide) at the terminator can be seen. On the right a visualisation of both the tropopause and the polar vortex can be seen, both are important mixing barriers.

This approach fits in well with the Sensor Web Concept (http://sensorwebs.jpl.nasa.gov/). A Sensor Web consists of a group of sensors (satellites, UAVs, aircraft, ground stations) set up to collect various kinds of information, communicate with other sensors in the web. NASA is already taking the first steps toward Internet-like connectivity among its Earth sensing satellites. A system composed of multiple science instrument/processor platforms that are interconnected by means of a communications fabric for the purpose of collecting measurements and processing data for Earth or Space Science objectives. An example how these ideas have

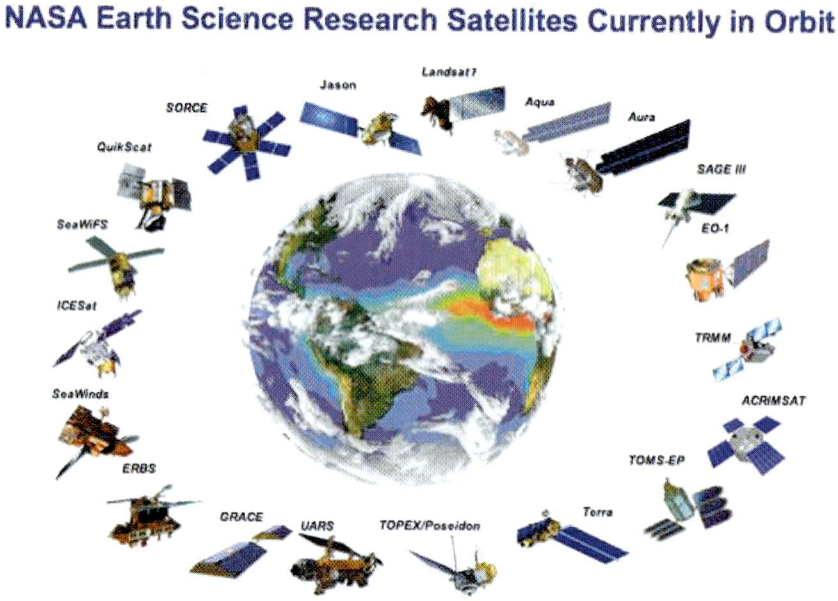

D. J. Lary & A. Koratkar

Fig. 2   **A schematic of the NASA Earth Science Research Satellites currently in orbit.**

already been used to track fires is available online at http://earthobservatory.nasa.gov/Newsroom/NasaNews/2003/2003072215047.html. Imagine, for example, if all the NASA earth science satellites currently in orbit (Fig. 2) had active communication with each other and with an intelligent observation direction system and global modelling tools. Such an arrangement would allow a much greater synergy than is currently possible and thus allow for an objectively optimised approach to earth observation.

NASA and ESA science missions have traditionally operated on the assumption that we can only manage scheduling priorities and scientific processing on the ground with significant human interaction, and that all scientific data must be downloaded and archived regardless of its scientific value. However, increases in onboard processing and storage capabilities of spacecraft, as well as increases in rates of data accumulation will soon force NASA operations staff and scientists to re-evaluate the assumption that all science must be done on the ground. In order to take advantage of these new in-flight capabilities, improve science return and contain costs, we must develop strategies that will help reduce the perceived risk associated with increased use of automation in all aspects of spacecraft operations. An important aspect of science operations is the ability to respond to science driven events in a timely manner. For such investigations, we must teach our observing platforms to intelligently achieve the scientists goals. The principles presented here are generic but a specific example will be taken from atmospheric chemistry. The assimilation system described is being used in the NASA Global Modeling and Assimilation Office to assist with EOS Aura data validation.

Throughout this chapter a central concept is that of state vectors. The first step in the mathematical formalisation of the system is the definition of the work space. The collection of numbers needed to represent the state of the system being studied is collected as a column matrix called the state vector, x.

2. Dynamic Data

The first and most important element is the concept of dynamic data. The dynamic data retrieval control system envisioned here dynamically adapts what measurements are made, where they are made, and when they are made. The dynamic adaption is performed online to maximise the information content, minimise the uncertainty in characterising the systems state vector, and minimise both the required storage and data processing time, and minimise the data heterogeneity minimisation within analysis grid cells. (It is conceivable that these ideas could be used in future to direct additional observations from unmanned automated sub-orbital platforms.)

The ability to develop a dynamic data retrieval control system for an objectively optimised earth observation system depends in large part on products made available when data assimilation is an integrated part of the earth observation system. Making data assimilation an integral part of the earth observation system is a prudent step since assimilation seeks to bring together heterogeneous information together with its associated uncertainty from a variety of sources (both observational and theoretical) in a self-consistent mathematical framework.

3. Science Goal Monitor

At the heart of the dynamic data retrieval control system is the Science Goal Monitor. The Science Goal Monitor (SGM; http://aaa. gsfc.nasa.gov/SGM) is a prototype software tool to explore strategies for implementing science goal driven operations for multiple sensors/platforms

[Koratkar *et al.* (2002)]. A space science SGM is being prototyped for dynamic automated reactions to intrinsically varying astronomical phenomenon using one of the Small and Moderate Aperture Research Telescope System (SMARTS) telescopes. An earth science prototype has been built for Earth Observing 1 (EO-1) to evaluate how multiple sensors can react dynamically to obtain rapid observations of evolving earth science events. Here we envision extending these previous prototypes to use objective metrics such as information content and system uncertainty so that SGM can schedule suitable observations that objectively optimise the use of our assets.

Higher level goals could also be specified. For example, malaria is a major international public health problem, causing 300–500 million infections worldwide and approximately 1 million deaths annually. If we have developed a risk model to predict the occurrence of malaria and its transmission intensity and its mapping to satellite-derived and meteorological data we could ensure that our earth observing system makes observing such conditions a priority. This would then facilitate the early identification of potential breeding sites of major vector species before a disease outbreak occurs and identify the locations for larvicide and insecticide applications in order to reduce costs, lessen the chance of developing pesticide resistance, and minimise the damage to the environment. Such projects already exist, for example, the NASA healthy planet project on Mekong Malaria and Filariasis, http://healthyplanet.gsfc.nasa.gov/project3.html.

4. Information Content and State Vector Uncertainty

As a dynamic system evolves with time not all of the state variables within the state vector contain equal amounts of information (information content), and not all state variables are known to the same precision. It is therefore clearly desirable that the observations made both contain the maximum information content possible with a given observing platform capability and allow the systems state to be characterised with a minimum uncertainty.

Information content is a broad term that could be quantified in any number of ways depending on the system or problem being studied. Therefore, although we propose to use a specific measure of information content for the atmospheric chemistry system, these measures could easily be substituted with alternative measures that may be more suitable depending on the given objectives of an investigation. Although we describe a specific example from atmospheric chemistry, the principle is clearly more general. The key new concept in this approach is that information content and system uncertainty are used in determining: What should be measured, when and where, thus providing a cost effective strategy for using resources and minimising the data storage required to characterise a system with a given level of precision.

One measure of information content/ranking that could be used is described by [Khattatov *et al.* (1999)] coupled with the so-called goal attainment algorithm to provide the information content ranking. The chemical assimilation system will provide analyses of the state vector together with an associated uncertainty. The information content/ranking software uses the analyzed state vector to provide the information content ranking. This information is then passed to the SGM to allow it to objectively determine the following days observation schedule.

5. Automatic Code Generation

The complexity of atmospheric chemistry varies tremendously with location: From the relatively simple chemistry of the mesosphere involving primarily oxygen, hydrogen, and nitrogen containing species, to the more complex chemistry of the stratosphere also involving chlorine, bromine, iodine, and sulphur containing species and simple hydrocarbons such as CH_4 and other greenhouse gasses, to the very complex chemistry of the troposphere, which also involves volatile organic hydrocarbons (VOCs) and their host of oxidation products. Therefore, any tool that is going to be involved in implementing a dynamic objectively optimised earth observation strategy must be capable of dealing with these very different chemical regimes. Consequently, it is most desirable to have an automatic code generator that is capable of creating and reusing code for the deterministic models required to describe the chemistry of these different regimes together with the entire data assimilation infrastructure required (i.e. time derivatives, Jacobians, Hessians, adjoints, and information content). The AutoChem code generation and modeling/assimilation system has these capabilities and has already been validated in a range of studies (www.AutoChem.info/). Code validation is an important part of this process. The AutoChem system has been extensively validated against a wide variety of data from aircraft, balloons, space shuttle borne instruments such as ATMOS and CRISTA and satellite based observations.

6. Data Assimilation

The information content metrics and uncertainty characterisation will be supplied by the chemical assimilation system, AutoChem. AutoChem (www.AutoChem.info/) is an automatic code generation system, documenter and symbolic differentiator for atmospheric chemical modeling and data assimilation [Fisher and Lary (1995); Lary *et al.* (2003)]. An advantage of assimilation is that it propagates information from data-rich regions to data-poor regions. Data assimilation also offers a mathematical framework to check and quantify the chemical consistency of multispecies observations with one another and with photochemical theory through the use of objective skill scores. That is, the analysis can examine both the consistency between different instruments observing the same constituent, and the photochemical self-consistency between multiconstituent observations and photochemical theory.

7. Automatic Data Compression

After the raw radiance data observed by a satellite is processed higher-level one and two datasets are generated. These higher-level datasets are usually stored at a uniform precision, where the stored precision is usually significantly greater than the certainty with which the level one and two data are known. For example, the data may be stored with eight significant figures when we are only confident in the first three or four. If the total data volume is small then this does not have significant cost implication. However, when we are dealing with very high data volumes this does have a significant cost implication for storage and/or data transfer. For many years now a variety of data compression techniques have been used that could be adapted to reduce the amount of space required for data storage and time for data transmission. The degree of data compression can be chosen to make the compression non-lossy for the accuracy characterised by the assimilation system, i.e. to three significant figures if that is how well we know the variable instead of eight or sixteen significant figures if we do not know the variable to that precision. If it is found at a later date that reprocessing is required then this can still be done as the raw radiance data is stored to the full machine precision. Automatic data compression uses the dynamic data concept in the addition of value added products without incurring prohibitive space requirements.

8. Machine Learning

The whole approach described depends in large part on the integration of a data assimilation system. When considering data assimilation of atmospheric chemistry, one of the computationally most expensive tasks is the time integration of a large and stiff set of ordinary differential equations (ODEs). However, very similar sets of ODEs are solved at adjacent grid points and on successive days, so similar calculations are repeated many thousands of times. This is the type of application that benefits from adaptive, error monitored, machine-learning technology. Our ODE solver already employs adaptive time stepping with error monitoring, if this is extended to an adaptive use of machine learning then there are literally massive potential savings in computational expense. A prototype code has been developed that we would like to extend here for use within the ODE solver. Early work seems promising that such an approach would work [Lary *et al.* (2004); Lary and Mussa (2004)]. A success in this area would mean a dramatic reduction in the computational cost of assimilation and hence of the entire dynamic data retrieval control system.

9. Automatic Analysis and Web Site Creation

To facilitate the analysis and scientific usefulness of the modeling and assimilation system and the dissemination of the data products the system includes an automatic web site generator called CDACentral (for **C**hemical **D**ata **A**ssimilation Central). An example is available online at www.CDACentral.info/. CDACentral creates a full cross-linked web site that presents not only the assimilated analyses, the associated uncertainties, detailed analysis of the uncertainties, assimilation skill scores, but also a break down of all the continuity equations and the contribution of each individual term to the overall continuity equation. It is easy to navigate to a given time period or constituent by using the site's javascript navigation bars. This allows detailed mechanistic studies to be performed. For example, the next subsection describes how the system has been recently used to show the often unrecognised role of halogen chemistry in the free troposphere.

9.1. *A case study: Chlorine oxidation of methane in the free troposphere*

Atmospheric methane is a key greenhouse gas. Methane and hydrocarbon oxidation are some of the most significant atmospheric chemical processes.

The hydroxyl radical (OH) is an important cleansing agent of the lower atmosphere, in particular, it provides the dominant sink for CH_4 and HFCs as well as the pollutants NO_x, CO and VOCs. Once formed, tropospheric OH reacts with CH_4 or CO within seconds. It is generally accepted that the local abundance of OH is controlled by the local abundances of NO_x, CO, VOCs, CH_4, O_3, and H_2O as well as the intensity of solar UV; and thus it varies greatly with time of day, season, and geographic location [Houghton and Ding (2001)].

Methane oxidation is usually initiated by hydrogen abstraction reactions such as

$$OH + CH_4 \longrightarrow CH_3 + H_2O, \tag{1}$$

$$O(^1D) + CH_4 \longrightarrow CH_3 + OH, \tag{2}$$

$$Cl + CH_4 \longrightarrow CH_3 + HCl, \tag{3}$$

$$Br + CH_4 \longrightarrow CH_3 + HBr. \tag{4}$$

However, the halogen initiation and catalysis of hydrocarbons is not usually considered in global chemistry models. This is not due to a lack of kinetic knowledge but rather an assumption that halogens play a minor role outside of the boundary layer [Vogt *et al.* (1996); Sander and Crutzen (1996); Richter *et al.* (1998); Dickerson *et al.* (1999); Sander *et al.* (2003); von Glasow and Crutzen (2004)] and stratosphere [Johnston and Podolske (1978); Cicerone *et al.* (1983); Farman *et al.* (1985)]. Figure 4(b) shows that in the lower stratosphere and even in the free troposphere, halogen-catalysed, and halogen-initiated, methane oxidation can be important. Halogen-catalysed methane oxidation can play a significant role in the production of HO_x (= H + OH + HO_2) radicals [Lary and Toumi (1997)] in just the region where it is usually accepted that nitrogen-catalysed methane oxidation is one of the main sources of ozone [Houghton and Ding (2001)]. Aspects of methane oxidation by halogens has been previously mentioned by [Crutzen *et al.* (1992); Burnett and Burnett (1995)] and the mechanism specifically described by [Lary and Toumi (1997)].

Figure 4(a) shows the fraction of CH_3 production due to the reaction of methane with OH as a height time series at an equivalent PV latitude of 74°S, i.e. in the polar vortex edge region. The analyses was produced using the AutoChem chemical data assimilation package and observations of methane, ozone, nitric acid, and hydrochloric acid from the NASA upper atmosphere research satellite (UARS). The overlaid dashed red line shows the tropopause as diagnosed by the WMO lapse rate definition, the

Toward an Optimized Earth Observation System

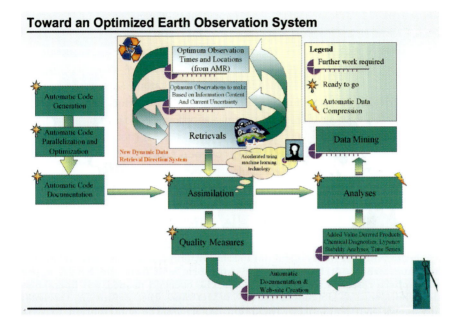

Fig. 3 Schematic overview.

solid line shows the temperature minimum. Although the contribution to CH_3 production by the reaction of Cl atoms with methane in the tropo-sphere (below the dashed red line) is usually considered to be unimportant the analysis produced by data assimilation shows that this is not true (panel b). Every spring the production of CH_3 due to the reaction of Cl with methane can contribute up to 80% of the total rate of CH_3 production. Likewise, the hydrolysis of $BrONO_2$ alone can contribute more than 35% of the HNO_3 production rate in the free-troposphere [Lary (2004)]. Comprehensive results from the chemical assimilation are available online at www.CDACentral.info/. Reaction (2) is most significant in the tropical upper-troposphere where it contributes up to 7% to the initiation of methane oxidation for much of the year as can be seen in the analysis presented in the CDACentral website. Reaction (4) plays a negligible role and is just included for the sake of completeness.

In this study sulphate aerosol observations from SAGE II [Ackerman *et al.* (1989); Oberbeck *et al.* (1989); Russell and McCormick (1989); Thomason (1991, 1992); Bauman *et al.* (2003)] and HALOE [Hervig *et al.* (1993); Hervig *et al.* (1996); Hervig *et al.* (1998); Massie *et al.* (2003)]

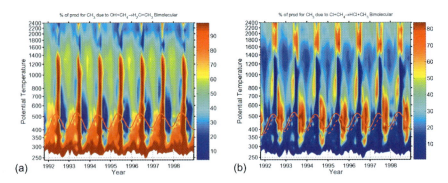

Fig. 4 Atmospheric methane is a key greenhouse gas. The main loss of methane occurs through the reaction of methane with OH to produce CH_3. Panel (a) shows the fraction of CH_3 production due to the reaction of methane with OH as a height time series at an equivalent PV latitude of 74°S, i.e. in the polar vortex edge region. The analyses was produced using the AutoChem chemical data assimilation package and observations of methane, ozone, nitric acid, and hydrochloric acid from the NASA upper atmosphere research satellite (UARS). The overlaid dashed red line shows the tropopause as diagnosed by the WMO lapse rate definition, the solid line shows the temperature minimum. Although the contribution to CH_3 production by the reaction of Cl atoms with methane in the troposphere (below the dashed red line) is usually considered to be unimportant the analysis produced by data assimilation shows that this is not true (panel b). Every spring the production of CH_3 due to the reaction of Cl with methane can contribute up to 80% of the total rate of CH_3 production. It can be seen that the active synergy between observations and modeling via data assimilation can facilitate scientific insights. If this synergy is extended to include a dynamic direction of observations based on objective measures routinely produced by data assimilation it can be seen how we have a sound strategy for focussing on the key scientific issues.

were used, ozone observations from UARS [Reber *et al.* (1993)] MLS v6 [Froidevaux *et al.* (1996); Waters (1998)], HALOE v19 [Russell *et al.* (1993)], POAM, ozone sondes and LIDAR, nitric acid observations from UARS MLS v6 [Santee *et al.* (1997, 1999)], CLAES, ATMOS, CRISTA [Offermann and Conway (1999)], ILAS [Wood *et al.* (2002)] and MOZAIC aircraft [Marenco *et al.* (1998)], hydrochloric acid observations from UARS HALOE and ATMOS, water observations from UARS MLS v6, HALOE v19, and ATMOS, methane observations from UARS HALOE v19, ATMOS and CRISTA were used. All though the bulk of these observations were in the stratosphere a significant number of satellite observations were available for the free troposphere down to 5 km, and from sondes and aircraft data is also available below 5 km.

The major uncertainty in the calculations just presented is the exact chlorine loading of the free-troposphere. UARS/HALOE did not make a significant number of measurements in the free-troposphere, and even when it did the altitude resolution is only 3 km. In the type of objectively optimised earth observations system envisioned here this type of information can be fed back to the earth observing system via the SGM to direct further observations to be made, for example by sub-orbital platforms such as the UAVs (unmanned arial vehicles). It can be seen that the active synergy between observations and modelling via data assimilation can facilitate scientific insights. If this synergy is extended to include a dynamic direction of observations based on objective measures routinely produced by data assimilation it can be seen how we have a sound strategy for focussing on the key scientific issues.

The example we have chosen is deliberately a little controversial. The point being that 'conventional wisdom' can make assumptions that do not square against a large body of observations. The example chosen was from earth observation but aircraft data and the beautiful ATMOS data set show exactly the same thing. In addition, one of the purposes of using assimilation is to validate the model, especially when using high quality in-situ data such as from aircraft. For example, one cannot explain the precise shape of the OH and HO_2 diurnal cycles observed from aircraft in the upper troposphere and lower stratosphere if halogen chemistry is not used. ATMOS and satellite data also strongly point to the same end. In other words observations from aircraft, ATMOS, and more than a decade of earth observation agree with the model used based on well established laboratory kinetics and disagree with the conventional wisdom that says halogens do not play a role in the free troposphere. The data speaks strongly against this 'conventional wisdom'.

10. Conclusion

A schematic overview of the objectively optimised earth observation system envisioned is shown in Fig. 3. The elements of the dynamic data retrieval control system can help in objectively planning mission goals, in the cost effective operation of future optimised earth observing systems, and for scientific analysis and dissemination. During the planning stage the objective measures of information content are invaluable in determining what the instrument capabilities should be. During the operation of future earth observing systems the dynamic data retrieval control system could dynamically adapt what measurements are made, where they are made, and when they are made, in an online fashion to maximise the information content, minimise the uncertainty in characterising the systems state vector, and minimise both the required storage and data processing time.

The same technology could be applied to the analyses and design of ground based pollution monitoring networks to provide regular pollution analyses. These could then be used for epidemiological studies in the precise quantification on the impacts of pollution on human health. For example, it was noted by Shallcross (personal communication) that high levels of benzene were associated with high hospital admissions of cardiovascular conditions.

At a more basic level the idea of symbiotic communication and dynamic data could be used in many applications to optimise monitoring and observing systems. The ideas of automatic code generation and automatic documentation to facilitate system implementation on a variety of hardware is also of quite general applicability. As is the concept of automatic data compression to minimise the required cost of both storage and dissemination.

Higher level goals could also be specified such as the remote identification of sites of likely malaria outbreaks. By facilitating the early identification of potential breeding sites of major vector species before a disease outbreak occurs and identifying the locations for larvicide and insecticide applications. This would reduce costs, lessen the chance of developing pesticide resistance, and minimise the damage to the environment.

Acknowledgements

It is a pleasure to acknowledge: NASA for a distinguished Goddard Fellowship in Earth Science and for research support; The Royal Society for a Royal Society University Research Fellowship; The government of Israel for an Alon Fellowship; NASA, NERC, EU, and ESA for research support.

References

Ackerman, M., Brogniez, C., Diallo, B. *et al.* (1989) European validation of SAGE II aerosol profiles. *J. Geophys. Res.* **94**(D6), 8399–8411.

Bauman, J. J., Russell, P. B., Geller, M. A. & Hamill, P. (2003) A stratospheric aerosol climatology from SAGE II and CLAES measurements: 1. Methodology. *J. Geophys. Res. (Atmos.)* **108**(D13), AAC 6–1 AAC 6–3.

Burnett, E. & Burnett, C. (1995) Enhanced production of stratospheric OH from methane oxidation at elevated reactive chlorine levels in Northern midlatitudes. *J. Atmos. Chem.* **21**(1), 13–41.

Cicerone, R. J., Walters, S. & Liu, S. C. (1983) Non-linear response of stratospheric ozone column to chlorine injections. *J. Geophys. Res. (Atmos.)* **88**(NC6), 3647–3661.

Crutzen, P. J., Müller, C., Brühl, R. & Peter, T. (1992) On the potential importance of the gas-phase reaction $CH_3OO + ClO \longrightarrow ClOO + CH_3O$ and the heterogeneous reaction $HOCl + HCl \longrightarrow H_2O + Cl_2$ in ozone hole chemistry. *Geophys. Res. Lett.* **19**(11), 1113–1116.

Dickerson, R. R., Rhoads, K. P., Carsey, T. P., Oltmans, S. J., Burrows, J. P. & Crutzen, P. J. (1999) Ozone in the remote marine boundary layer: A possible role for halogens. *J. Geophys. Res. (Atmos.)* **104**(D17), 21385–21395.

Farman, J. C., Gardiner, B. G. & Shanklin, J. D. (1985) Large losses of total ozone in antarctica reveal seasonal ClO_x/NO_x interaction. *Nature* **315**(6016), 207–210.

Fisher, M. & Lary, D. (1995) Lagrangian 4-dimensional variational data assimilation of chemical-species. *Q. J. R. Meteorol. Soc.* **121**(527 Part A), 1681–1704.

Froidevaux, L., Read, W. G., Lungu, T. A., Cofield, R. E., Fishbein, E. F., Flower, D. A., Jarnot, R. F., Ridenoure, B. P., Shippony, Z., Waters, J. W., Margitan, J. J., McDermid, I. S., Stachnik, R. A., Peckham, G. E., Braathen, G., Deshler, T., Fishman, J., Hofmann, D. J. & Oltmans, S. J. (1996) Validation of UARS microwave limb sounder ozone measurement. *J. Geophys. Res. (Atmos.)* **101**(D6), 10017–10060.

Hervig, M., Russell, J., Gordley, L., Drayson, S., Stone, K., Thompson, R., Gelman, M., McDermid, I., Hauchecorne, A., Keckhut, P., McGee, T., Singh, U. & Gross, M. (1996) Validation of temperature measurements from the halogen occultation experiment. *J. Geophys. Res.* **101**(D6), 10277–10285.

Hervig, M. E. & Deshler, T. (1998) Stratospheric aerosol surface area and volume inferred from HALOE, CLAES, and ILAS measurements. *J. Geophys. Res. (Atmos.)* **103**(D19), 25345–25352.

Hervig, M. E., Russell, J. M., Gordley, L. L., Park, J. H., & Drayson, S. R. (1993) Observations of aerosol by the HALOE experiment onboard UARS — a preliminary validation. *Geophys. Res. Lett.* **20**(12), 1291–1294.

Houghton, J. & Ding, Y., eds. (2001) *Climate Change 2001: The Scientific Basis*, IPCC, UNEP.

Johnston, H. S. & Podolske, J. (1978) Interpretation of stratospheric photochemistry. *Rev. Geophys.* **16**, 491.

Khattatov, B., Gille, J., Lyjak, L., Brasseur, G., Dvortsov, V., Roche, A. & Waters, J. (1999) Assimilation of photochemically active species and a case analysis of UARS data. *J. Geophys. Res. (Atmos.)* **104**(D15), 18715–18737.

Koratkar, A., Grosvenor, S., Jones, J. E., Memarsadeghi, A. & Wolf, K. R. (2002) Science goal driven observing: A step towards maximizing science returns and spacecraft autonomy. *SPIE* **4844**, 250.

Lary, D. (2004) Halogens and the chemistry of the free troposphere. *Atmospheric Chemistry and Physics Discussion* **4**, 5367–5380.

Lary D. & Toumi, R. (1997) Halogen-catalyzed methane oxidation. *J. Geophys. Res.* **102**(D19), 23421–23428.

Lary D. J. & Mussa, H. Y. (2004) Using an extended Kalman filter learning algorithm for feed-forward neural networks to describe tracer correlations. *Atmospheric Chemistry and Physics Discussions* **4**, 3653–3667.

Lary, D. J., Khattatov, B. & Mussa, H. Y. (2003) Chemical data assimilation: A case study of solar occultation data from the Atlas 1 mission of the atmospheric trace molecule spectroscopy experiment (atmos). *J. Geophys. Res. (Atmos.)* **108**(D15).

Lary, D. J., Muller, M. D. & Mussa, H. Y. (2004) Using neural networks to describe tracer correlations. *Atmospheric Chemistry and Physics* **4**, 143–146.

Marenco, A., Thouret, V., Nedelec, P., Smit, H., Helten, M., Kley, D., Karcher, F., Simon, P., Law, K., Pyle, J., Poschmann, G., Von Wrede, R., Hume, C. & Cook, T. (1998) Measurement of ozone and water vapor by airbus in-service aircraft: The MOZAIC airborne program, an overview. *J. Geophys. Res. (Atmos.)* **103**(D19), 25631–25642.

Massie, S., Randel, W., Wu, F., Baumgardner, D. & Hervig, M. (2003) Halogen occultation experiment and stratospheric aerosol and gas experiment II observations of tropopause cirrus and aerosol during the 1990s. *J. Geophys. Res. (Atmos.)* **108**(D7).

Oberbeck, V. R., Livingston, J. M., Russell, P. B., Pueschel, R. F., Rosen, J. N., Osborn, M. T., Kritz, M. A., Snetsinger, K. G. & Ferry, G. V. (1989) SAGE-II aerosol validation — selected altitude measurements, including particle micromeasurements. *J. Geophys. Res. (Atmos.)* **94**(D6), 8367–8380.

Offermann, D. & Conway, R. R. (1999) Crista/mahrsi — preface. *J. Geophys. Res. (Atmos.)* **104**(D13), 16309–16310.

Reber, C. A., Trevathan, C. E., Mcneal, R. J. & Luther, M. R. (1993) The upper-atmosphere research satellite (UARS) mission. *J. Geophys. Res. (Atmos.)* **98**(D6), 10643–10647.

Richter, A., Wittrock, F., Eisinger, M. & Burrows, J. P. (1998) Gome observations of tropospheric bro in northern hemispheric spring and summer 1997. *Geophys. Res. Lett.* **25**(14), 2683–2686.

Russell, J. M., Gordley, L. L., Park, J. H., Drayson, S. R., Hesketh, W. D., Cicerone, R. J., Tuck, A. F., Frederick, J. E., Harries, J. E. & Crutzen, P. J. (1993) The Halogen Occultation Experiment. *J. Geophys. Res. (Atmos.)* **98**(D6), 10777–10797.

Russell, P. B. & McCormick, M. P. (1989) SAGE-II aerosol data validation and initial data use — an introduction and overview. *J. Geophys. Res. (Atmos.)* **94**(D6), 8335–8338.

Sander, R. & Crutzen, P. J. (1996) Model study indicating halogen activation and ozone destruction in polluted air masses transported to the sea. *J. Geophys. Res. (Atmos.)* **101**(D4), 9121–9138.

Sander, R., Keene, W. C., Pszenny, A. A. P., Arimoto, R., Ayers, G. P., Baboukas, E., Cainey, J. M., Crutzen, P. J., Duce, R. A., Hönninger, G., Huebert, B. J., Maenhaut, W., Mihalopoulos, N., Turekian, V. C. & Van Dingenen, R.

(2003) Inorganic bromine in the marine boundary layer: A critical review. *Atmospheric Chemistry and Physics* **3**, 1301–1336.

Santee, M. L., Manney, G. L., Froidevaux, L., Read, W. G. & Water, J. W. (1999) Six years of UARS microwave limb sounder HNO_3 observations: Seasonal, interhemispheric, and interannual variations in the lower stratosphere. *J. Geophys. Res. (Atmos.)* **104**(D7), 8225–8246.

Santee, M. L., Manney, G. L., Froidevaux, L., Zurek, R. W. & Waters, J. W. (1997) MLS observations of ClO and HNO_3 in the 1996-97 arctic polar vortex. *Geophys. Res. Lett.* **24**(22), 2713–2716.

Sujit, P. B. & Ghose, D. (2004) Search using multiple UAVS with flight time constraints, *IEEE Trans. Aero. Electronic Sys.* **40**(2), 491–509.

Thomason, L. W. (1991) A diagnostic stratospheric aerosol size distribution inferred from SAGE-II measurements. *J. Geophys. Res. (Atmos.)* **96**(D12), 22501–22508.

Thomason, L. W. (1992) Observations of a new SAGE-II aerosol extinction mode following the eruption of Mt. Pinatubo. *Geophys. Res. Lett.* **19**(21), 2179–2182.

Trego, L. (1994) Unmanned aerial vehicles. *Aero. Eng.* **14**(3), 15.

Vogt, R., Crutzen, P. J. & Sander, R. (1996) A mechanism for halogen release from sea-salt aerosol in the remote marine boundary layer. *Nature* **383**(6598), 327–330.

von Glasow, R. & Crutzen, P. J. (2004) Model study of multiphase DMS oxidation with a focus on halogens. *Atmospheric Chemistry and Physics* **4**, 589–608.

Waters, J. W. (1998) Atmospheric measurements by the MLS experiments: Results from UARS and plans for the future. *CIRA Part III Reference Atmospheres — Trace Constituent Models — Comparison with Latest Data, Advances in Space Research* **21**, 1363–1372, Elsevier.

Wood, S. W., Bodeker, G. E., Boyd, I. S., Jones, N. B., Connor, B. J, Johnston, P. V., Matthews, W. A., Nichol, S. E., Murcray, F. J., Nakajima, H. & Sasano, Y. (2002) Validation of version 5.20 ILAS HNO_3, CH_4, N_2O, O_3, and NO_2 using ground-based measurements at Arrival Heights and Kiruna. *J. Geophys. Res. (Atmos.)* **107**(D24).

Zelenka, R. E., Smith, P. N., Coppenbarger, R. A., Njaka, C. E. & Sridhar, B. (1997) Results from the NASA automated nap-of-the-earth program. *J. Am. Helicopter Soc.* **42**(2), 107–115.

Afterword

There can be little doubt that these are exciting times for the earth sciences, and in many ways contemporary research in the field can be considered as marking a golden age in the history of the science. The Earth Sciences currently occupy a position firmly at the heart of today's critical issues; tackling climate change, managing and mitigating natural hazards, prospecting for essential new resources and addressing future energy needs. At the same time, earth scientists find themselves at the cutting-edge of discovery and exploration; within the Earth's interior, in the abyssal depths of the oceans, and on the surfaces of our sibling planets and their satellites.

Without question the new millennium has brought with it a range of problems that the earth sciences are perfectly placed to address, and many of which are examined by contributors to this volume. In relation to increasing vulnerability and exposure to natural hazards, a growing role for satellite monitoring is critical for disaster risk reduction, using radar interferometry and other means to monitor and forecast hazards such as landslides, earthquakes and volcanic eruptions. Combined with this, improving our understanding of the mechanisms that underpin landslide and earthquake formation, enable better predictions to be made about future hazards and the threats they pose to life and property. With both observation and modelling highlighting an acceleration in the rate of climate change this century and beyond, learning more about the atmospheric carbon budget and the effects of particulate loading, and about how we can cost mitigation policies to ensure we tackle rising greenhouse gas emissions in the most effective manner, is critical. While clearly disturbing, taking a retrospective look at the causes of past extinction level events (ELEs) is also vital if we are to start to understand how abrupt climate change might affect our planet's complex ecosystems.

Inevitably, as our population expands and more and more nations strive for industrial economies to raise living standards, there is increasing demand for ever more resources, both in the form of hydrocarbons and metallic and non-metallic ores. At the same time, increasing pollution arising directly from enhanced resource usage is creating problems of environmental pollution and contamination that must be addressed.

While there has never been a time when the application of the earth sciences to contemporary economic and social issues has been so great, the purer aspects of the field are also prospering. New data are providing us with a greater knowledge of the near-earth environment, the cryptic processes that operate deep within and beneath volcanoes — and even deeper, at the Earth's core — and the mechanisms that underpin the slow dance of the tectonic plates across our planet's surface.

We are perhaps at a crossroads in the growth and development of our civilisation, providing options that may see our race and our civilisation bloom or suffer a knock-back as climate change and a potential energy crisis conspire to make life increasingly difficult for our children and grandchildren. The buoyant field of Earth Sciences may, however, help to provide some of the answers to the problems we undoubtedly face, ensuring for future generations a better life, rather than one fraught with danger and difficulty.

Bill McGuire
Benfield Professor of Geophysical Hazards and
Director of the Benfield Hazard Research Centre
University College London, London, England

INDEX